Disposal of Weapon Plutonium

NATO ASI Series

Advanced Science Institutes Series

A Series presenting the results of activities sponsored by the NATO Science Committee, which aims at the dissemination of advanced scientific and technological knowledge, with a view to strengthening links between scientific communities.

The Series is published by an international board of publishers in conjunction with the NATO Scientific Affairs Division

A	**Life Sciences**	Plenum Publishing Corporation
B	**Physics**	London and New York
C	**Mathematical and Physical Sciences**	Kluwer Academic Publishers
D	**Behavioural and Social Sciences**	Dordrecht, Boston and London
E	**Applied Sciences**	
F	**Computer and Systems Sciences**	Springer-Verlag
G	**Ecological Sciences**	Berlin, Heidelberg, New York, London,
H	**Cell Biology**	Paris and Tokyo
I	**Global Environmental Change**	

PARTNERSHIP SUB-SERIES

1.	**Disarmament Technologies**	Kluwer Academic Publishers
2.	**Environment**	Springer-Verlag / Kluwer Academic Publishers
3.	**High Technology**	Kluwer Academic Publishers
4.	**Science and Technology Policy**	Kluwer Academic Publishers
5.	**Computer Networking**	Kluwer Academic Publishers

The Partnership Sub-Series incorporates activities undertaken in collaboration with NATO's Cooperation Partners, the countries of the CIS and Central and Eastern Europe, in Priority Areas of concern to those countries.

NATO-PCO-DATA BASE

The electronic index to the NATO ASI Series provides full bibliographical references (with keywords and/or abstracts) to more than 50000 contributions from international scientists published in all sections of the NATO ASI Series.
Access to the NATO-PCO-DATA BASE is possible in two ways:

– via online FILE 128 (NATO-PCO-DATA BASE) hosted by ESRIN, Via Galileo Galilei, I-00044 Frascati, Italy.

– via CD-ROM "NATO-PCO-DATA BASE" with user-friendly retrieval software in English, French and German (© WTV GmbH and DATAWARE Technologies Inc. 1989).

The CD-ROM can be ordered through any member of the Board of Publishers or through NATO-PCO, Overijse, Belgium.

Series 1: Disarmament Technologies – Vol. 4

Disposal of Weapon Plutonium

Approaches and Prospects

edited by

Erich R. Merz
Forschungszentrum Jülich,
KFA,
Jülich, Germany

and

Carl E. Walter
Lawrence Livermore National Laboratory,
University of California,
Livermore, California, U.S.A.

Kluwer Academic Publishers

Dordrecht / Boston / London

Published in cooperation with NATO Scientific Affairs Division

Proceedings of the NATO Advanced Research Workshop on
Disposal of Weapon Plutonium – Approaches and Prospects
St. Petersburg, Russia
May 14–17, 1995

A C.I.P. Catalogue record for this book is available from the Library of Congress.

ISBN 0-7923-3841-3

Published by Kluwer Academic Publishers,
P.O. Box 17, 3300 AA Dordrecht, The Netherlands.

Kluwer Academic Publishers incorporates the publishing programmes of
D. Reidel, Martinus Nijhoff, Dr W. Junk and MTP Press.

Sold and distributed in the U.S.A. and Canada
by Kluwer Academic Publishers,
101 Philip Drive, Norwell, MA 02061, U.S.A.

In all other countries, sold and distributed
by Kluwer Academic Publishers Group,
P.O. Box 322, 3300 AH Dordrecht, The Netherlands.

Printed on acid-free paper

Table of Contents

PREFACE

This NATO Advanced Research Workshop on Disposal of Weapons Plutonium is a follow-up event to two preceding workshops, each dealing with a special subject within the overall disarmament issue: "Disposition of Weapon Plutonium", sponsored by the NATO Science Committee.

The first workshop of this series was held at the Royal Institute of International Affairs in London on 24-25 January 1994, entitled "Managing the Plutonium Surplus, Applications, and Options". Its overall goal was to clarify the current situation with respect to plutonium characteristics and availability, the technical options for use or disposal, and their main technical, environmental, and economic constraints.

In the immediate term, plutonium recovered from dismantled nuclear warheads will have to be stored securely, and under international safeguards if possible. In the intermediate term, the principal alternatives for disposition of this plutonium are: irradiation in mixed-oxide (MOX) fuel assemblies in existing commercial light-water reactors or in specially adapted light-water reactors capable of operation with full cores of MOX fuel and irradiation in future fast reactors. Another option is to blend plutonium with high-level waste as it is vitrified for final disposal in a geologic repository. In both cases, the high radioactivity of the resulting products provides "self-shielding" and prevents separation of plutonium without already-developed and available sophisticated technology. The so-called "spent fuel standard" as an effective protection barrier is acquired in either case.

A second workshop took place at the Institute of Physics and Power Engineering (IPPE) in Obninsk, Russia, on 16-19 October 1994, which concentrated its objective upon the application of "Mixed Oxide Fuel (MOX) Exploitation and Destruction in Power Reactors". After achieving a typical fuel burnup level the spent fuel would be disposed of without further reprocessing in a deep underground geologic repository.

This time, in the third workshop, the non-reactor disposition option of immobilization of plutonium in glass or other durable waste ceramics, followed by suitable geologic disposal deep underground was thoroughly discussed and the pros and cons evaluated. Immobilization of plutonium can be done either during vitrification of high-level liquid waste, after adding weapon plutonium as solid plutonium dioxide or in a plutonium nitrate solution, or in a specially developed process by mixing plutonium oxide or an acid solution of plutonium nitrate with glass- or ceramic-forming components. In both cases this puts the plutonium into a highly chemically stable matrix that hinders its re-extraction and that also keeps it out of the environment with a fair degree of reliability. However, the second option does not fulfill the so-called spent fuel standard, as it lacks the extant radiation barrier. Thus, the second option would require, as a substitute, another effective additional barrier, e.g. depth, as in the deep-borehole approach.

Clearly, vitrification like any of the other proposed methods for making weapon plutonium secure, must satisfy a number of conditions, the main ones being:

- engineering feasibility
- cost
- possibility of verification
- absence of damage to the environment
- long-term storage
- public acceptance of the given method.

In assessing vitrification from the standpoint of engineering feasibility, consideration must be given both to the possibility, in principle, of accomplishing the given method, and to the possibility of utilizing a production facility for solidification of high-level waste that already exists. An important characteristic is the capacity of glass for homogeneous dissolution of radioactive admixtures.

The third workshop, organized in collaboration with Professor Aleksander Karelin, Director General, and his team of the V.G.Khlopin Radium Institute, St.Petersburg, was held on 14-17 May 1995. The purpose of this workshop was to:

- exchange information on existing knowledge and relevant experience in the field of waste vitrification. The idea of mixing plutonium directly into high-level waste glass seems to be technically feasible, but stringent safeguard and criticality-avoidance measures are required.

- discuss the different approaches from Eastern and Western countries on the optimum techniques and choices of credible alternative media for immobilizing transuranics and fission products for their ultimate disposal.

- introduce new disposal techniques, in particular the idea of embedding plutonium in cement or some other matrix in boreholes several kilometers deep. This approach was first proposed by Swedish researchers and endorsed for serious examination in the U.S. National Academy of Sciences study. Preliminary examinations by U.S. and Russian laboratories have come out in favor of in-depth investigation of this option.

- assess the implications stemming from safeguards requirements.

- identify common research areas, where defined research tasks in cooperative partnerships can be initiated. For example, at this time the maximum amount of plutonium that can be added to the vitrified glass logs has not been well established.

About 35 participants (43% Russian) from nine countries took part in the third workshop. The high international level demonstrates the relevance and importance of the workshop items. Twenty-six oral presentations gave an impressive overview of the state of the art. Additionally, several representatives from related institutions observed the workshop.

The content of the presented scientific papers can be grouped into five topical areas:

- Disposal Strategies
- Product Forms
- Vitrification Technology
- Disposal in Deep Boreholes
- Long-Term Safety Aspects.

Although Russia has experience with vitrification, largely with phosphate glass rather than borosilicate glass, considerable reluctance prevails in Russia with regard to disposition of weapon plutonium via the vitrification route. While there is strong Russian interest in further pursuing the use of vitrification for high-level waste disposal, Russia shows no interest in using vitrification as a means for disposition of weapon plutonium. Ideas and views expressed during the sometimes excited discussions were not for attribution - lively debate was encouraged, and no consensus was sought.

With the intention of broadening the base of information input for the dialog of pros and cons of plutonium disposition alternatives and the related safety areas, some critically minded experts from Russia were invited to participate in the workshop. Unfortunately, only one of the three accepted the invitation.

On May 17th, a technical tour of the Gatchina Radiochemical Department of the Radium Institute was arranged. The very informative visit to the Hot Cell Complex was guided by Dr.Boris A.Galkin, Head of the Technological Department.

It appears difficult to gain Russian acceptance for a major vitrification facility for Russian weapon plutonium at this time. On the other hand, vitrification is one of the options under consideration in the United States. In addition to the plutonium removed from warheads, both the U.S. and Russia have substantial amounts of residual scrap plutonium, for which, for technical and economic reasons, vitrification appears to be a preferable solution. Thus, Russia may modify its earlier reluctance to the vitrification route as a suitable means to dispose of any plutonium.

After these workshops and their resulting perceptions, together with the provision of additional information from outside sources, the question arises whether a concluding Advanced Research Workshop comprising a small number of experts would be useful. Its subject might be focused on two main goals:

1. Scientific and technical appraisal of the most promising plutonium disposition options on the basis of the most advanced knowledge available.
2. Recommendations on further necessary research and development within NATO`s Science Programs, e.g. Linkage and Collaborative Research Grants. Political implementations and consequences would generally not be considered.

A decision is pending.

Prof.Dr.Erich Merz
Co-Director, NATO Advanced
Research Workshop Programme

ACKNOWLEDGEMENTS

The editors are most grateful to NATO for providing financial support, and to the hosts of the workshop, the staff of the V.G.Khlopin Radium Institute, St.Petersburg, Russia, for their generous hospitality.

It is a pleasure to thank Dr.Michail M.Moshkov, Head of the International Department of the V.G.Khlopin Radium Institute, and all his collaborators, for their contributions to the success of the meeting and for managing all logistical arrangements.

The organizing committee and the editors would like to express their thanks to the speakers, commentators, and all others who attended the workshop and contributed to the discussions; they all made this publication possible.

We are grateful to Mrs. Nel de Boer for editorial assistance and to the staff of the V.G.Khlopin Radium Institute for organizational help.

LIST OF SPEAKERS AND PARTICIPANTS

Aloy, Albert S.
Head of Laboratory
V.G.Khlopin Radium Institute
194021 St.Petersburg
Russia

Anderson, Evgeniy B.
Deputy Director General
V.G.Khlopin Radium Institute
194021 St.Petersburg
Russia

Baltes, Bruno
Gesellschaft für Anlagen-und
Reaktorsicherheit, GRS mbH
Schwertnergasse 1
D-50667 Cologne
Germany

Bashmakova, Anna Y.
Ministry of the Russian Federation
on Atomic Energy, MINATOM
Committee for External Relations
109180 Moscow
Russia

Belot, Claude
Colonel
Secrétariat Général de la Défense
Nationale
51 bld de Latour-Maubourg
Paris
France

Burakov, Boris E.
Mineralogical Group
V.G.Khlopin Radium Institute
194021 St.Petersburg
Russia

Demonie, Maurits
Deputy Department Head
Belgoprocess
Gravenstraat 73
B-2480 Dessel
Belgium

Dworschak, Heinz
Institute Safety Technology
Joint Research Centre Ispra
European Union
I-21020 Ispra (Varese)
Italy

Ewing, Rodney C.
Professor, Department of
Earth & Planetary Sciences
University of New Mexico
Albuquerque, NM 87131
USA

Furuya, Hirotaka
Professor, Department of Nuclear
Engineering, Faculty of Engineer-
ing, Kyushu University
Fukuoka 812
Japan

Galkin, Boris Ya.
Head of Technological Department
V.G.Khlopin Radium Institute
194021 St.Petersburg
Russia

Gopalo, Tatyana A.
Leading Scientific Worker
All-Russian Research and Design
Institute of Production Engineer-
ing, VNIPIPPromtechnologii
Kashirskoje shosse 33
115409 Moscow
Russia

Hicken, Enno F.
Professor, Institute of Safety
Research and Reactor Technology
Research Center Jülich, KFA
D-52425 Jülich
Germany

Jardine, Leslie J.
Lawrence Livermore National
Laboratory, P.O.Box 808
University of California
Livermore, CA 94551
USA

Jouan, Antoine
CEA Marcoule, DCC/DIR
BP 171 Marcoule
F-30207 Bagnols sur Ceze
France

Kagramanyan, Vladimir S.
Head of Laboratory
State Scientific Centre, Institute of Physics and Power Engineering
Bondarenko sq.1
249020 Obninsk
Russia

Karelin, Aleksander I.
Director General
V.G.Khlopin Radium Institute
194021 St.Petersburg
Russia

Korolyov, Evgeny
Deputy Head of Nuclear Physics Department
V.G.Khlopin Radium Institute
194021 St.Petersburg
Russia

Ledebrink, Friedrich W.
Section Head
Siemens AG, KWU-BW
P.O.Box 110060, D-63434 Hanau
Germany

Lyubtsev, Rem I.
Head of Section
V.G.Khlopin Radium Institute
St.Petersburg
Russia

Marples, J.Anquin C.
Materials and Chemistry Group
AEA Technology
220, Harwell
Didcot, Oxon OX 11 ORA
England

Matzke, Hans Joachim
Section Head
Institute for Transuranium Elements, Joint Research Centre
European Union, P.O.Box 2340
D-76125 Karlsruhe
Germany

Matyunin, Yuri I.
Scientist
A.A.Bochvar All-Russian Scientific Research Institute of Inorganic Materials, Rogov st.5
123060 Moscow
Russia

Menschikov, Valery Federovich
Consultant
Interagency Commission on Ecological Safety, The Council on Safety of Russian Federation
Staraya pl.4
103132 Moscow
Russia

Merz, Erich R.
Professor Emeritus
Research Center Jülich, KFA-ISR-3
D-52425 Jülich
Germany

Moshkov, Michail
Head of International Department
V.G.Khlopin Radium Institute
194021 St.Petersburg
Russia

Nadykto, Boris A.
Head of Department
Russian Federal Nuclear Centre
All-Russian Scientific Research Institute for Experimental Physics
VNIIEF
Arzamas-16
Russia

Pavlov, Alexander B.
Head of Department
All-Russian Research and Design Institute for Energy Technology
VNIPIET
195248 St.Petersburg
Russia

Rimsky-Korsakov, A.A.
Deputy Director
V.G.Khlopin Radium Institute
194021 St.Petersburg
Russia

Schulte, Nancy
Programme Director
NATO Scientific Affairs Division
Disarmament Technologies
B-1110 Brussels
Belgium

Shpunt, Lev B.
Head of Department
V.G.Khlopin Radium Institute
194021 St.Petersburg
Russia

Sutcliffe, William B.
Senior Scientist
Lawrence Livermore National
Laboratory, CSTS, L-019
University of California
P.O.Box 808
Livermore, CA 94551
USA

Swahn, Johan
Institute of Physical Resource
Theory, Chalmers University of
Technology and Göteborg University
S-41296 Göteborg
Sweden

Teule, Jacques
Secrétariat Général de la Défense
Nationale
51 bld de Latour-Maubourg
Paris
France

Trauwaert, Etienne
Deputy General Manager
Belgoprocess
Gravenstraat 73
B-2480 Dessel
Belgium

Walter, Carl E.
Senior Scientist
Lawrence Livermore National
Laboratory, L-125
P.O.Box 808
University of California
Livermore, CA 94551-9900
USA

Wicks, George G.
Senior Scientist
Westinghouse Savannah River
Company, Savannah River Site
SRTC Building No.773-A, B-129
Aiken, SC 29808
USA

WEAPONS PLUTONIUM MANAGEMENT: THE PRESENT SITUATION, THE METHODS AND THE PROSPECTS

Alexander I. Karelin
General Director
V.G.Khlopin Radium Institute
194021 St.Petersburg, Russia

In Russia investigations on the management of weapons fissile materials involve scientists of quite a few scientific research institutes, including scientists of the V.G.Khlopin Radium Institute.

In solving the problem of the management of weapons plutonium and uranium it is necessary to take into consideration many factors: the amount of plutonium and uranium in both the military and the civilian sectors, the possibility for using plutonium and uranium, taken out of the arsenals, again for military purposes, the reliability of the accounting and the physical protection of the safety of fissile materials, the political situation in the world, technical possibilities, economical conditions, etc.

It is known that as a result of reduction of military programmes in each of the countries: the USA and Russia, not less than 50 tons of plutonium and several hundred tons of highly-enriched uranium will be released.

In developing an optimal concept of fissile materials management the main criteria should be the safety, the ecological protection of the working personnel, the population and the environment from the radiation impact, as well as reducing financial expenses for the solution of the problem. The main part in the choice of this concept should belong to scientists and specialists, not to populist declarations of politicals. In this respect a great harm and damage in working out the public opinion was caused by the much clamoured article in "The New York Times" of August 19, 1994: "Russia preserves plutonium while the US wants to destroy it". Such a contrariety does not promote the solution of the problem.

How can a fissile material be destroyed by mixing it with wastes or glass? A fissile material may be destroyed only as a result of nuclear transformations. And all the rest is a palliative, a half-measure, a self-deception and a misleading of the public opinion.

An assessment of the amounts of fissile materials shows that the world stock of civil plutonium is much greater than that of military one, and grows much more rapidly, being on the average 60 - 70 tons per year. And this stock will grow annually, irrespective whether Russia, the European Community or other countries will develop programmes of fast breeder reactors. For example, quite recently a store repository for 50 tons of plutonium dioxide has been put into operation by the BNFL company at the THORP plant in Sellafield, UK, which corresponds to the requirements of the Safeguards of the Euratom and the IAEA. Why should the isotopically purer military plutonium

E.R. Merz and C.E. Walter (eds.), Disposal of Weapon Plutonium, 1-7.

be mixed with high-activity raffinate and incorporated into glass, and the high-background plutonium be extracted from spent nuclear fuel from the same raffinate and stored as a material of value? As, according to the statement of Mr. Yohn Gibbons, a scientific councillor of the White House, "plutonium essentially has a negative economical value". Or is Sellafield already in Russia?

CIS- scientists and specialists steadfast adhere to the position that fissile materials, military plutonium and uranium have in them a high-energy potential and should be used for electric power generation for peaceful purposes. Main Russian companies, former producers of military plutonium, refused to participate in the work of this symposium, when they learnt that the NATO scientists propose the mixing of weapons plutonium with high-radioactivity raffinates and their vitrification as advanced technologies in weapons fissile materials management. In Russia, scientific investigations are conducted for the development of efficient ways of dissolution of plutonium metal and plutonium dioxide, for searching new matrices based on glass and ceramics for incorporation into them of the high-background plutonium unsuitable for further use, new methods for equipment decontamination, and other projects. They will be reported at this sysmposium.

It also seems to us that scientific approaches and positions of different countries in weapons fissile materials management are getting closer, and already now one may speak about working out joint recommendations on weapons fissile materials management.

Thus, the Committee for International Security and Arms Control (CISAC), formed in 1980 at the USA Academy of Science, carried out and published an investigation of the problems of the management of fissile materials and released as a result of nuclear arms reduction ("Management and Disposition of Excess Weapons Plutonium", Washington D.C., National Academic Press, 1994). As a result of it, criteria were developed on which basis the CISAC worked out recommendations on the long-term management of weapons and fissile materials. Russian scientists and specialists agree with most stated recommendations, and constantly insist on their observance.

1. Establishing in the USA and Russia a new mutually agreed regime of weapons and fissile materials management, which includes:

- a declaration about stocks of nuclear weapons and all fissile materials;
- joint arrangements for revealing and confirming all the stocks declared;
- an agreement about stoppage of fissile materials production for weapons. It is impossible to agree to this recommendation, because one cannot strictly distinguish fissile materials as to their usage - they are two-purpose. Otherwise the development of nuclear power engineering should be stopped;
- an agreement about the control of the decrease of fissile materials from the declared stocks and the control of the production of fissile materials by international system of safeguards;

- a transfer of surplus fissile materials to the non-military sector to civil reactor fuels. Besides, both existing and new reactors can be used.

2. Guanranteed storage of fissile materials and control so that the materials stored should be taken only for non-military purposes. To guarantee high standards of safety and registration of nuclear materials at all stages of the process.

It is necessary to improve the system of safeguards and to ensure the physical safety of all the forms of plutonium and high-enriched uranium on the world scale.

As plutonium incorporated in MOX fuel or glass does not exclude the risk of its use in nuclear weapons, further steps should be taken for decreasing the long-term risk of its application for military or diversion purposes. This may be fuel cycles more resistant to the risk of nuclear arms proliferation, including such options that could almost fully use the world stocks of plutonium.

3. Russian scientists and specialists still reject as an unacceptable option the fissile materials management by vitrifying them together with high-radioactive wastes of radiochemical plants. Plutonium disposal into deep geological formations may be recommended only in exclusive cases when it is completely unsuitable for further use.

The fulfilment of the above mentioned recommendations will require certain financial expenses that, by our estimations, may be fully recompensed due to using the fissile materials not for military purposes, but as for electric power generation.

However, the main result of managing weapons and fissile materials from their reduction is national and international security.

APPENDIX

to an article by William I. Brod in New York Times International from November 21st, 1994, "Nuclear Roulette for Russia" concerning a Russian Paper entitled:

"Disposal of Liquid Radwaste in Russia: Nuclear Roulette or an Analog of American Technologies?"

Commentators: Dr. Nikolai N. Egorov
Dr. Andrei I. Rybal`chenko
Dr. Elena V. Zakharova

This article and the opinions of some U.S. experts commented in the article have very much surprised Russian participants in the International Symposium on Scientific and Engineering Aspects of Deep Injection Disposal of Hazardous and Industrial Wastes, 10-13 May, 1994, Berkeley, CA, USA.

As follows from this article, the reports presented at this symposium by Russian scientists about injection of liquid radwastes into deep geological formations have become a revelation for American experts, although this technology has been used in Russia for more than 30 years. Reports on this topic have been presented at the symposia of the International Atomic Energy Agency since 1967, and pertinent publications are available from Russian scientific journals. In addition, a large monography on this topic was published in 1994.

In William I. Brod`s article, the International Symposium on the Scientific and Engineering Aspects of Deep Injection of Hazardous and Industrial Wastes, which was sponsored by the U.S. Environmental Protection Agency and the U.S. Department of Energy is called "a small group of western experts". "A small group of western scientists carefully selected for receiving information", etc. This is not actually so.

Similar symposia have been periodically held in the U.S.A., because the technology of deep well injection of toxic liquid wastes has been widely used in the North American continent. This technology of treatment of industrial non-radioactive wastes, which has been employed in the U.S.A. and other countries for more than half a century, has become a prototype for the development of the facilities designed for the disposal of liquid radioactive wastes in Russia. Naturally, the original technology was significantly modified and improved for this specific purpose.

The method of treatment of radwastes, used at the repository sites in Russia, is basically different from that used in Hanford, WA, U.S.A. In Russia, liquid radwastes were injected through drilled wells into permeable layers situated at a depth of several hundred metres and more. These layers are isolated from the surface with layers of water-resistant clay rocks. In Hanford, liquid radwastes were injected directly into soils and surface grounds, which resulted in the contamination of vast areas. Injection well technologies were used in Idaho in unsuitable geological environment (Snake River valley) and New Mexico (Anaconda uranium processing company).

The injection technologies used in Russia for the disposal of liquid radwastes are similar to the technologies used in the U.S.A. and Canada for the disposal of non-radioactive toxic wastes, but the volume of injected radioactive wastes is significantly lower and there is a wide system of control over the state of the Earth`s interior. Injected liquid radwastes are situated in collector layers within the predetermined boundaries of the injection zones. As a result of physicochemical processes, radioactive nuclides are retained on the rocks of these collector layers. There are no reasons to believe that the behavior of injected radwastes is unpredictable.

The article by William I. Brod contains a number of deliberate or casual inaccuracies. Thus, the figure with the scheme of the location of the radwaste repository site in Tomsk differs from that presented in the Russian papers at the Symposium in Berkeley, and is a work of the author`s imagination. In addition, the results of the experimental studies of the distribution of radwastes in collector layers, presented in the Russian papers, are interpreted by the author as data on an unpredictable migration of water.

We believe that the publication of William I. Brod`s article in the New York Times International does not contribute to the development of the cooperation between Russia and the U.S.A. in the important field of nuclear waste management. Russia can give any international experts all necessary information concerning the safety aspects of the Russian technology of deep well injection of liquid radwastes.

About 100 experts from eight countries took part in the Symposium in Berkeley. More than forty reports, mainly by U.S. and Canadian scientists and engineers, on a deep injection of toxic wastes were presented. Russian scientists presented five reports on the disposal of liquid radioactive wastes. From these reports it follows that Russia has three facilities for deep well injection of liquid radioactive wastes.

Extensive studies of the geological structures and properties of host rocks, as well as of wastes were carried out prior to the development and construction of these facilities. These studies were performed by geological organizations and the National Academy of Sciences under the supervision of well-known scientists and highly-qualified experts. The work on the injection of liquid radioactive wastes has been performed in full conformity with the recommendations of the International Atomic Energy.

Thirty-year experience of the operation of the facilities for a deep injection of liquid radwastes has demonstrated a high reliability of this technology, which enabled Russia to minimize the hazardous environmental and human impacts of radwastes. Therefore, the concern expressed by some U.S. scientists about the disposal of liquid radwastes in Russia is not justified.

During discussion at the Symposium in Berkeley, the technologies of deep well injection of radwastes used in Russia, did not arouse any radical criticism. The participants in this Symposium showed a strong interest in these technologies, and proposals were made for joint studies of the Russian experience in this field.

Therefore, William I. Brod is wrong when saying that the Russian technology is contradictory to the generally accepted rules of the disposal of nuclear wastes. Such rules by the way do not exist. Nevertheless, Russia has many times urged that a concept of the solidification of radwastes before their ultimate storage or disposal should be adopted, as required by the International Atomic Energy Agency, and has been working toward this goal. However, the adoption of this concept would not mean that all existing technologies ensuring a reliable protection of people from radioactive wastes should be refused. The more so as the development of waste solidification systems takes much time and resources.

As a supplement to Karelin`s statements the following diagram (Fig.1) illustrates the different stragegies of plutonium disposition.

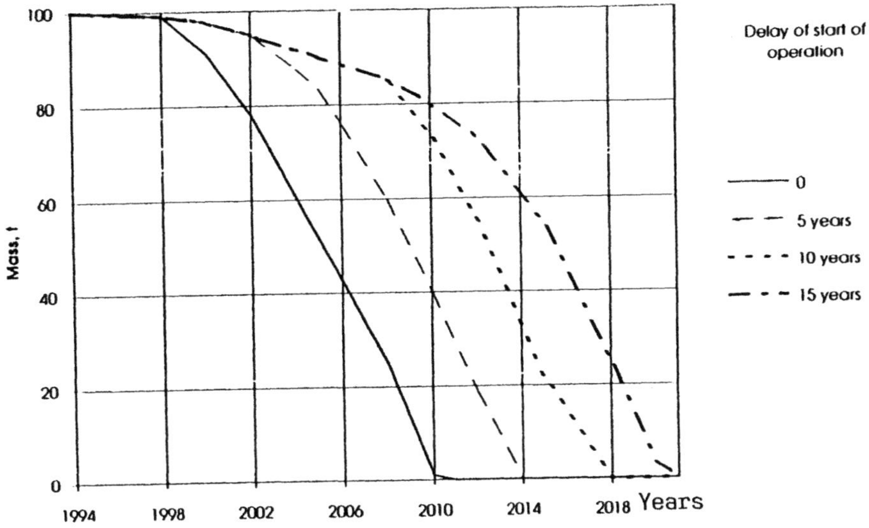

Fig. 1: The evolution of the stock of Pu for a mixed scenario with the delay in the introduction of fast reactors of the types BN-600 and BN-800

RETRIEVABILITY AND SAFEGUARDS CONCERNS REGARDING PLUTONIUM IN GEOLOGICAL REPOSITORIES

Dr. Johan Swahn
Institute of Physical Resource Theory
Chalmers University of Technology/Göteborg University
S-412 96 Göteborg (Gothenburg), Sweden

Abstract. In the paper the following conclusions of relevance for policy-making for plutonium and nuclear waste management are made:

- *There are three main lines of thinking when it come to managing surplus separated plutonium:*
  1. *The plutonium should be converted into a form that satisfies the "spent fuel standard", i.e., the plutonium is to be made as inaccessible for weapons purposes as plutonium contained in un-reprocessed spent fuel from nuclear reactors. The two principal methods for doing this are a) using separated plutonium as MOX fuel in standard nuclear reactors and b) "anti-reprocessing" separated plutonium. In both cases the resulting waste is proposed to be disposed of in mined geologic repositories with similar methods as those being developed for direct final disposal of spent nuclear fuel. A mined geologic repository is a repository at depths of approximately 500 m, where the intention is to keep the content of the repository away from the surface environment for very long time periods, i.e., on the order of 100 000 to 1 million years.*
  2. *The plutonium should be stored while waiting for the possibility to use it for energy supply in a fast breeder plutonium economy.*
  3. *The plutonium should be made irretrievable for all future use.*
- *Regardless of whether we chose to use surplus military and civil plutonium as MOX-fuel or if we decide to "anti-reprocess" it, the result is a material of "spent fuel standard" when it comes to proliferation risks.*
- *When we "convert" separated plutonium into a "spent fuel standard" material we arrive at the same long-term safeguards and retrievability problem that we already have when discussing the final direct disposal of spent fuel in mined geologic repositories.*
- *The costs, risks, and delays of going beyond the "spent fuel standard" to eliminate excess weapons plutonium completely or nearly so would not offer substantial additional security benefits unless society were prepared to take the same approach with the global stock of civil plutonium.*
- *Spent nuclear fuel and "spent fuel standard" material in a mined geologic repository has to be kept under safeguard and be surveilled for very long time periods.*
- *If the fast breeder plutonium economy does become part of a large-scale global energy supply system, there will be no long-term safeguards problem with the plutonium in the waste.*

E.R. Merz and C.E. Walter (eds.), Disposal of Weapon Plutonium, 9-22.

- *There are, however, very difficult nuclear weapon proliferation problems to solve if the plutonium economy is to be realised because the long-term energy carrier is separated high-quality military plutonium.*
- *It is not implausible that the plutonium energy economy never becomes a reality.*
- *If the plutonium economy never becomes a reality, then separated plutonium will be regarded as a waste product that needs to be disposed of or destroyed and we have to decide if we should dispose of it in a way that leaves long-term safeguard concerns or not.*
- *If we decide that we want to avoid long-term safeguard and retrievability concerns and we decide that we are prepared to accept the costs, risks, and delays of disposing all plutonium in such a manner, there exist methods for irretrievable plutonium disposition.*
- *The main contenders for such disposition are 1) transmuting the plutonium, and perhaps all the high-level waste, in "plutonium burners" and 2) putting all the high-level nuclear waste, including plutonium, in very deep boreholes.*
- *If costs for long-term safeguards are included in the costs for disposal in mined geologic repositories, the costs for disposal in very deep boreholes might be found attractive.*
- *The very deep borehole disposal method might be found acceptable also by those who want the plutonium to be retrievable for use in a possible future plutonium economy, as the plutonium would remain retrievable in the borehole until it was decided that it was time to seal the holes up once and for all with "locks".*

1. Introduction

The objective of this paper is to discuss retrievability and safeguard concerns for long-term management of plutonium, especially with regard to final disposal of plutonium in geologic repositories. In order to be able to do this in a systematic manner it is necessary to first establish some basic points.

Firstly, there exists at present an excess of both separated military and civil plutonium in the world. The military surplus is due to far-reaching cuts in the global nuclear weapon arsenals. The civil surplus is due to the non-appearance of the fast breeder plutonium economy that was earlier incorrectly foreseen to be needed today.

Secondly, all forms of plutonium should be regarded with equally serious concern when it comes to risks for nuclear weapons proliferation [1]. This means that it is not meaningful to discuss disposition of military plutonium without also looking at the civil plutonium dilemma.

Thirdly, after the successful conclusion of the 1995 Non-Proliferation Treaty extension conference it is no longer arguable that there can be special "low-safeguard" plutonium disposition solutions in countries that at present have nuclear weapons. The ultimate global disarmament goal, as reconfirmed by the conference, is a nuclear-weapons-free world. In the long term there will thus only exist non-military plutonium in non-nuclear-weapon states and thus no special "low-safeguard" solutions would be acceptable.

Fourth, when discuss "irretrievable" disposal of plutonium we are always, unless the plutonium is actually physically destroyed in a nuclear process, discussing retrievability in relative terms. When we use the word "irretrievable" in this paper, however, we mean that the material is difficult enough to retrieve so that safeguard measures can be removed [2]. Just how high a level of irretrievability is needed in practice depends on alternative means of getting access to fissile material for nuclear explosives. These alternative means may be highly varying in both time and place and are highly dependent upon the presence or lack of large-scale use of military or civil nuclear technology in a future society.

Fifth, there are three main lines of thinking when it come to managing surplus separated plutonium:

1. The plutonium should be converted into a form that satisfies the "spent fuel standard", i.e., the plutonium is to be made as inaccessible for weapons purposes as plutonium contained in un-reprocessed spent fuel from nuclear reactors. The two principal methods for doing this are a) using separated plutonium as MOX fuel in standard nuclear reactors and b) "anti-reprocessing" separated plutonium. In both cases the resulting waste is proposed to be disposed of in mined geologic repositories with similar methods as those being developed for direct final disposal of spent nuclear fuel. A mined geologic repository is a repository at depths of approximately 500 m, where the intention is to keep the content of the repository away from the surface environment for very long time periods, i.e., on the order of 100 000 to 1 million years.
2. The plutonium should be stored while waiting for the possibility to use it for energy supply in a fast breeder plutonium economy.
3. The plutonium should be made irretrievable for all future use.

When analysing long-term retrievability and safeguard concerns for plutonium disposition these methods lead to very different, in fact fundamentally different, results.

In the paper we will discuss all of the three lines of thinking described above. Firstly, we will discuss, at some length, important safeguard concerns for the disposal in mined geologic repositories of plutonium satisfying criteria for the "spent fuel standard". Secondly we will shortly discuss, while elaborating on possible long-term options for global energy use, the safeguard concerns of the fast breeder plutonium economy. Thirdly we will briefly examine some of the possibilities to avoid long-term safeguards of plutonium by irretrievable disposal of the plutonium. In this final section an attempt is made to make a synthesis with conclusions from the previous sections.

In each of the sections we will towards the end summarise the conclusions of interest for policy-making for plutonium and nuclear waste management.

2. Safeguarding "spent fuel standard" plutonium in mined geologic repositories

When discussing military plutonium disposition, it is vital that the present state of the military *and* civil nuclear complex is taken into account. Such an analysis easily leads to the conclusion that there are two main ways to dispose of the military plutonium because they, in principle, do not drastically differ from how plutonium is handled in today's civil nuclear fuel cycles. These two options are:

1. Use the plutonium as MOX fuel in the present civil nuclear reactors, and then dispose of the spent MOX fuel in mined geologic repositories, or,

2. Treat the plutonium as waste, "anti-reprocess" it, and then dispose of the resulting waste in mined geologic repositories.

These options are seemingly each others opposites in the way plutonium is viewed, as a waste or as a resource. Which of these methods is preferably is an extremely infected issue [3]. However, in both cases the result is a highly radioactive waste containing plutonium, and the plutonium has been "converted" into a form that satisfies the "spent fuel standard". The "spent fuel standard" was introduced in a U.S. National Academy of Science report on plutonium management and disposition in 1994 and is defined as "making the plutonium roughly as inaccessible for weapons use as the much larger and growing stock of plutonium in civilian spent fuel" [4]. The plutonium is thus to be made as inaccessible for weapons purposes as plutonium contained in un-reprocessed spent fuel from nuclear reactors. The report goes on to state that:

> "Options that left the weapons plutonium more accessible would mean that this material would continue to pose a unique safeguards problem indefinitely. Conversely, the costs, risks, and delays of going beyond the spent fuel standard to eliminate excess weapons plutonium completely or nearly so would not offer substantial additional security benefits *unless society were prepared to take the same approach with the global stock of civil plutonium* [emphasis in original].

This seems reasonable. One can discuss whether the plutonium should be used in MOX fuel or whether it should be "anti-reprocessed", but the long-term result is that both methods leave us with a material that is similar to spent fuel when it comes to long-term retrievability and safeguard concerns. What are those concerns?

2.1. PROPOSED SAFEGUARD MEASURES FOR FINAL GEOLOGIC REPOSITORIES FOR CIVIL SPENT FUEL

There are several countries that have opted not to reprocess spent nuclear fuel, but to instead dispose of the spent fuel in mined geologic repositories. A mined geologic repository is a repository at depths of approximately 500 m, where the intention is to keep the content of the repository away from the surface environment for very long time periods, i.e., on the order of 100 000 to 1 million years. Apart from long-term environmental security concerns, there are also long-term concerns that the plutonium in such a repository shall be used for the construction of nuclear weapons. Safeguards are applied to the nuclear fuel when it is in the reactor and while it is stored awaiting final disposal. Can safeguards be removed after the mined geologic repository has been sealed shut?

Let us first study what the present safeguard regime under auspices of the International Atomic Energy Agency (IAEA) has to say on the subject. There are two documents that give the structure and content of the safeguard agreements that are entered between the IAEA and separate countries. The first, INFCIRC/66 [5], was first written in 1966 and is still used by some countries not party to the Non-proliferation Treaty (NPT) and the second, INFCIRC/153 [6], was written in 1970 to be used by countries that have signed the NPT. Concerning the possibility of removing safeguarded material from safeguards, i.e. the termination of safeguards, INFCIRC/66 states in paragraph 26 (c):

> "Nuclear material shall no longer be subject to safeguards after The Agency has determined that it has been consumed, or has been diluted in such a way that it is no longer usable in any nuclear activity relevant from the point of view of safeguards, or has become practicably irrecoverable."

INFCIRC/153 states, on the same issue, in paragraph 11 that:

> "The Agreement should provide that safeguards shall terminate on nuclear material subject to safeguards thereunder upon determination by the Agency that it has been consumed, or it has been diluted in such a way that it is no longer usable for any nuclear activity relevant from the point of view of safeguards, or has become practicably irrecoverable."

We, and others, use the term irretrievable instead of practicably irrecoverable. If the plutonium in a spent fuel repository is considered irretrievable then no safeguards need to be applied. If not, then some form of safeguards will need to be applied "indefinitely".

Construction of the first mined repository for spent fuel will not be started until in a decade's time or more. Nevertheless the IAEA Department of Safeguards has started to examine the possibility of removing safeguards from a closed mined geologic repository [7]. Two IAEA reports on the subject have been produced as a result of an advisory group and a consultants meeting [8]. The result of these reports can be summarised by quoting an interview with Bruno Pellaud, the IAEA's deputy director general for safeguards [9]:

> "Safeguards on spent fuel, Pellaud said, obviously cannot ever be terminated. He said: 'A spent fuel repository has been likened to a plutonium mine'."

This does not mean that one needs to have intrusive monitoring. The use of remote sensors and satellite monitoring may be possible. But, evidently some sort of surveillance has to be carried out for a very long time. The half-life of plutonium-239 is 24 100 years, so if we are discussing a reduction in plutonium concentration by a factor of a thousand (approximately ten half-lives) we are talking about time scales comparable to those involved with regard to environmental security.

We have thus found that long-term safeguards need to be applied to plutonium in spent fuel in mined geologic repositories. This is then, of course, also true for other nuclear waste that conforms to the "spent fuel standard". This means that both spent MOX fuel and "anti-reprocessed" plutonium that is stored in mined geologic repositories need long-term safeguarding.

2.2. LONG-TERM CHANGES IN THE CHARACTERISTICS OF "SPENT FUEL STANDARD" MATERIAL IN GEOLOGIC REPOSITORIES

The characteristics of the "spent fuel standard" material that is put in a final repository changes with time. Some of these changes are relevant when discussing nuclear explosives proliferation concern.

Firstly, there is a decay of the radiation in the high-level waste. This "radiating shield" is often considered to be an important barrier against proliferation. Figure 1 shows how the gamma source rate decrease with time in spent nuclear fuel. Spent MOX fuel and "anti-reprocessed" plutonium would show similar characteristics. One can see that the radiation levels fall fast. After one thousand years, the source rate has fallen by a factor of 10 000. The radiation protection needed to process such material is very much less difficult to achieve compared to processing fresh spent nuclear fuel.

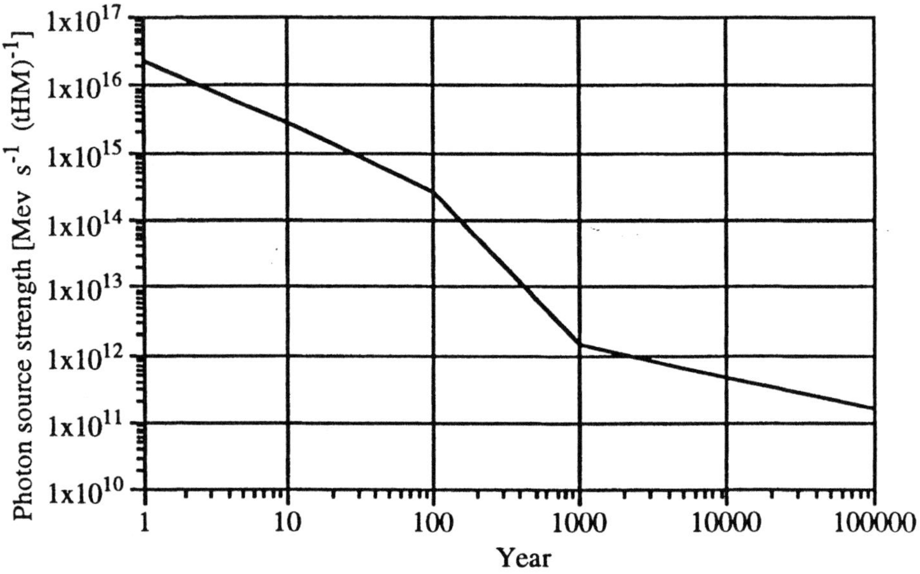

Figure 1. Photon source strength in spent nuclear power reactor fuel

Secondly, the characteristics of the plutonium in the repository changes in ways that are beneficial to the design of a nuclear explosive. The different plutonium isotopes have different half-lives and the heat production in and surface dose from a specific mass of plutonium decreases with time.

2.3. THE ETHICAL DILEMMA: SHOULD WE INSPECT "FOREVER"?

The ethics of nuclear waste disposal has been discussed at length. In Sweden the KASAM principles were developed in the late 1980s and can be summed up as follows [10]:

> "A repository should be constructed so that it makes controls and corrective measures unnecessary, while at the same time not making

> controls and corrective measures impossible. In other words, our generation should not put entire responsibility for maintenance of repositories on coming generations; however, neither should we deny coming generations the possibility of taking control."

In short, we should not leave any burdens on future generations, nor should we make it difficult for them mend or improve the repository. This has since then become the mainstream thinking in Sweden. In Sweden the present planning is to dispose of the spent fuel in mined geologic repositories at a depth of about 500 m in granite rock. The spent fuel rods are to be encapsulated in copper and steel and then enclosed in bentonite clay before sealing the repository.

However, it must be clearly stated that *only* environmental issues had been discussed when these principles were developed. This has later proven to be a problem. After all, the necessity to carry out long-term surveillance due to a safeguard concerns contradicts these principles. Likewise one can discuss whether it is really a positive aspect that future generations can carry out maintenance on the repository if this also allows them to become nuclear weapon proliferators.

2.4 POLICY CONSIDERATIONS DUE TO SAFEGUARD AND RETRIEVABILITY CONCERNS FOR DISPOSAL OF "SPENT FUEL STANDARD" MATERIAL IN MINED GEOLOGIC REPOSITORIES

The following policy considerations can be made:

- Regardless of whether we chose to use surplus military and civil plutonium as MOX-fuel or if we decide to "anti-reprocess" it, the result is a material of "spent fuel standard" when it comes to proliferation risks.
- When we "convert" separated plutonium into a "spent fuel standard" material we arrive at the same long-term safeguards and retrievability problem that we already have when discussing the final direct disposal of spent fuel in mined geologic repositories.
- The costs, risks, and delays of going beyond the "spent fuel standard" to eliminate excess weapons plutonium completely or nearly so would not offer substantial additional security benefits unless society were prepared to take the same approach with the global stock of civil plutonium.
- Spent nuclear fuel and "spent fuel standard" material in a mined geologic repository has to be kept under safeguard and be surveilled for very long time periods.

3. The fast breeder plutonium economy: Safeguard concerns and prospects

When we in the introduction described the three main lines of thinking when it comes to disposing of surplus separated plutonium we said that the second line was that the plutonium should be stored while waiting for the possibility to use it for energy supply in a fast breeder plutonium economy. Strong nuclear proponents often point out that in their view it is a great pity to see plutonium be converted into waste, whether it be done as a result of decisions to opt for direct disposal of spent fuel or whether it is a result of using separated plutonium in MOX fuel or of "anti-reprocessing" it. They would like to see the plutonium used as a long-term energy source and see the fast breeder plutonium economy as the only reasonable large-scale and long-term nuclear future. This argumentation comes from the realisation that the natural uranium that fuels the present generation of reactors will relatively soon be exhausted, especially if nuclear energy supply is rapidly increased from today's level of approximately 5% of the global energy supply.

If the fast breeder plutonium economy does become part of a large-scale global energy supply system, there will not be a long-term safeguards problem with plutonium in waste, as the plutonium will instead be used as fuel. There are, however, difficult nuclear weapon proliferation problems to solve when implementing the plutonium economy, much larger difficulties than those encountered in today's nuclear activities. After a transition period when reactor-grade plutonium from the present nuclear reactors would be recycled, the use of fast plutonium breeder reactors will introduce plutonium of high military quality into the cycle and in a long-term system only plutonium of high military quality will be used as the energy carrier [11]. Proponents of a long-term plutonium energy economy are thus advocating the use of separated high-quality military plutonium as an energy carrier in a long-term global energy system. This is an issue to be seriously concerned about when one considers nuclear proliferation concerns of long-term energy futures.

However, it is not at all implausible that the plutonium energy economy never becomes a reality. The present developments within the global energy system points in other directions [12]. There is a definite shift from increasing energy supply towards increasing efficiency in use of energy. The economic competitiveness of renewable energy sources is improving at a fast rate [13]. If external costs for environmental degradation were included in the costs for using fossil and nuclear fuels, the comparative economics of using renewables would improve even faster [14].

Towards the middle of the next century, fossil fuels will start to become more scarce and prices for energy will start to rise. However, drastic reductions in use of fossil fuels may be necessary much earlier in order to avoid catastrophic climatic change. Renewable energy sources will in either case compete with nuclear energy as a replacement for fossil fuel use. The first scenarios, albeit very preliminary, of a global energy system using only renewable energy resources are just emerging [15]. In such scenarios no nuclear energy, and thus no plutonium, is used. Another scenario can be developed where renewable energy sources are complemented with non-plutonium recycle nuclear energy cycles, for instance the Th-232/U-233 breeder nuclear fuel cycle [16].

There are thus at least two possible scenarios where the plutonium energy economy does not develop as a viable long-term energy system. If these scenarios are realised, then separated plutonium will be regarded as a waste product that needs to be disposed of or destroyed. And, we are back to where we were in the previous section. Should we dispose of it in a way that leaves long-term safeguard concerns or not?

The following policy considerations can thus be made:

- If the fast breeder plutonium economy does become part of a large-scale global energy supply system, there will be no long-term safeguards problem with plutonium in waste.

- There are, however, very difficult nuclear weapon proliferation problems to solve if the plutonium economy is to be realised because the long-term energy carrier is separated high-quality military plutonium.
- It is not implausible that the plutonium energy economy never becomes a reality.
- If the plutonium economy never becomes a reality, then separated plutonium will be regarded as a waste product that needs to be disposed of or destroyed and we have to decide if we should dispose of it in a way that leaves long-term safeguard concerns or not.

4. Long-term irretrievable plutonium disposition

The third main line of thinking regarding plutonium disposition options that were presented in the introduction was that the plutonium was to be made irretrievable for all future use. It is in this direction we should look if we do not believe that safeguarding mined geological repositories for very long time-scales is an attractive image of the future. However, we should keep in mind that we earlier concluded that these methods should not be used if we did not decide the costs, risks, and delays of disposing *all* plutonium in this manner were acceptable.

There are two principal methods for achieving irretrievable plutonium disposition. The first is to physically convert the plutonium into something else. The second is to place the plutonium somewhere where it would be so hard to retrieve that it can be regarded as irretrievable.

Physically destroying the plutonium can be done by transmutation of the plutonium in a "plutonium burner". A "plutonium burner" is a nuclear reactor that fissions plutonium without the creation of new plutonium. An example is the operation of a fast breeder reactor with a plutonium core and without a breeding mantle for new plutonium production [17]. Another example are accelerator-driven reactors that also allow the burning of plutonium without the creation of new plutonium [18].

A common proposal for placing plutonium in a place where it would be difficult to retrieve it from is to send it into space. The plutonium would be launched into earth orbit and later placed in a solar orbit. This proposal, albeit for nuclear waste, has been evaluated by NASA in two studies in the beginning of the 1980s [19]. The nuclear waste would be loaded into a special container that is designed to withstand the consequences of launch accidents. The load would then be launched into low earth orbit and later transferred to a solar orbit. Even though it may sound tempting to suggest that the plutonium should be sent into the sun itself, where it would be destroyed, this would require between two to three times as much energy. There are presently severe environmental problems to overcome before the space disposal proposal can be realised, mainly due to the risk for accidents. The use of the space disposal method relies to a large extent on the reliability of the launch vehicles, even if some care is taken to minimise environmental impact of a launch failure. Today, the reliability of launch vehicles is much too low.

A second proposal for making the separated plutonium irretrievable is to disposed of it in very deep drilled boreholes, at depths exceeding 2-3 kilometres [20]. The principle behind this proposal is that it would be possible to make the re-drilling of the hole very difficult and retrievability would be low. This is, however, not the case unless special measures are taken to prevent re-drilling. Re-drilling of a plugged hole that had not been specially prepared would be fairly straightforward. But, if techniques were developed to use wedges, rock fragmentation and post-closure concealment to make retrievability more difficult, the method might meet high standards for irretrievability. A schematic drawing of a deep borehole repository is shown in figure 2, where it is compared to a basic mined final repository for high-level nuclear waste or spent nuclear fuel.

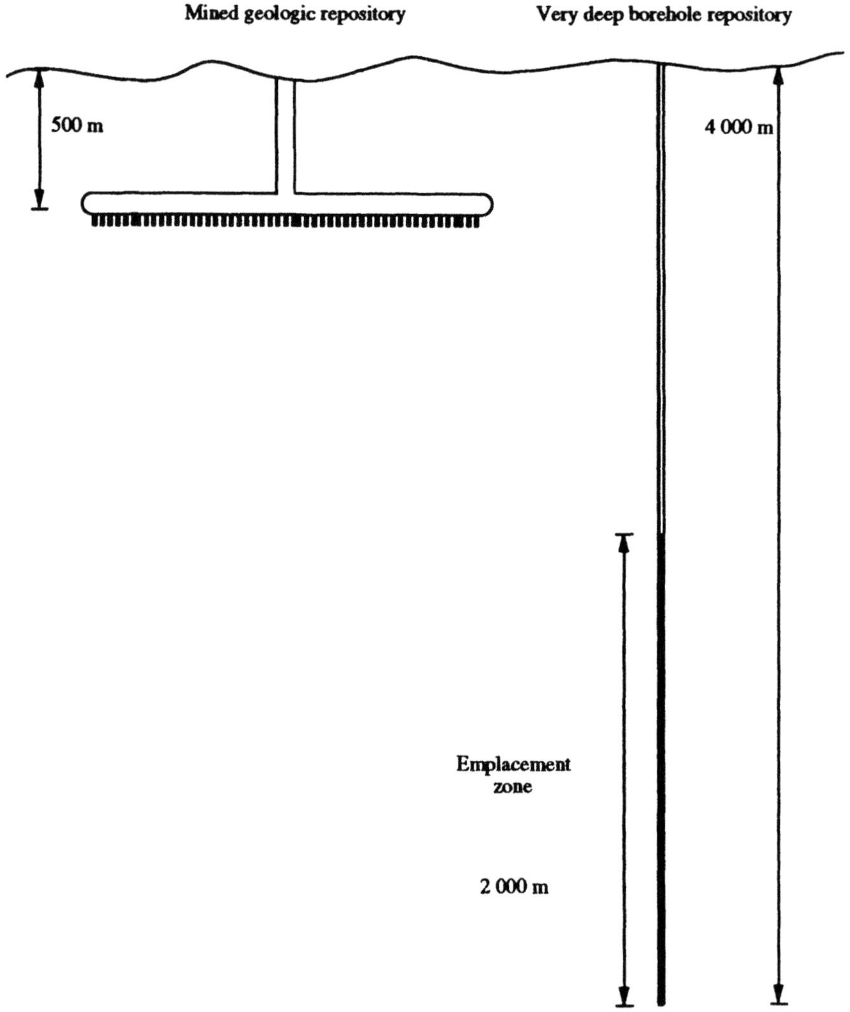

Figure 2. Very deep borehole repository.

When it comes to irretrievable disposal of plutonium we have earlier emphasised that these methods only become interesting if we are prepared to go one step further than meeting the "spent fuel standard" and disposing of the material in mined geologic repositories. Let us now imagine that this will at some time in the future be the case. IN such a future we may have been convinced that it is not a good idea to leave a safeguards burden on future generation by leaving plutonium in mined geologic repositories. We may have a nuclear-weapon-free world and a world where plutonium is not used in nuclear fuel cycles and we wish to eliminate the future global long-term safeguard costs for plutonium once and for all.

We then have the choice of transmuting the plutonium, and perhaps all the high-level waste, in plutonium burners or putting all the high-level nuclear waste including plutonium, in very deep boreholes [21]. Building transmuting reactors may be possible. So may be the use of very deep boreholes.

Are there any especially attractive features with the deep borehole method? There may be. Firstly, the method may prove to be relatively inexpensive. It is possible that costs may not be much higher than disposal in mined geologic repositories and that this higher cost may easily be absorbed when comparing with the very high future costs of surveillance of a mined repository. Secondly, the method could be found acceptable also by those who want the plutonium to be retrievable for use in a possible future plutonium economy. We have previously stated that using the very deep boreholes method does not make an irretrievable disposal option unless active methods are employed to prevent re-drilling. The plutonium would remain retrievable in the borehole until it was decided that it was time to seal the holes up once and for all by activating these "locks".

The following policy considerations can thus be made:

- If we decide that we want to avoid long-term safeguard and retrievability concerns and we decide that we are prepared to accept the costs, risks, and delays of disposing *all* plutonium in such a manner, there exist methods for irretrievable plutonium disposition.
- The main contenders for such disposition are 1) transmuting the plutonium, and perhaps all the high-level waste, in "plutonium burners" and 2) putting all the high-level nuclear waste, including plutonium, in very deep boreholes.
- If costs for long-term safeguards are included in the costs for disposal in mined geologic repositories, the costs for disposal in very deep boreholes might be found attractive.
- The very deep borehole disposal method might be found acceptable also by those who want the plutonium to be retrievable for use in a possible future plutonium economy, as the plutonium would remain retrievable in the borehole until it was decided that it was time to seal the holes up once and for all with "locks".

Notes

1. The possibility of using different grades of plutonium for the construction of nuclear devices is now widely acknowledged. For more insight into this subject the reader is foremost referred to Swahn (1992), Mark (1993), Garwin (1994), DOE (1994), Hiruo & Marshall (1994) and Cochran & Paine (1994). The existence of laser isotope separation technology for converting any form of plutonium to high-purity military plutonium only aggravates the problem (Marshall 1989).

2. The International Atomic Energy Agency (IAEA) uses the term "practicably irrecoverable" as a standard for allowing removal of safeguards from fissile material.
3. An analysis of this issue can be found in Bunn (1994). For an analysis of the use of plutonium in light water reactors, see Murogov et al. (1994), Salvatores (1994), Novikov et al. (1994) and Merz (1994). For an analysis of direct disposal options for separated plutonium, see Swahn (1994).
4. NAS (1994, p. 34).
5. The document was later revised twice and is now called INFCIRC/66/Rev. 2 (IAEA 1968).
6. IAEA (1972).
7. Linsley & Fattah (1994).
8. IAEA (1988) and IAEA (1991).
9. Seneviratne (1994).
10. From KASAM (1987, p. 91) as translated in SKN (1988B, p. 15).
11. In the long run fast plutonium breeder reactors will use as fuel plutonium of the isotope composition of the breeding blanket. A Superphenix type radial blanket irradiated for 39 months of full power contains 96.7% Pu-239, 3.2% Pu-240 and 0.1% Pu-241 (Salvatores 1994).
12. See Grubb (1994) for an analysis of these developments.
13. See, for example, Johansson et al. (1993), WEC (1994) and Flavin & Lenssen (1994).
14. See, for example, Hohmeyer (1992).
15. See Lazarus et al. (1993) and Flavin & Lenssen (1994)
16. For a discussion of this fuel cycle, which has been analysed as being more proliferation-resistant then the plutonium breeder fuel cycle, see Williams & Feiveson (1990).
17. See, for example, MacLachlan (1993).
18. See, for example, Bowman (1994).
19. Rice & Priest (1980) and NASA (1982).
20. The very deep borehole disposal method has been examined as an alternative for Swedish spent fuel disposal (SKB 1989).
21. If one were to use space disposal we would be very tempted to bring down costs by only sending out the plutonium into space and dispose of the high-level reprocessing waste below the Earth's surface.

References

Bowman, Charles D. 1994. "Weapons and Commercial Plutonium Ultimate Disposition Choices – 'Destroy' Completely or Store 'Forever'?", pp. 125-138 in Garwin, R.L., Michael Grubb & Emma Matanle, eds. *Managing the Plutonium Surplus: Applications and Technical Options*. Kluwer Academic Publishers.

Bunn, Matt 1994. "Management and Disposition of excess Weapons Plutonium", pp. 59-75 in Garwin, R.L., Michael Grubb & Emma Matanle, eds. *Managing the Plutonium Surplus: Applications and Technical Options*. Kluwer Academic Publishers.

Cochran, Thomas B. & Christopher E. Paine 1994. *The Amount of Plutonium and Highly-Enriched Uranium Needed for Pure Fission Nuclear Weapons*. August 22. Washington D.C.: Natural Resources Defense Council.

DOE 1994. *Additional Information Concerning Underground Nuclear Weapon Test of Reactor-Grade Plutonium*. DOE Facts. Openness Press Conference Fact Sheets. June 27. U.S. Department of Energy.

Flavin, Nicholas & Nicholas Lenssen 1994. *Power Surge: A Guide to the Coming Energy Revolution*

Garwin, Richard L. 1994. "Explosive Properties of Various Types of Plutonium", pp. 15-22 in Garwin, R.L., Michael Grubb & Emma Matanle, eds. *Managing the Plutonium Surplus: Applications and Technical Options*. Kluwer Academic Publishers.

Grubb, Michael 1994. "Energy from Nuclear Fission Over the Very Long Term – U-238 and Th-232: A Commentary", pp. 121-124 in Garwin, R.L., Michael Grubb & Emma Matanle, eds. *Managing the Plutonium Surplus: Applications and Technical Options*. Kluwer Academic Publishers.

Hiruo, Elaine & Pearl Marshall 1994. "Reactor-Pu Used in 1962 Bomb Test Highlights Proliferation Concerns", *Nucleonics Week*, June 30, pp. 12-13.

Hohmeyer, O. 1992. "Renewables and the Full Costs for Energy", *Energy Policy* , vol. 20, pp. 365-375.

IAEA 1968. *The Agency's Safeguards System (1965, As Provisionally Extended in 1966 and 1968)*. Information Circular INFCIRC/66/Rev. 2. 16 September. International Atomic Energy Agency.

IAEA 1972. *The Structure and Content of Agreements Between the Agency and States Required in Connection with the Treaty on the Non-proliferation of Nuclear Weapons*. INFCIRC/153. June. International Atomic Energy Agency.

IAEA 1988. *Advisory Group Meeting on Safeguards Related to Final Disposal of Nuclear Material in Waste and Spent Fuel (AGM-660)*. System Studies Section. Department of Safeguards. Report STR-243 (Revised). December. International Atomic Energy Agency. Vienna.

IAEA 1991. *Consultants' Report on Safeguards for Final Disposal of Spent Fuel in Geological Repositories*. Department of Safeguards. Report STR-274, International Atomic Energy Agency. Vienna.

Johansson, T.B., Kelly, H., Reddy, A.K.N. & Williams, R.H. (eds.) 1993. Renewable *Energy: Sources for Fuels and Electricity*, Washington D.C.: Island Press.

Lazarus M., et al. 1993. Towards *a Fossil Free Energy Future: The Next Energy Transition*. A technical analysis for Greenpeace International by the Stockholm Environmental Institute – Boston Centre. Amsterdam: Greenpeace.

KASAM 1987. *Kunskapsläget på kärnavfallsområdet 1987 [The State of Knowledge on the Nuclear Waste Front 1987, in Swedish]*. Samrådsnämnden för kärnavfallsfrågor [Consultative Committee for Nuclear Waste Management], Stockholm.

Lindslay, Gordon & Abdul Fattah 1994. "The Interface Between Nuclear Safeguards and Radioactive Waste Disposal". *IAEA Bulletin*. vol. 36, no. 2, pp. 22-26.

MacLachlan, Ann 1993. "Superphenix backed as Actinide Burner, Phenix Gets Power Hike, *Nucleonics Week*, January 7, pp. 1, 10-11.

Mark, J. Carson 1993. "Explosive Properties of Reactor-Grade Plutonium", *Science & Global Security*, vol. 4, no. 1, pp. 111-128.

Marshall, Eliot 1989. "Bomb Factories of the 21st Century", *Science*, vol. 243, no. 4889, January 20, p. 305.

Merz, Erich R. 1994. "Disposition of Plutonium from Nuclear Weapons", pp. 97-104 in Garwin, R.L., Michael Grubb & Emma Matanle, eds. *Managing the Plutonium Surplus: Applications and Technical Options*. Kluwer Academic Publishers.

Murogov, V.M., V.S. Kagramanian & N.S. Rabotnov 1994. "The Use of Weapon and Reactor Plutonium in WWER and BN Type Reactors", pp. 77-84 in Garwin, R.L., Michael Grubb & Emma Matanle, eds. *Managing the Plutonium Surplus: Applications and Technical Options*. Kluwer Academic Publishers.

NAS 1994. *Management and Disposition of Excess Weapons Plutonium*. Committee on International Security and Arms Control. National Academy of Sciences. Washington D.C.: National Academy Press.

NASA 1982. *Analysis of Space Systems for the Space Disposal of Nuclear Waste: Follow-on Study, Final Report*. 2 vol. NASA-CR-161991 & 161991. National Aeronautics and Space Administration. Boeing Aerospace Company.

Novikov, A.N., V.V. Saprykin, A.A. Suslov & A.P. Lazarenko 1994. "Use of MOX (R-Pu and W-Pu) Fuel in VVER-1000", pp. 93-96 in Garwin, R.L., Michael Grubb & Emma Matanle, eds. *Managing the Plutonium Surplus: Applications and Technical Options*. Kluwer Academic Publishers.

Rice, E.E. & C.C. Priest 1980. *Analysis of Nuclear Waste Disposal in Space -- Phase III, Final Report*. 2 vol. NASA-CR-161418 & 161419. National Aeronautics and Space Administration, Batelle Columbus Laboratories.

Salvatores, M. 1994. Commissariat à l'energie atomique (CEA). Cadarache. Personal communication.

Salvatores, M. 1994. "Pu Utilisation: Recycling in PWRs and Fast Neutron Reactor Perspective", pp. 85-92 in Garwin, R.L., Michael Grubb & Emma Matanle, eds. *Managing the Plutonium Surplus: Applications and Technical Options*. Kluwer Academic Publishers.

Seniviratne, Gamini 1994. "AEA Developing Safeguard Conditions for Vitrified Waste, Repositories", *Nuclear Fuel*, September 12, pp. 9-10.

SKB 1989. *Storage of Nuclear Waste in Very Deep Boreholes: Feasibility Study and Assessment of Economical Potential*. Technical Report 89-39. December. Svensk Kärnbränslehantering AB [Swedish Nuclear Fuel and Waste Management Co.]. (In English)

SKB 1993. *SKB Annual Report*. Swedish Nuclear Waste and Management Co. Stockholm.

SKN 1988. *Ethical Aspects on Nuclear Waste – Some salient points discussed at a "Seminar on Ethical Action in the Face of Uncertainty"*, Stockholm, September 8-9, 1987. SKN Report 29. April. National Board for Spent Nuclear Fuel [Statens Kärnbränslenämnd]. (English translation of Chapter 6 of KASAM (1987).)

Swahn, Johan 1992. *The Long-term Nuclear Explosives Predicament: The Final Disposal of Militarily Usable Fissile Material in Nuclear Waste from Nuclear Power and from the Elimination of Nuclear Weapons*. Institute of Physical resource Theory. Chalmers University of Technology/Göteborg University. Göteborg.

Swahn, Johan 1994. "Direct Disposal Options for Separated Plutonium", pp. 47-57 in Garwin, R.L., Michael Grubb & Emma Matanle, eds. *Managing the Plutonium Surplus: Applications and Technical Options*. Kluwer Academic Publishers.

WEC 1994. *New Renewable Energy Resources: A Guide to the Future*. World Energy Council. London: Kogan Page.

Williams, R.H. & H.A. Feiveson 1990 "Diversion-resistance criteria for future nuclear power"; *Energy Policy*, vol. 18, no. 6, July/August, pp. 543-549.

MULTIBARRIERS: THEIR EFFICACY AGAINST INTENTIONAL INTRUSION INTO A GEOLOGICAL REPOSITORY TO RETRIEVE DISPOSED PLUTONIUM

Erich R. Merz
Forschungszentrum Jülich GmbH, KFA
P.O.Box 1913, D-52425 Jülich, Germany

1. INTRODUCTION

One of the new and urgent challenges in nuclear waste management is the disposal of plutonium recovered from nuclear weapons. Under the first and second Strategic Arms Reduction Treaties(START I/II), as well as unilateral pledges made by both the United States and Russia, several thousand nuclear weapons will be dismantled. This will result in an estimated 100 metric tons of weapons plutonium that will require long-term disposal. The disposal strategy should not only protect the public and the environment, but should also ensure that the plutonium is not readily recoverable for use in weapons.

The primary goal of any plutonium disposition option is to exclude or at least to minimize the possibility that discarded ex-weapons plutonium could be reused as strategic arms. Indefinite storage of the recovered plutonium from the dismantled warheads in strongly secured storage facilities indeed represents an acceptable option. However, its reliability can only be guaranteed for several decades at the maximum, because indefinite storage is encumbered with a number of grave liabilities.

It is not just the risk of breaking-in and theft on a short time scale where the urgency is obvious; in the long term either a physical destruction of the plutonium by fission or transmutation is required, or its disposal in a form and location where it is intended to remain indefinitely.

This latter approach represents the concept of so-called minimized accessibility in which physical, chemical or radiological barriers reduce the availability of plutonium for weapons use. The most promising concept seems to be the vitrification of the dissolved plutonium after mixing it with a highly radioactive fission product solution stemming from spent fuel reprocessing and the disposal of the resulting glass canisters in a deep underground geological repository.

E.R. Merz and C.E. Walter (eds.), Disposal of Weapon Plutonium, 23-35.

2. DISPOSITION ALTERNATIVES

Since the plutonium recovered from nuclear weapons is endangering national and global security, two alternative approaches have been itentified as viable solutions to tackle the problem:

1. using the plutonium as fuel (MOX) for nuclear electric power plants and subsequent direct disposal of the spent fuel elements;
2. vitrification the plutonium together with the high-level wastes that inevitably accompany its production, with subsequent burial.

In both cases, the high radioactivity of the resulting materials is like "self-shielding", and prevents separation of plutonium without developed technology.

It is necessary to take a closer look at these two options whether they represent a real alternative in a genuine sense. The schematic scheme, depicted in Figure 1 reveals that they fulfill the pretension only partially.

After dismantling of the warheads the recovered plutonium has to be stored safely for an interim period of time in both cases before either the MOX-route or the vitrification-route comes into force. The resulting products are of comparable quality with regard to their effective protection barriers against a succeeding misuse. In either event, the succeeding steps are characterized by a second interim storage period prior of an ultimate disposal in a well-suited deep underground geological repository, because a suitable underground repository may not be available before the year 2025.

The associated risks for the two options show only technical differences for a time frame in the range of 20 to 40 years. All other spaces of time, in particular the very long term geological disposal with its dominant overall risk contribution, are more or less equal.

The choice of the preferred route (MOX versus vitrification) must be decided upon employing additional critera, e.g. resource preservation, availability of techniques, maturity, time element, cost.

Genuine alternative disposition options may only be realized following either the futuristic options:

- Multiple reprocessing and isotpe transmutation;
- Production of very resistant Pu-ceramics with subsequent very deep borehole disposal or launching into outer space.

Neither one of these options will be mature before a rather long time. Much research and development work as well as a practical demonstration are required until their feasibility may be proven.

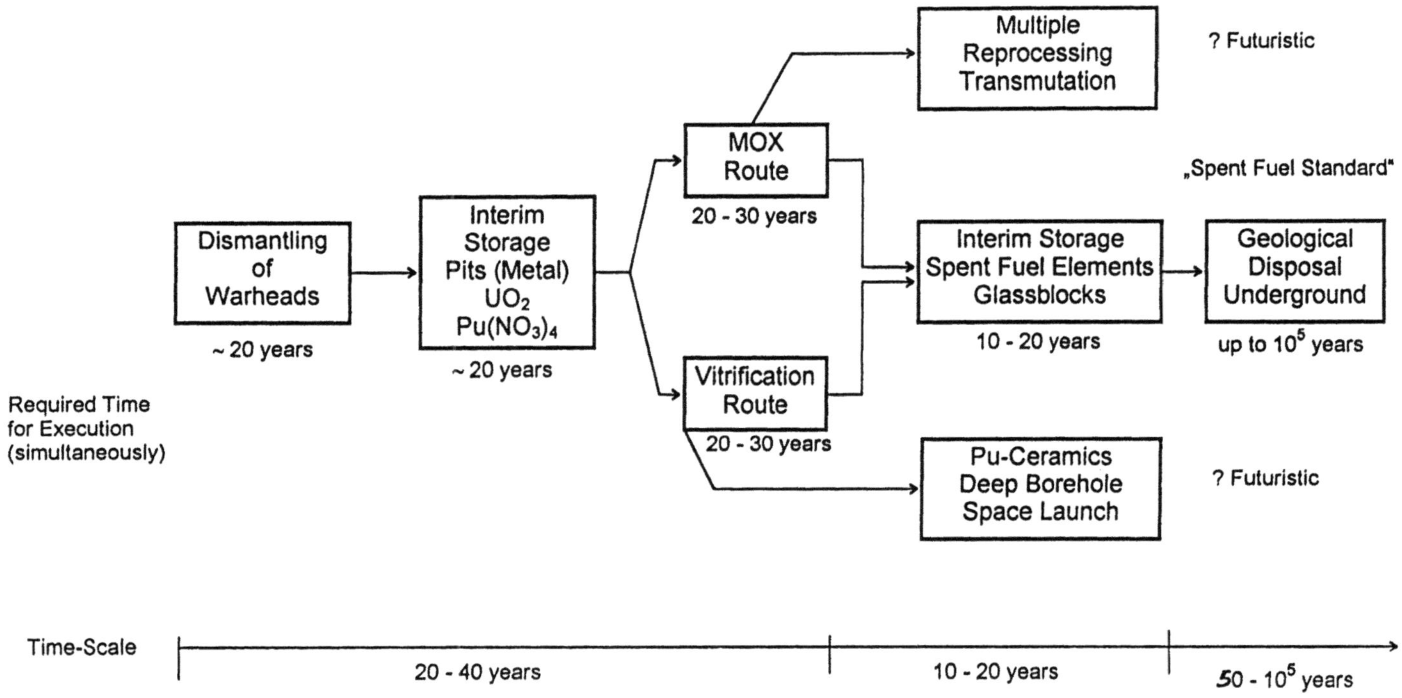

<u>Fig. 1</u>: Plutonium Disposition Alternatives

3. THE MULTIBARRIER PRINCIPLE

Mined repositories in deep underground geologic formations have been identified as the preferred disposal method for spent reactor fuel, solidified high-level waste, and transuranic elements. A major consideration of such a disposal concept is to supply protective barriers against the diffusion of radioactivity products and radiation. The idea of having a series of barriers for the purpose of protecting burried radioactive waste has been called the "multibarrier concept". In the majority of conceptual underground repository designs the multibarrier aim is achieved by a series of four components as illustrated in Figure 2.

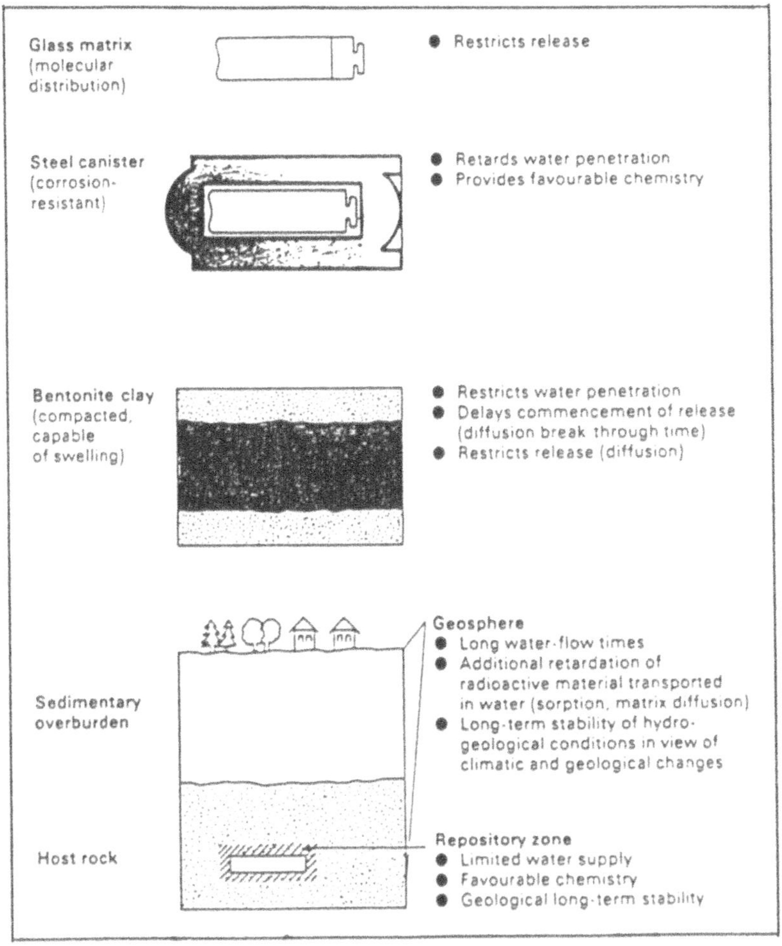

Fig. 2: Schematic illustration of the multibarrier system for waste containment in an underground repository

In the case of containment of only radioactive wastes, preventing the return of burried radioactive material into the biosphere is the prime protection objective. For the sake of secure and ultimate plutonium disposition on the other hand, another safety criterion is of decisive importance, namely the efficacy of the various barriers to illicit retrieval of the disposed plutonium.

Thus, the performance of a multiple barrier should be considered only to the extent necessary to obtain a realistic evaluation of each component, made up of the following:

1. The waste product itself, which is designed to retain integrity and leach resistance for long times under the anticipated heat, radiation, and chemical conditions that will occur in the repository. It should be borne in mind that any conceivable product, once retrieved, cannot withstand in principle an intentional chemical reprocessing for the purpose of plutonium recovery. However, by appropriate measures, it can be repressed considerably.

2. The waste packaging that is chosen to protect the waste form against physical and chemical attacks as well as interactions between the emplaced waste and the host rock and any water that may be present.

3. The backfill material which may fulfill a number of roles, including minimizing of groundwater access to the waste package, alteration of groundwater chemistry, and retardation of solute transport. For the purpose under consideration here, the prevention of plutonium retrieval, a well-devised backfill concept can greatly strengthen the protection goal.

4. The repository deserves the highest esteem. There are several reasons for this statement:
   - the long distance from the surface provides shielding not only from radiation exposure, but primarily in the case in hand from intentional intrusion;
   - the long distance and the methods used to seal any excavated areas reduce the probability of inadvertent contact with the waste in the future and make intentional removal much more difficult;
   - appropriately sited repositories will have little, if any, water in contact with the waste, therefore the potential leaching of plutonium and transport of plutonium via the groundwater path is minimized;
   - geological phenomena occur over long timescales compared to the length of time waste isolation is required.

4. POST-CLOSURE SURVEILLANCE OF THE REPOSITORY, MONITORING

Ensuring that the waste remains isolated from the biosphere and human grasp requires that the underground repository is backfilled and sealed as perfectly as possible. The repository must be sited, designed, used, operated, and closed to achieve long-term stability of the disposal site and to eliminate, to the extent practicable, the need for ongoing active maintenance of the disposal site following closure so that only surveillance, monitoring, or minor custodial care are required. However, components of the site environmental monitoring program and the site security plan that will be continued after closure must be developed and put into practice. Of utmost importance are measures to control the integrity of the sealed repository. The site aboveground must be controlled permanently by a government agency.

A schematic of a closed and backfilled repository in a rock salt formation is shown in Figure 3.

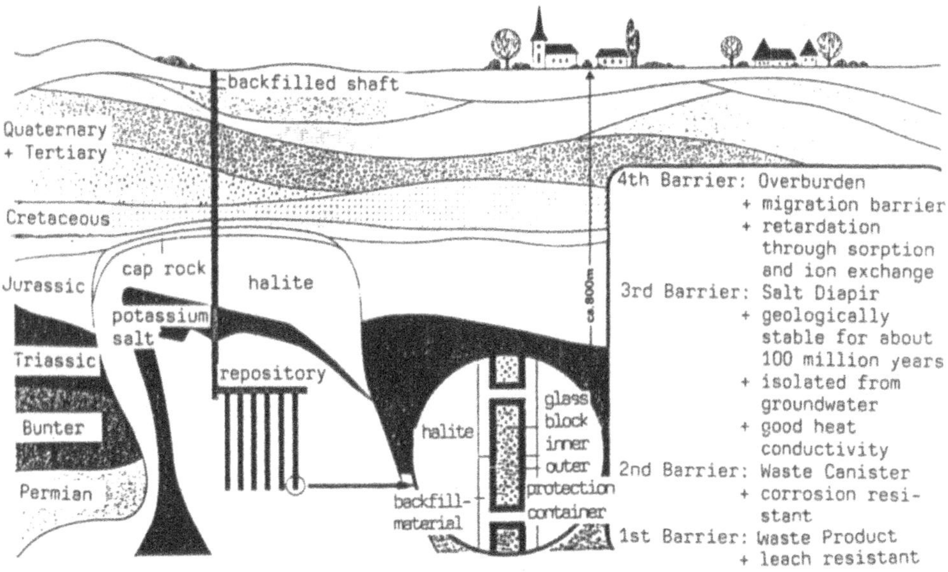

Fig. 3: Emplacement technique and repository closure schematics of a repository in a salt dome

With the emplacement of the direct disposal of spent fuel elements as well as the disposal of large-scale plutonium waste packages, the hitherto well established safeguards measures in nuclear materials accountancy require a new approach. Nuclear materials accountancy will be difficult, if not impossible without changing the whole safeguards philosophy so far applied.

A periodical nuclear inventory verification is no longer feasible due to the inaccessibility of the disposed and buried waste containers. Safeguards measures "at the material" by the authorities are restricted to item counting and identification at the repository entrance and to the verification that all containers arriving at the repository are in fact buried in the geological repository. Additional containment and surveillance measures implemented in the future have to ensure that no material is removed from the repository without being discovered.

The most effective countermeasure to intentional disruptive actions is active institutional control of the surface above and for some distance around the disposal site. Surveillance via satellite observation of the aboveground area may help to tackle the problem. However, active institutional control cannot be relied upon over the timescales for which the plutonium waste presents a potential hazard.

5. INTENTIONAL INTRUSION

The greatest risk associated with plutonium waste disposal may arise from intentional intrusion into the repository and thus retrieval of the plutonium waste canisters for the purpose of reclaiming the plutonium for reuse as weapons material. Disruptive human actions can be divided into those in which the barrier system is intentionally disrupted and those in which it is inadvertently disrupted.

Only the former event will be further evaluated since this event is the incident of interest. The various engineered and natural barriers should render intruders' actions harmless.

Effective countermeasures should be:

- Siting the repositories away from areas which present difficulties for remote control by international bodies.
- Isolation of the waste in deep underground repositories, the very depth of burial is an important feature mitigating potentially disruptive human actions.
- Reinforced physical barriers to deter attempted intrusion as well as the simple chemical recovery of plutonium.
- Dilution of the plutonium with an awkward matrix which makes chemical recovery of plutonium very difficult.
- Strong radiation of the waste product.

Conservation and communication of information under international cooperation about the location, contents and hazards of the repository can help reduce the likelihood of disruptive human activities. This implies the development of an international archive of plutonium waste repositories. Also, the development of a suitable marker system for such repositories is required.

6. PROTECTION EFFECTIVENESS OF INDIVIDUAL BARRIERS

Physical protection measures are necessary to protect plutonium from theft and misuse, especially by terrorists, but also by other sub-national or foreign organizations. Physical protection can also be regarded as one of the measures which reduces the risk of proliferation and may overlap with some safeguards measures verified by the IAEA such as containment and surveillance. It is not a function of international safeguards nor of the IAEA to protect nuclear material, and thus plutonium, from theft or sabotage either within a country or in international transit.

Physical protection measures are partly the responsibility of the owner of the plutonium, or the organization in whose possession it might be held, and partly the responsibility of the country concerned. The intensity of physical protection measures varies depending on the strategic value of the nuclear material and the perceived threat. Without doubt, plutonium falls into the highest risk category.

The various barriers are elements of an overall protection system. Their effectiveness depends on a number of factors, which are discussed in some detail in the following sections.

7. THE WASTE FORM BARRIER

The chemical form of plutonium plays an important role with regard to its suitability as a strategic material. Best qualified is the metallic form, followed by the carbide, oxide or nitrate. In order to make the plutonium less sensitive and accessible for direct use in weapons it should be transformed into a resistant chemical compound of very poor neutronics, highly diluted with other elements, and preferably incorporated into a very insoluble matrix in which plutonium is a partner in a mineral lattice structure.

Another favored waste form of a similar kind is a vitrified product in which the plutonium is dissolved in an amorphous glass matrix. In either case the isolation of pure plutonium from such materials becomes very difficult and requires great efforts.

Borosilicate glass is today the only waste form used commercially to immobilize radioactive waste. Production on an industrial scale is mature. The capacity of borosilicate glass for the incorporation of larger amounts of plutonium has not been investigated in detail as yet. However, it should be possible to produce durable glass products with plutonium contents in the range of one to a few percent. The quality and chemical behavior of a well-designed borosilicate glass composed of a suitable glass frit as the matrix together with high-level radioactive fission products and 1 - 4 % of plutonium should be about the same compared with highly irradiated spent oxide fuel. Both materials exhibit equal barrier protection quality.

With the objective of improved immobilization of radioactive waste, alternatives to borosilicate glass have been the subject of research and development. These alternatives have included other types of glass, non-vitrous or crystalline materials.

The main alternative glasses have higher silicon or aluminium content than borosilicate glass to improve their leach resistance. They melt at higher temperatures and are therefore more difficult and expensive to produce. Crystalline materials applied to the encapsulation of waste include special cements, super-calcines, and synthetic minerals, e.g. Synroc. In particular, so-called cement glasses could represent a suitable and advantageous host matrix for plutonium. Efforts are being made to produce such superior concrete ceramics by curing the liquid waste-cement mix under pressure and at temperatures of a few hundred degrees Celsius and also by using cements with a high alumina content.

Additional R&D would be required in this field if the plutonium disposal route should be favored. Various crystalline and ceramic waste forms have been proposed and many are being tested at present. Mineral assemblages can be chosen which are chemically very stable under chemical attack. In particular the flexibility of various aluminate and titanate matrix types to variable waste compositions and loadings is worthwhile mentioning. However, loading with quantities of plutonium has not yet been investigated.

The choice of the preferred solidification process depends primarily on the properties required for the final waste form, in particular its stability and resistance to the back-dissolution of plutonium. Of comparable importance are considerations of the risks involved in the solidification process and the ability to carry it out on a large scale. Most of these questions have not yet been resolved.

8. THE WASTE CONTAINER BARRIER

Encapsulation of solidified plutonium waste involves sealing it directly into containers in a form suitable for burial in a deep geological repository. Container materials of high resistance to mechanical and chemical attack are thick-walled models of cast carbon or stainless steel, copper, and some special alloys, possibly with multi-shells. Encapsulation is a well established technique.

The function of a high-integrity waste canister is to protect the waste during transport, interim storage, and emplacement in the repository. The mechanical performance required to ensure that it will withstand these operations and subsequent pressure re-equilibrium in the filled boreholes generally leads to the choice of the canister design. Also, shielding requirements must be considered. Lid seals, welds and seams are all potential sources of weakness and likely sites for enhanced, localized corrosion.

With regard to protection against the theft or misuse of plutonium, the canister barrier is of no value. Its excellent integrity over a certain time span may even be an advantage if retrieval of the buried plutonium waste is attempted.

9. THE BACKFILL MATERIAL BARRIER

The backfill material may fulfill a number of roles, including minimization of groundwater access to the waste package, alteration of groundwater chemistry, and retardation of solute transport. In the case under consideration here, the prevention of the exhumation of buried plutonium waste canisters is an important additional aim. Its prime ability should be to provide a strong mechanical shield with a high power of resistance to human intrusion.

In customary waste repository equipment the backfill material may display plastic properties for the purpose of filling any gaps formed during and after emplacement of the waste canisters. A reliable and widely advocated backfill material is bentonite clay compacted to a high density prior to emplacement. Bentonite displays the ability to swell as it takes up water from the surrounding rock, and in doing so would squeeze into fissures around the borehole wall and fill any other cavities, while exerting a considerable swelling pressure on the waste package itself. It also represents a very effective migration barrier for dissolved radioisotopes. However, bentonite buffer backfills do not exhibit high mechanical resistance to exhumation measures.

Several other buffer materials which also display very low permeabilities have been proposed, such as cement grouts or concretes, but with much better mechanical stabilities than bentonite. What is needed are filler materials with a very fine grained and impermeable cement-like hydraulic reaction product, which might then seal and protect the waste packages from easy retrieval.

On the basis of purely theoretical considerations one could advocate a containerless in-situ solidification of plutonium for example in a large very deep-sited salt cavern together with high-level radioactive waste. The plutonium would be mixed with a large amount of inert buffer material which could undergo continuous chemical reactions until special selected backfill materials and the plutonium are totally altered into stable products with strong mechanical resistance. Besides the mechanical barrier, the high dilution of plutonium in a complex matrix would render a plutonium re-isolation very difficult. However, the evaluation and proof of long-term safety seem to be rather problematic.

Another daring proposal for an improved exhumation backfill material barrier is the application of a cement grout injection soaked with a highly toxic chemical substance stable in the long term which would be released and dispersed during exhumation drilling and mining activities. The effectiveness of resistance would be due to serious injury to the health of workers and possibly the people living in the direct vicinity of the repository.

In summary, the strength of the backfill material barrier to retrieval of the buried plutonium waste is not overwhelming. It can reduce the ease of recovering disposed plutonium but not prevent it.

10. THE GEOLOGICAL BARRIER

Deep underground geological repositories are designed to provide long-term isolation of radioactive waste. Long-lived radioactive wastes, including plutonium, require almost complete isolation for time periods of many thousands of years. Since it is not conceivable that human society will be able or willing to maintain controls on repository sites for such long times, isolation systems are designed to be passive in nature. In other words, the safety of the system depends on the intrinsic properties of the geological isolation barriers and not on the existence of surveillance and maintenance procedures.

The host rock overburden barriers are of course outside the engineer`s control and can only be modified to a limited extent. They are designed to counteract or minimize as much as possible the natural process which could lead to release of the waste and in some cases also to minimize the effects of human intrusion into the disposal zone.

Many conceptual designs exist for geological waste repositories. The most advanced one is deep burial in a purpose-build or modified existing mine in a depth of about 500 to 1000 metres. In contrast, a very deep borehole disposal has lately been receiving serious attention in Australia, Denmark, Sweden and the USA. Generally speaking, deep boreholes are projected to 2 to 4 kilometres deep. This concept involves lowering waste packages down wide (60 to 100 cm∅) single boreholes drilled from the surface, with subsequent sealing of the holes. In theory boreholes can penetrate many kilometres into the earth`s crust where little water movement is thought to occur. However, a retrieval of the plutonium is also not impossible in such a concept.

The deep borehole concept is still in an exploratory stage of development. The risks belonging to it are complex and difficult to evaluate. Cost estimates in the order of 100 million US dollars per hole have been produced in Sweden, but they are rather questionable with regard to their trustworthiness.

At present the most advanced and only mature technique is the geological disposal in mined repositories. The required costs are attainable, but most likely altogether not below the expenditures for the MOX disposition route.

Principal control on the rock barrier is exercised when the site is selected, which is the reason why so much basic geological research is required to ensure an adequate understanding of the properties of rock types and environments considered potentially suitable for disposal purposes. These aspects are beyond the scope of this paper and the relevant literature should be consulted.

The protection of man against the risk associated with the dissemination of radioactive substances must be provided in all situations considered. This includes intentional exhumation of the waste canisters and thus theft and misuse of plutonium. In the long term the geological barrier represents the most effective and reliable barrier. An important presupposition is therefore that for a period of at least 10,000 years the stability (covering limited and practicable evolution) of the geological barrier must be demonstrated and proved. The method used to assess the quality of the site in terms of safety is of the deterministic type.

Since beyond this period of stability of the geological barrier and uncertainty concerning the evolution of the repository increases progressively with time, consideration may be given to using the risk concept to allow for the probability of the occurrence of a human intrusion into the repository.

11. CONCLUSIONS

The advantage of the multibarrier concept is that the reliability of the disposition concept under consideration to prevent the theft and misuse of disposed plutonium does not depend on a single barrier whose failure might seriously compromise the preset protection goal. The barriers are complementary in this respect, in the course of which the waste form and the backfill of the repository are of less efficacy compared to the geological barrier. The geological barrier of the repository and the shaft sealing materials must be able to ensure containment alone in the long term.

Absolute certainty that an intentional human intrusion into an underground repository can be ruled out is not achievable. Clearly, however, it would not be possible for anyone to retrieve the plutonium without the permission of the host country, as long as political control in the host country remained intact. Moreover, because necessary mining activities would be highly visible, the host country could not retrieve the plutonium without detection.

Figure 3 (see following page) illustrates in summary and in a readily understandable form the characteristics of the three effective barriers preventing an easy plutonium retrieval.

The repository should be safeguarded without compromising safety features. Since excavation for a sealed repository could not be carried out in a short time, nor made invisible, an obvious approach would be through the analysis of periodically obtained satellite images. Additionally, the aboveground site of the former repository could be subject to periodic inspection by international inspectors. Such a safeguards surveillance would also reduce the possibility of inadvertent intrusion into the repository by humans.

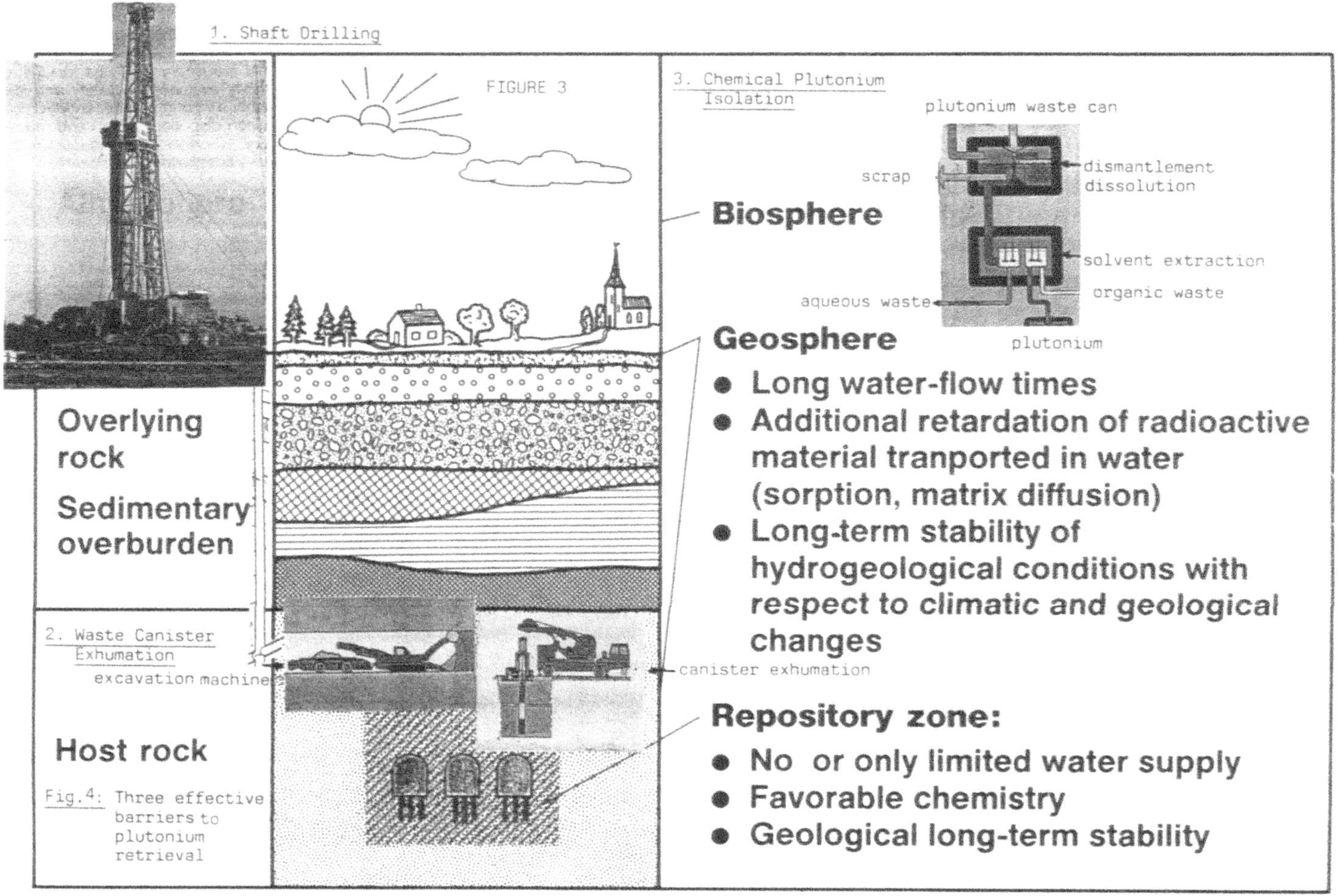

Fig.4: Three effective barriers to plutonium retrieval

PROSPECTS FOR USING PLUTONIUM AS A FUEL IN RUSSIA

V.N. MIKHAILOV, V.V. BOGDAN
Ministry for Atomic Energy of Russian Federation (MINATOM), Moscow, Russia

V.M. Murogov, V.S. Kagramanian
Institute of Physics & Power Engineering
Obninsk, Russia

1. INTRODUCTION

The planned release of large amounts of fissile materials resulting from nuclear disarmament has created interest in the matter not only for power engineers and weapon manufacturers, but also for environmentalists, political scientists, economists, politicians, and the public around the world /1/.

"Civil" plutonium accumulated from spent uranium fuel in power reactors also shares this spotlight. Mainly produced in light water reactors (LWRs), over 1100 tons of plutonium have already accumulated all over the world. The annual output is about 70 tons. The plutonium problem becomes even more critical when the desire of the world community to make nuclear disarmament irreversible is considered /2/.

Table 1 shows the projected total plutonium that will exist in Russia. Weapons-grade plutonium is depicted. All of it will arise as a result of disarmament activities. Civil plutonium is estimated as that which will accumulate from the spent nuclear fuel of reactors now operating. The accumulation is expected over an operating life time of 30 years for each nuclear power plant.

Table 1: Amount of plutonium in irradiated nuclear fuel of modern Russian reactors after 30 years in service

Reactor type	Power [$GW_{t(e)}$]	Pu* [t]	Pu** [t]
VVER-440	3	27 /3.7/	20 /16/
VVER-1000	7	55 /9.1/	40 /33/
RBMK-1000	11	76 /8.3/	45 /37/
BN-600	0.6	12 /0/	11 /11/
Weapons Pu	-	100 /0/	95 /95/
In total		270 /21/	211 /193/

* in brackets mass of Pu-241+Am-241

** in brackets counting on Pu-241 decay after 30 years storage

E.R. Merz and C.E. Walter (eds.), *Disposal of Weapon Plutonium*, 37-49.

Our development and experience shows that usage of plutonium to close the nuclear fuel cycle not only increases our national power resources but also lowers the radiotoxicity of long-lived nuclear wastes of the future /1,2,3/.

We clearly understand that the fuel cycle technology should be sufficiently mastered and assimilated in order to:

- follow non-proliferation policies
- assure the safety and environmental succession of plutonium in the fuel cycle
- be economically viable.

Using these criteria, plutonium can be considered a national property. Otherwise it represents a permanent general threat.

The aim of this presentation is to clarify our point of view concerning the plutonium problem.

2. EXPERIENCE GAINED WITH PLUTONIUM

2.2 MOX Fuel Fabrication and Reactors

Russia began using plutonium as a nuclear fuel in the second half of the 50ties. In 1957 a core was fabricated of a metallic plutonium alloy for the pulsing fast reactor IBR-30. In 1959-65 plutonium dioxide fuel was made for the BR-5 and IBR-2 reactors.

Systematic study of plutonium fuel began in 1970 in the BOR-60 reactor. This work eventually conformed to the fast breeder reactor using plutonium as the most efficient way to expand nuclear power fuel resources.

Two cores of weapons-grade plutonium have been tested in the BR-10 experimental fast reactor. Large batches of MOX fuel pins made of different technologies with plutonium of various isotopic composition have been tested in the BOR-60 research reactor. This reactor has been operated for many years recycling its own plutonium.

With continued successful MOX fuel tests, we expanded the scope of these tests to the prototype fast reactors BN-350 and BN-600. Both reactors have been fueled from the very beginning with enriched uranium. The semi-industrial plant "Paket" capable of producing up to 10 plutonium fuel subassemblies annually for these reactors was built at the PO Mayak at Chelyabinsk to experimentally substantiate the technology of plutonium utilization in fast reactors. In the BN-350 reactor the tests have been performed with the subsequent investigation and chemical reprocessing of test fuel subassemblies with MOX fuel (350 kg of weapons-grade plutonium). More than 2000 such fuel elements have been built and tested.

Post-reactor investigation showed that fuel element endurance was not exhausted with a burn-up of 9-11% h.a. MOX fuel fabrication technologies that produced less dust than the mechanical mixing of oxides were studied and developed to improve radiation characteristics. The sol-gel fabrication process of granulated MOX fuel was used for the subsequent pellet pressing. However, this process has not provided very high or stable pellet quality. Therefore, a method of ammonia coprecipitation of uranium and plutonium was developed in parallel. Most of these twelve fuel subassemblies manufactured for the BN-600 reactor passed in-pile tests.

Vibro-compacted fuel has been studied using granulated fuel obtained through various means. Among them is a method of electrochemical uranium oxide and plutonium oxide coprecipitation. All fuel elements loaded into the BOR-60 reactor have been built using this technology. Also, subassemblies have been tested in the BN-350 and BN-600 reactors.

MOX fuel development for BN-type fast breeder reactors has progressed to the initial design and construction of 3-4 units of BN-800 power fast reactors at the South Urals and Beloyarskaya sites. Construction of a MOX fuel plant called Complex-300 is under way. It will produce up to 900 subassemblies per year as fuel for these reactors. The reactors are in the initial stage of construction while construction of Complex-300 is 50% complete.

Two pilot installations at the Mayak site called Granat and Paket will be backfitted to meet the latest safety and environmental requirements of our state nuclear surveillance commission, Gosatomnador.

As with regard to plutonium utilization in thermal reactors, Russia is presently experimenting with and analyzing these methods. The studies compare the advantages and disadvantages of this direction in plutonium utilization to that of fast reactors taking into account Western European experience. A critical assembly, called SUPR, is being constructed at IPPE in Obninsk. It will be used to study safety characteristics of LWRs using plutonium, including weapons-grade plutonium. General information on MOX fuel irradiation and production in Russia is presented in tables 2 and 3 /4/. Information on spent uranium fuel disposition is given in table 4 /4/.

Table 2: MOX fuel usage in reactors

Reactor	BR-10	BOR-60	BN-350	BN-600
Pelletized fuel	PuO_2	$PuO_2 + UO_2$		
Number of fuel pins	3287	370	1778	1524
Maximum burnup %h.a.	14.1	12.6	10.8	10.5
Max.linear rating,kW/m	17	50	48	48
Vibro-acked fuel	$PuO_2 + UO_2$			
Number of fuel pins	-	12800	254	762
Maximum burnup %h.a.	-	26	7.2	9.8
Max.linear rating,kW/m	-	54	48	48

Table 3: MOX fuel production capacity in Russia

Plant / Facility	Reactor	Annual Capacity	Production in 1992
"Paket"at Mayak Chelyabinsk	BN-350 BN-600	10-12 Fuel Assem. 300 kg MOX	4 Fuel Assemblies 100 kg MOX(~20%Pu)
"Paket" modified since 1993	BN-600	40 Fuel Assemblies 1 ton MOX	
Facility at RIAR (Dimitrovgrad)	BOR-60 BN-600	1 ton MOX (Vibropack)	600 kg MOX
Plant at Chelyabinsk complex (50-60% complete)	BN-600 BN-800 (VVER-1000)	60 tons of heavy metal	
Plant in Krasnoyarsk	VVER-1000	planned for the future	

Table 4: Industrial nuclear fuel and fuel cladding production

Plant	Type of Production		Annual Capacity [tons]	Production in 1992 [tons]
	Reactor	Final Product		
Electrostal (near Moscow)	VVER-440	fuel assemblies	700	230
	RBMK	fuel assemblies	570	570
	BN	fuel for reactor core	20	
	BN	fuel for reactor breeding blanket	15	
Novosibirsk	VVER-1000	fuel assemblies	1000	210
Ust-Kamenogorsk (Kazakhstan)	VVER	fuel pellets	2650	220
	RBMK			570
Glazov	VVER RBMK	zirconium alloy tubing	2000 6000 km/a (tubing)	2000 km/a (tubing)

2.2 Reprocessing and Plutonium Storage

Figure 1 demonstrates the scheme of the U-Pu closed nuclear fuel cycle being implemented in Russia. Shown in the figure are industries and installations in operation as well as partially implemented and planned ones /5/.

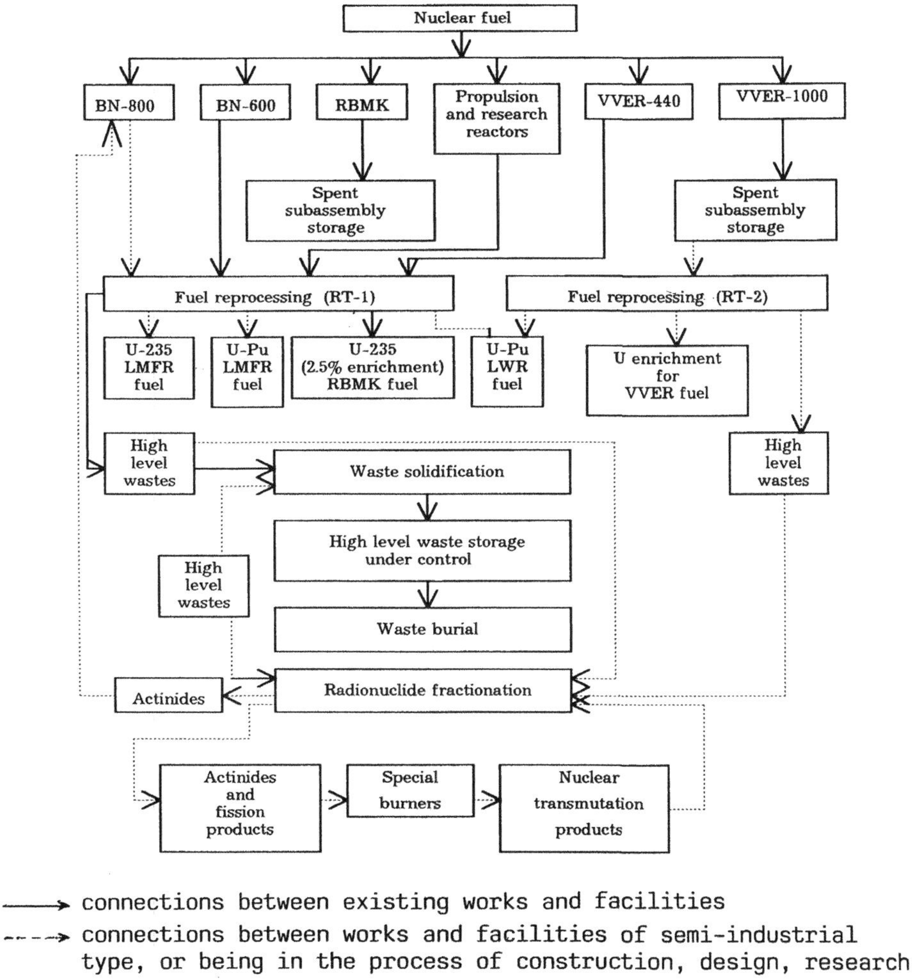

Fig. 1: Schematic diagram of Nuclear Fuel Cycle in Russia

The Russian closed nuclear fuel cycle was first realized with the commissioning of the RT-1 reprocessing plant in 1976 at Mayak near Chelyabinsk. RT-1 is a multi-purpose enterprise that reprocesses spent fuel from VVER-440 reactors, fast converter reactors BN-350 and BN-600, icebreaker and submarine transport units, research reactors, and other power units. The plant capacity is 400 tons/year with respect to the main fuel type, that of the VVER reactor. Therefore, it has the capacity to reprocess fuel not only from Russian nuclear power plants but also from foreign reactors of the same type. To date about 3000 tons of spent fuel have been reprocessed in the plant. Since 1978 reprocessed uranium (with U-235 isotope content conditioning by mixing) has been recycled into BN-350 and BN-600 reactors. Later, uranium was reprocessed into RBMK reactors, and recently into VVER-1000 reactors.

At the present time the Russian nuclear fuel cycle is closed for VVER and BN reactors for only uranium. Without simultaneous use of uranium and plutonium, the closed fuel cycle is neither economically nor ecologically justified.

The main product of the RT-1 spent fuel reprocessing is enriched uranium. However, the plant can also extract as much as 2.6 tons of civil plutonium per year. Last year`s plutonium production dropped to 0.6 tons.

The total quantity of extracted civil plutonium stored at the Mayak site is about 30 tons. This plutonium is stored reliably in the form of dioxide to be used in fast reactors. The main problem of this forced, extended storage of civil plutonium is decay of short-lived fissile isotope plutonium-241 into americium-241. The latter is very environmentally troublesome. About 10% of the discharged plutonium is initially plutonium-241.

A new radiochemical plant RT-2 is under construction not far from Krasnoyarsk. Its principle objective will be to reprocess VVER-1000 spent fuel. Its first phase, to be put in operation after 2005, will reprocess 1500 tons of fuel per year. The plant cooling pond already contains 1100 tons of VVER-1000 spent fuel assemblies, while about 1000 tons of this fuel are currently in storage at the nuclear power plants.

RBMK-1000 and RBMK-1500 reactor spent fuel is temporarily delivered to nuclear power plant storage areas. The resolution concerning the monitored long-term storage of several decades is in force for this fuel category.

The accumulation of the 30 tons of plutonium at Mayak, the plutonium to be extracted from 2100 tons of spent fuel RT-2, along with the 100 tons of ex-weapons plutonium expected to be released aggravate the problem of fast reactor development. In our opinion it should be accelerated not only from the standpoint of power engineering and economics, but for non-proliferation and safety. There is no doubt that in the near-term management measures must be taken for reliable plutonium storage. In addition to the storage of civil plutonium, storage for released ex-weapons plutonium should also be arranged. At the same time such storage should not be too long for economic, environmental and political reasons.

General information on spent fuel management in Russia is given in table 5 /4/.

Table 5: Reprocessing and spent fuel management in Russia

Facility	Reactor	Capacity	Product
RT-1 reprocessing plant at Mayak complex in Chelyabinsk (since 1971)	VVER-440 Fast and transport reactors	400 tons of heavy metal per year (t HM/y)	Reprocessed uranium returned to fuel fabrication for RBMK Plutonium stored in dioxide form
RT-2 reprocessing plant in Krasnoyarsk	VVER-1000	1st line: 1500 t HM/y 2nd line: 1500 t HM/y Total:3000 tHM/y	Reprocessed fuel will be returned to fuel fabrication for VVER and BN
Storage facilities at each RBMK nuclear power plant site	RBMK	2000 t HM	
Storage facility at RT-2 plant	VVER-1000	6000 t HM (3000 t HM in operation)	

Note: The RT-2 reprocessing plant is under construction, however, the date for its completion is still under discussion.

3. MANAGEMENT OF SEPARATED CIVIL AND EX-WEAPONS PLUTONIUM

The Ministry of the Russian Federation for Atomic Energy has developed a basic concept for the management of extracted civil and released ex-weapons plutonium based on the following principles:

- Russian experience of plutonium management should be maximized;
- Plutonium diversion resistance should be considered an important criteria;
- Plutonium management should be ecologically and economically acceptable and lead to the development of a long-term fuel cycle option.

Separated plutonium utilization based on a nuclear center at Mayak, including the Shop-300 plant for MOX fuel manufacturing and an NPP with three BN-800 units together with the operating RT-1 for chemical processing of uranium spent fuel, fully meets these criteria /6/.

Construction of this center has already been foreseen by the Russian program. The main purpose of these reactors was to utilize civil plutonium accumulating at Mayak, the product of chemical processing uranium reactor spent fuel. The BN-800 reactor design is described by a breeding ratio (BR) of about 1. That is, it was expected that these reactors will use plutonium from thermal reactors as fuel for their first loading only, changing to their own plutonium in the future. For this purpose it is foreseen to organize BN-800 spent fuel processing.

Estimates show that not only the civil plutonium accumulated at the RT-1 but also all released ex-weapons plutonium could be handled by a nuclear center at Mayak with the addition of one BN-600 reactor built on the Beloyarskaya site. It would only be necessary to postpone chemical reprocessing of the BN-800 reactor spent fuel until a considerable portion of already released civil and weapon-grade plutonium will have been involved in the fuel cycle. An improved BN-800 reactor design without blankets decreases plutonium content in the spent fuel and the costs of spent fuel subassembly storage.

Fissionable material nonproliferation is addressed first by limiting transportation of released civil and weapons-grade plutonium within the militarized zone on the Mayak site. Secondly, a barrier in the form of radioactive fission products present in the spent fuel would act as a barrier for nonproliferation.

Annual weapons-grade plutonium consumption for the fabrication of fuel subassemblies for one BN-800 reactor is about 1.6 tons. If the BN-800 reactors were put in operation according to the program, all civil plutonium accumulated at Mayak and all released weapons-grade plutonium could be used in the once-through mode (that is, transformed into spent fuel) during the first two or three decades of the next century (see Figures 2 and 3) /6/.

Thus, weapons-grade plutonium utilization in the context of the basic concept is inseparably linked with the realization of the program on fast reactors using traditional MOX fuel.

In principle, the initial stage of plutonium utilization could also be based on the construction of a new generation of MOX fuel LWRs within the framework of Mayak, at Tomsk or at Krasnoyarsk defense sites. However, this decision is unacceptable for Russia for several reasons.

First, the analysis of such plutonium utilization is still basically the study of advantages and disadvantages of this direction. Therefore, no Russian thermal reactor has been designed to use MOX fuel. With the limitation of one third of the core loaded with MOX

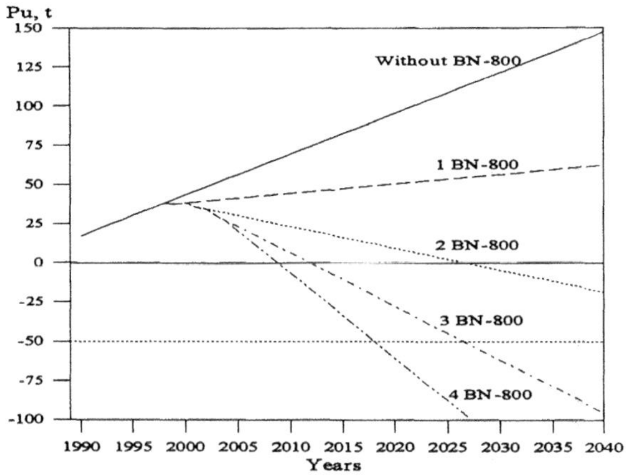

Fig. 2: Pu balance at the PO "Mayak" as a function of BN-800 reactors number (once through cycle, RT-1 capacity 2.6 t Pu/a)

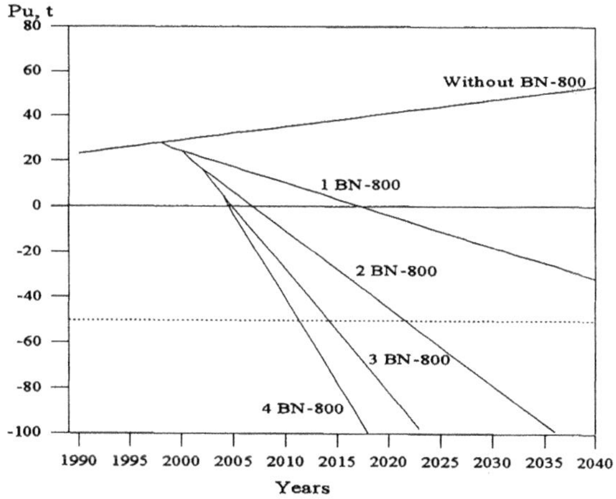

Fig. 3: Pu balance at the PO "Mayak" as a function of BN-800 reactors number (once through cycle, RT-1 capacity 0.6 t Pu/a)

fuel subassemblies, as in Western Europe, constructing VVER reactors in place of the three BN-800 units at the South Urals NPP would only ensure 1 ton of weapons-grade plutonium consumption per year. Thus, 15-20 LWR units would have to be built at Mayak to dispose of all accumulated and released plutonium even using an open cycle. This is virtually impossible.

Second, the issue of spent fuel radiotoxicity is also essential. Long-lived isotopes in spent fuel such as americium, neptunium, and curium complicate both MOX-fuel recycling and long-term waste disposal. To a great extent this is attributed to the accumulation of plutonium-241 in spent fuel. The specific radiotoxicity of plutonium-241 is 40 times higher than that of the basic isotope plutonium-239. During storage the plutonium-241 isotope is transformed into the toxic americium-241 with a half-life of 433 years. This isotope is the main contributor to the radiotoxicity of transuranium elements in spent fuel after the decay of short-lived fission products. During the operation of light water reactors with uranium, about 250 kg/GWt(e)/year of reactor grade plutonium are built up. About 30 kg of this mass is plutonium-241. Disarming or burning-up weapons-grade plutonium in thermal reactors increases three-fold the annual build-up of plutonium-241, compared with burning uranium-based fuel in VVERs. If this spent fuel is stored for the long-term, a considerable amount of the plutonium-241 will be transformed into americium-241, highly complicating further utilization of plutonium and waste disposal /2/.

Third, the burning of weapons-grade plutonium at VVERs may produce several times more minor actinides as compared with the VVERs using uranium-based fuel. The burning of the basic isotope plutonium-239 during weapons-grade plutonium utilization at Mayak facilities would result in an accumulation of nuclides with a total radiotoxicity three times higher than that of the VVER-uranium variant /2/. The situation is quite different however, if we use BN-type reactors to utilize weapons-grade plutonium. The radiotoxicity of disarmed plutonium does not, for all practical purposes, exceed that of the initially loaded plutonium (see Figure 4).

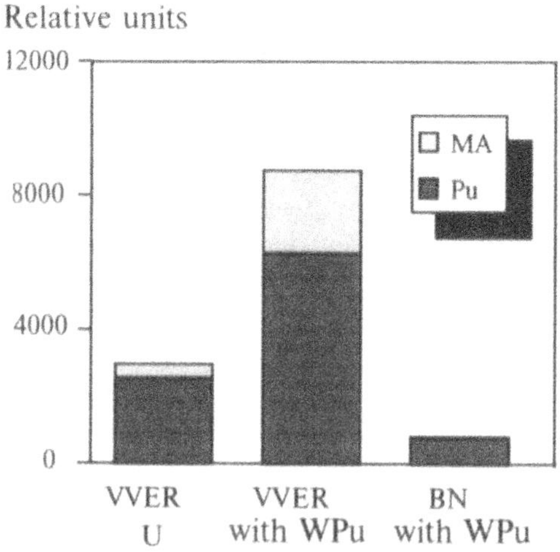

Fig. 4: TRU radiotoxicity relative evolution after 1st recycle of weapon Pu (WPu) in VVER and BN type reactors

And fourth, the economic efficiency of plutonium utilization in thermal reactors is connected in many countries with the relatively high cost of fast reactors. In Russia, the electricity generated by the first BN-600 is 40% more expensive than that of a VVER-1000. But the experience of BN-600 was taken into account in the BN-800 project and the metal consumption factor is only 80% of the BN-600 project. The improved economic characteristics of the BN-800 fuel cycle have been reached through the transition from uranium (which is ineffective for fast reactors) to MOX fuel and the further improvement of the burn-up factor /6/.

The economic characteristics of the fast and thermal reactors have also been compared, based on the increased safety standards required in our country after Chernobyl. Inherent characteristics of fast reactors, as well as new technical solutions, enabled us to upgrade the BN-800 project to a new generation of world-level nuclear power plant safety. It is the only project which has passed all, including ecological, necessary certifications. Local authorities have already approved the project and have reached an agreement with MINATOM /6/.

But work is not completed on new projects with new, safer uranium reactors such as the VVER-500 and VPBER-600. Preliminary estimates show that average capital investment in these projects is higher than in the VVER-1000, but corresponds approximately to the same investment necessary to build the three BN-800 units at the South Urals site /6/.

Regarding BN-type reactors, Complex-300 at Mayak is 50% completed. With sufficient investment it can be commissioned by the end of the century, before the commissioning ot the first BN-800 unit. After putting three BN-800 units into operation, Complex-300 will reach its maximum capacity in MOX fuel fabrication.

It is evident now that plutonium use in thermal reactors is a forced decision and is limited by the number of recycles because of radiotoxic minor actinide accumulation. Fast reactors will complete the fuel cycle for thermal reactors, solving the problems of plutonium isotopic composition and the burning of minor actinides.

The aim of further system studies is to determine the optimal parameters of thermal and fast reactor cores and the optimal composition of the nuclear power grid. Criteria for fuel cycle efficiency should encompass all aspects of economics, ecology, and fissile material non-proliferation.

One key direction of research in the field of plutonium utilization is the destruction of plutonium surplus and radiotoxic minor actinides stored in thermal reactors. For these purposes new fast reactor cores with MOX fuel, and more plutonium in the fuel without blankets are under consideration. Also, new fuel composition on the basis of an inert matrix instead of uranium are being developed /6/.

Along with the traditional uranium-plutonium cycle, fuel cycles with thorium and uranium-233 are under consideration. These will allow us to decrease the quantity of plutonium and minor actinides in wastes of the nuclear industry to a minimum/7,8/.

In the distant future the use of uranium-233 blended with uranium-238 will enable civil nuclear power plants to operate outside the plutonium centers on low-enriched uranium fuel only. Now we use U-235 + U-238 fuel, in the future we can use the secondary nuclear fuel U-233 + U-238 (denatured fuel cycle) /7,8/.

In the Figure 5 different options of plutonium management are presented as a scheme which has been realized in the countries with developed and hypothetical plutonium technology.

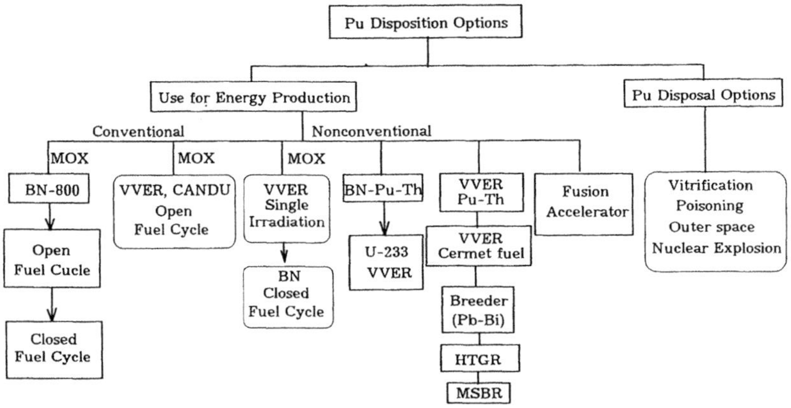

Fig. 5: Plutonium Disposition Options
Comparison criteria: economics, safety, ecology, non-proliferation
Boundary conditions: technical evolution, adaptation, reality in the frames of traditional nuclear industry development

4. CONCLUSIONS

1. The Russian concept of plutonium management (both civil and weapons) is based on the postulate of the outer fuel cycle closure, necessity to enhance fuel efficiency, and decreasing radioactivity of disposed long-lived wastes.
2. In view of plutonium utilization of existing fast reactors being restricted, safe and reliable storage of separated civil plutonium at PO Mayak and ex-weapons plutonium is required.

3. Longer range plutonium disposition options are based on development of a Nuclear Power Center at PO Mayak (RT-1, Complex-300 and 3 BN-800 reactors) to use accumulated civil and ex-weapons plutonium in fast reactors.
4. Long-term management options, which call for fulfillment of research, envisage burning of excess plutonium and minor actinides in fast reactors with new core compositions.
5. Investigation is underway to estimate plutonium utilization options possibly using foreign technology including those on light-water reactors in the framework of defense nuclear center concepts.
6. International cooperation with the aim not only to develop the current technical policy but also to determine the optimum long-term disposition is required.

5. REFERENCES

/1/ V.N.Mikhailov, Returning Weapon Plutonium to Nuclear Power. GLOBAL`93, Intern. Conf., September 1993, Seattle, WA, USA

/2/ V.M.Murogov et al., Strategy of Nuclear Weapon Material Utilization in Nuclear Power Engineering. A Concept of Uranium-Plutonium Fuel Cycle Implementation, IPPE-2303, Obninsk, Russia (1993)

/3/ V.N.Mikhailov, V.M.Murogov, M.L.Solonin et al., Plutonium Usage in Nuclear Power in Russia, Atomic Power, Vol.76, issue 4 (1994)326

/4/ Yu.K.Bibilashvili, F.G.Reshetnikov, Russia`s Nuclear Fuel Cycle-An Industrial Perspective, IAEA Bulletin, Vol.35,No.3 (1993) Vienna, Austria

/5/ M.Solonin, B.Zakharkin, Technical and Ecology Aspects of Plutonium Involvment in Atomic Power, Second Plutonium Forum, March 1995, USA

/6/ N.I.Ermakov, V.M.Murogov, V.M.Poplavski, Role of Fast Reactors in the Future of Power Engineering, the Fuel Supply and the Environment, Proceedings of Intern. Topical Meeting "Sodium-Cooled Fast Reactor Safety", Obninsk, Russia (1994)7

/7/ G.N.Kazantsev, V.M.Murogov, F.P.Raskach et al., Nuclear Fuel Cycle on the Basis of Th and U-233, IPPE-2183, Obninsk, Russia (1993)

/8/ V.N.Mikhailov, V.V.Bogdan, V.M.Murogov et al., Utilization of Plutonium in Nuclear Power Industry of Russia, Intern. Policy Forum, Management and Disposition of Nuclear Weapon Material (1994) USA

PLUTONIUM HANDLING AND VITRIFICATION: MAIN PROCESS STEPS AND THEIR COST EVALUATION

Etienne TRAUWAERT, Maurits DEMONIE
Belgoprocess
Gravenstraat 73
B-2480 Dessel (Belgium)

1. Introduction

One of the key questions in the nuclear non-proliferation strategy is what to do with the fissile material that becomes available from the dismantling of nuclear weapons. It appears clearly that some destination must be given to this material in a near foreseeable future, if one is to avoid that this material sooner or later finds its way to the wrong hands.

Two "proven" solutions appear today readily applicable providing only slight adaptations to existing techniques. Depending whether one wants to consider this weapons fissile material as an energy source or as waste, one can use it for nuclear fuel production or condition it to high-level radiological waste packages.

For uranium material this question is most easily answered as the uranium fuel fabrication process is well developed in most of the concerned countries. The only particular step to be added is to reduce the grade of enrichment which for weapons uranium is presumably around 93 % and for reactor uranium generally around 5 %. This de-enrichment can easily be obtained by mixing the highly enriched uranium with natural or depleted uranium. This process step can be performed in gaseous, liquid or solid powder phase. In this approach the uranium conserves its positive economic value which in the case of its processing as waste would become negative.

Things are not so straightforward with weapons plutonium, which probably constitutes the largest part of the weapons fissile material.

It is well known that the economical most suitable way to burn plutonium is to use it as fuel for fast breeder reactors (FBR). However, this route is presently under development in only a few countries (Russia, France, Japan, UK) and it is very clear that in the short term no new developments of this route will take place.

This is why a number of countries (Belgium, France, Germany, Switzerland) started to use their plutonium in light water thermal reactors (LWR), in the form of mixed oxide (MOX) fuel. Up to now the largest fabrication capacity (35 THM per year)

E.R. Merz and C.E. Walter (eds.), Disposal of Weapon Plutonium, 51-57.

for MOX is in Dessel, Belgium. This installation will probably be superseded by the Melox plant (120 THM per year) of Marcoule in France before the end of the year.

However this technology is not universally accepted and the available capacity is anyway very limited. An alternative way to handle the plutonium is to consider it as high-level waste (HLW) and therefore to condition it in a glass matrix within a stainless steel canister with a view to its long term final disposal. This is the option that will mainly be analysed in this paper.

In the comparison between the two options it is quite clear that the fuel option will be economically more rewarding. At the expense of fuel fabrication costs one gets a highly valuable energy source. It could even be cheaper than the classical fuel cycle cost. Reprocessing is no longer required but weapons dismantling costs have to be added.

When the plutonium is considered as waste the conditioning costs are not compensated by any return, which makes this option undoubtedly more expensive.

However, cost considerations alone are not enough to decide what technology to use. Many other arguments on strategic, safeguarding and political grounds are in favour of the burial of plutonium. Therefore it is almost certain that at least a part of the available weapons grade plutonium in the world (200 t) will follow the vitrification route. This will certainly be the case with small batches of plutonium, contaminated with chemical or fission poisons. These are difficult to remove. The fissile material becomes unsuitable for fuel production.

The possibilities and the limitations of this vitrification route are analysed. The description of the process will serve as a basis for a conceptual cost evaluation. This estimate is one of the elements in the comparison to be made with the MOX-fuel option.

2. The Constraints on the Plutonium Content in Glass

The only significant industrial experience on introduction of plutonium in a glass matrix is that of traces of plutonium in the high-level waste streams from reprocessing. Depending on the irradiation level and the type of fuel the content of plutonium can vary in the range of 0.01 to 0.10 g in the waste stream and between 0.01 and 0.20 g /kg in the glass product, which is very low.[1] The actinide content of glass products amounted to 5 g/kg within a waste oxide content of 200 g/kg.

The question with plutonium vitrification is whether the totality of the actinide content could be plutonium. In other words, is a plutonium content of 5 g /kg in the glass product achievable ? And what about 10 g /kg or 20 g /kg plutonium in the glass product ?

To answer these questions one must consider the qualities that are expected from the glass product and the vitrification process that can be adopted.

One of the primary characteristics of the glass is its chemical stability and its low leachability. Borosilicate glasses generally present a high chemical resistance and can accomodate a relatively large variety of chemical elements and compounds. [2] This ability stems from the capacity of the glass to solubilize these elements. The solubility of Pu within borosilicate glass may be larger than in other glasses. Nevertheless it is limited. 5 wt % is quoted by some authors as the upper limit for Pu solubility in borosilicate glass [3]. Furthermore this figure is likely to be adversely influenced by the presence of rare earths in the high level waste. Higher concentrations of Pu - between 7 wt % and 15 wt % - are mentioned by other researchers but only as a result of experiments on laboratory scale. [4]. However other factors than the solubility will also limit the amount of Pu that can be incorporated in the borosilicate glass.

Indeed, α-radiating isotopes like Pu can considerably damage the matrix from inside and form He-gas inclusions capable of embrittling the glass blocks. [5] In classical vitrified waste the α-activity is limited to about 10 Ci/kg glass. Weapons-grade plutonium with low content of ^{238}Pu and ^{241}Pu has an activity of less than 1 Ci/g Pu. This would correspond to about 10 to 20 g Pu/kg glass. The main material characteristics influenced by an increased radiation are the reduction of the material hardness, the modification of Young's modulus and the enhanced fracture toughness. It is not sure that all these modifications are necessarily negative. It is the instability of the material that makes the problem. The change in characteristics could involve volume modifications or pressure build-up.

The most stringent limitation on the plutonium content is probably the criticality risk. This risk must be considered both during waste conditioning and long term storage and disposal. In another presentation this limit is estimated at 16 g Pu/kg glass [1]. This limit has to be very carefully calculated with respect to the incommensurate consequences of trespassing a safe limit. It should further be appreciated that the presence of boron in the borosilicate glass, typically in the range of 10 to 15 wt %, is instrumental in reducing criticality problems once the Pu and the glass are properly mixed. This neutrophagus effect can also be obtained by the addition of some other neutron poisons such as Gd, that is often used as burnable poison in fuel designs.

The criticality problem conserves however its sharpness in the stage before homogenization of the Pu and the poisons and after possible segregation during the vitrification process or during intermediate or final storage.

Further constraints are dictated by the operational limitation on the electrical resistance and the viscosity of the molten glass. However this can be compensated within some limits by additives such as Sb_2O_3 and by a proper design of the melter [6].

3. Main Steps of the Vitrification

3.1. PRECONDITIONING OF THE WEAPONS-GRADE PLUTONIUM

The feed material for the melter should be chemically stable and compatible with the glass production process; therefore the most suitable form for the plutonium should be a nitrate solution form. As most of the dismantled weapons-plutonium will presumably be in metallic form, it should first be converted to a nitrate solution and then mixed with the high-level liquid waste.

Other compounds of Pu are also acceptable but will then possibly require a supplementary conversion step. The difficulty of this step should not be underestimated. It is well known that for instance pure PuO_2 is not easy to dissolve in nitric acid. Hence the preconditioning step must be considered carefully, as it can have important consequences on the overall economy of the project.

3.2. MELTING PROCESS

Worldwide mainly two types of vitrification processes are operational: a calcination followed by melting in an induction-heated metallic melter, as in the French AVM (Atelier de Vitrification de Marcoule) and a one-step melting in a joule-heated ceramic melter as in the Pamela installation at Belgoprocess. This paper is based on the latter type.

The high-level liquid waste is mixed with plutonium nitrate. This mixture is fed to the ceramic melter, simultaneously with the borocilicate glass frit. On top of the glass pool, a process zone is formed where drying calcination and melting reactions occur. The working temperature of the melting furnace is about 1150°C and the average retention time is about 10 h.

Critical for the operation of the melter is to have a constant composition of the feed material and hence of the molten glass in order to assure a stable joule heating effect between the electrodes. It is also important to avoid the sedimentation of noble metals such as Pd, Rh or Ru in the melter. Due to their lower specific electrical resistivity the heat dissipation between the electrodes would be reduced and consequently also the throughput of the melter. The accumulation of noble metals can be lowered by improving convection of the molten glass mass by a nitrogen bubbling system and by an appropriate melter design.

3.3. WELDING AND DECONTAMINATION OF THE CANISTERS

Canisters filled with glass are stored for some time to cool down. Once the equilibrium temperature is reached the rim of the canister wall is thoroughly cleaned before being covered with a lid. This lid is then tig-welded, assuring the leak thightness of the container. After a complete decontamination of the external surface of the canister it is ready for transport to the temporary storage.

3.4. OFF-GAS AND SECUNDARY WASTE

The melter off-gas is continuously routed through a dust scrubber, a condensor and a jet scrubber for removing the particulate elements. The nitric components are trapped in a NOx absorber. The final steps in the off-gas system consist of a glass fiber filter, followed by a HEPA filter.

The secundary liquid waste from the condensor and the scrubbers is treated by evaporation. The concentrate is recycled to the melter whereas the distillate is further treated as low-level liquid waste.

4. Cost Evaluation of the Plutonium Vitrification Process

4.1. ECONOMIC ANALYSIS

It is not possible, with all the unknown parameters and open options, to make a precise evaluation of what weapons plutonium vitrification will cost. The aim of this analysis is merely to specify and to define some principles and concepts, and to indicate the order of magnitude with rough figures which can be refined later on.

4.2. REFERENCE SIZE OF THE PROGRAMME

As mentioned before (§ 1) the total plutonium problem is of the order of 200 t but it is unlikely that the totality of the weapons grade plutonium will be treated by vitrification. Rather than trying to give an adequate figure for this quantity we will base our cost estimate on a small installation with a throughput of 100 t glass per year, which is comparable to the Pamela facility at Belgoprocess. Extrapolation of the figures may generate some scale economies. However if Pu is to be handled in small quantities, the scale economies will remain small. They will not exceed the overall precision of the present estimations.

4.3. INVESTMENT AND OPERATIONAL COST

The investment for such a vitrification installation in US dollars of 1995 would amount to $ 200 million. If one considers that this installation will have an economical life-time of 20 years and that the interest rate during this period will amount to 5 %, the annual depreciation rate (a) would be

$$a = \frac{i\,Vo\,(1+i)^n}{(1+i)^n - 1}$$

with $i = 5\ \%$
$Vo = \$\ 2\ 10^8$
$n = 20$ years

resulting in $ 16 million.

Operational costs can be estimated on a yearly basis at about $ 12 million for an operational crew of 50.

According to this evaluation the yearly vitrification cost amounts to $ 28 million for a production of 100 t of glass.

In order to refer this cost to the amount of plutonium vitrified the plutionium content of the glass product must be determined. To our experience the glass can contain 10 w% to 20 w% of waste oxides. Adopting the lowest proportion one gets a conservative cost per unit of waste oxide of $ 2.8 million per tonne or $ 2.8 per gramme.

As in this approach the vitrification serves the double role of conditioning both waste oxide and plutonium, and that within certain limits plutonium oxide and waste oxide can be exchanged with each other without changing the structure of the glass, the same unit cost is attributable to both waste oxide and plutonium oxide. This approach supposes however that apart from the plutonium to be conditioned, a corresponding HLLW volume can bear its part of the total cost of the vitrification operations. This evaluation shows that the cost depends on the total amount of waste oxides in the glass product, inclusive plutonium oxide.

4.4. INTERMEDIATE STORAGE AND FINAL DISPOSAL COST

Fixed tariffs for geological disposal of HLW are not yet known as no such installation is yet operational. There are however best estimates for different disposal options which should give a fair approximation of what the final cost will be.
The cost for intermediate storage and final disposal in deep clay layers, inclusive the R and D, is estimated at \$ 500,000 per m^3 of conditioned waste

Considering that 1 m^3 corresponds with 2.5 t of glass, the cost for one tonne of glass product is \$ 200,000. If the waste oxide content of the glass product is 10 % the cost for one tonne of waste oxide is \$ 2,000,000. If waste oxide can be exchanged with plutionium oxide, the cost per gramme of plutionium oxide would be \$ 2.

5. Conclusion

The vitrification of plutonium as high-level waste appears to be a feasible solution for at least part of the large amounts of weapons grade plutonium that will become available after dismantling of the nuclear warheads.

Mainly due to criticality problems, but also from the safeguarding point of view, plutonium should not be vitrified alone but mixed with other high-level waste.

Extensive development work is required to define optimal mixture proportions, glass composition and ways to precondition the plutonium and introduce it into the melter.

Providing the plutionium is vitrified together with high-level waste the overall unit cost per gramme of waste oxide and of plutonium oxide can be considered the same and amounts to approximately to \$5.

If however the expenses would have to be borne by the sole plutonium, the unit cost per gram plutonium oxide would become 5 to 10 times higher depending whether 2 % or only 1 % of plutonium oxide can be mixed with the glass.

6. References

1 Demonie M., Cuyvers H.; (1995) On the feasibility of vitrifying high-level liquid waste containing high amounts of plutonium; *NATO Advanced Research Workshop on Disposal of Weapons Plutonium - Approaches and Prospects; St. Petersburg, Russia, May 14 - 17, 1995.*

2 Schneider A., Li K.K. and Schumacher R.F.; (1991) Remote monitoring of molten radioactive glass; *The XV International Symposium on the Scientific Basis for Nuclear Waste Management; Strasbourg, France, November 4 - 8; 1991.*

3 Matzke Hj. and van Geel J.; (1995) Incorporation of Pu and other actinides in borosilicate glass and in waste ceramics; *Nato advanced Research Workshop, St. Petersburg, Russia, May 14 - 17, 1995.*

4 Wicks G.G. and al.; (1995) SRS Vitrification studies in support of the US program for disposition of excess plutonium; *Nato advanced Research Workshop, St.Petersburg, Russia, May 14 - 17, 1995.*

5 Inagaki Y., Furuya H. and Idemitsu K.; (1991) Microstructure of simulated high-level waste glass doped with short-lived actinides, ^{238}Pu and ^{244}Cm; *The XV International Symposium on the Scientific Basis for Nuclear Waste Management; Strasbourg, France, November 4 - 8; 1991.*

6 De A.K., Wiese H. and Demonie M.; (1990) Improvement of vitrification operation by optimization of the borosilicate glassfrit at the Pamela plant; *Spectrum '90; Knoxville, Tennessee, U.S.A.; September 30 - October 4, 1990.*

DISSOLUTION OF MATERIALS CONTAINING METAL PLUTONIUM AND OXIDIZED FORMS OF PLUTONIUM

G.P. NIKITINA, Yu.E. IVANOV, M.Yu. KIRSHYN,
A.A. LISTOPADOV, L.B. SHPUNT
Khlopin Radium Institute, S.-Petersburg, Russia

Rich wastes and reversed products of plutonium metallurgical production are enough various. Among them the considerable part belongs to the wastes containing metal plutonium: defective ingots, cutting, casting wastes, refining-melt wastes. Plutonium content in this wastes attains 990 g/kg of raw material. Other group of wastes contains dominantly oxidized forms of plutonium, to which it is assigned: substandard plutonium dioxide, calcination furnace sweepings, chamber sublimates, insoluble residues, ash of calenders, rags, paper combustion and so on.

Dissolution of materials containing metal plutonium is generally carried out by mechanically sectioning plutonium metal to small pieces, combustion them to plutonium dioxide with following PuO_2-dissolving in mixture of nitric and hydrofluoric acids [1]. Besides, it is possible the anodic dissolving of metal plutonium (with utilizing the "metal contact" principle or "liquid contact" one) [1,2].

Four general directions exist in dissolution techniques development of materials containing oxidized forms of plutonium:

- improvement of PuO_2-dissolving in mixtures nitric and hydrofluoric acids by introduction oxidizing agents (Ag_2O_2, Co(III), Ce(IV), Am(VI) [3] or considerable quantity of fissile materials (100÷200 g/l Pu or Pu+U) [4];
- introduction of stage preceding PuO_2-dissolving in nitric acid with following digestion of the product prepared at this stage by boiling nitric acid. As such stage it is proposed the preparation of the substoichiometric oxide $PuO_2 \rightarrow PuO_{1,8}$ by reduction with H_2 or CO at 1800°C [5]; different melting techniques; solid solution preparation of plutonium dioxide in UO_2 [6];
- improvement of autoclave dissolution techniques [7,8];
- redox-dissolving in nitric acid with using homogeneous catalysts [9-16], reoxidized with ozone or electrochemically [9-11,13-16].

E.R. Merz and C.E. Walter (eds.), Disposal of Weapon Plutonium, 59-64.

In France technology practice the electrochemical oxidizing PuO_2-dissolution used Ag-ions as catalyst has recently received wide acceptance [11,13-16].

Khlopin Radium Institute proposes the new concept of dissolving Pu-containing materials. In accordance with this concept it is proposed to bring about anodic dissolving the materials containing metal plutonium under conditions of "metal contact" and "liquid contact" principles combination (combined approach). Hereafter the formed sludge is dissolved electrochemically in dilute nitric acid (4 M) containing cerium with catalyst-additives.

In process based on "metal contact" principle dissolvable plutonium is physically attached to the anode (direct Pu-screwing on anode or using the principle of valve metal [1]). In the latter case plutonium is contained in valve metal basket which serves as both the anode and a container for dissolvable materials. At the attachment point of plutonium and valve metal there is additional feeding the oxide barrier (blocking) layer of valve metal with plutonium electrons, transforming this layer into electroconducting one and providing electric contact of plutonium with anode.

Dissolving in the current field ("liquid contact" principle) [1] is based on electrostatic induction. Physically insulated from electrodes dissolvable metal is placed in basket between cathode and anode. Electrons pass through the dissolvable metal at transfer from one electrode to another. In this case the plutonium side facing to anode turns into cathode, the contrary side becomes anode. Dissolving takes place on anode side of metal.

Both techniques suffer from many disadvantages. "Metal contact" process is characterized by the difficulty in maintaining a good high-conducting contact between the anode and dissolving material. "Liquid contact" process is characterized by complex construction of apparatus-dissolver, which is bound to provide the most short distance between cathode and anode with exception of electrodes shorting and the possibility of initiation arc which can break down the electrodes. Therefore, it is proposed to accomplish dissolving by combination of the both approaches - "metal contact" and "liquid contact" those.

In combined approach electrode-basket system contains insulated cathode and valve-metal anode basket connected by electric contact with Pt-anode (fig.1). As long as there is direct contact dissolvable metal with anode the dissolving is carried out in contact regime. On breaking this contact dissolution is continued in current field. For combined regime the current efficiency depends only from charging extent of basket and practically does not depend from the current density. So, at j_a=2 A/cm^2 and filled basket the current efficiency of stainless-steel anodic dissolving is equal ~90% from theoretical dissolving. On basket charging of 50% the current efficiency in combined regime significantly exceeds that in non contact regime and constitutes 70 and 50% from the current efficiency for filled basket respectively.

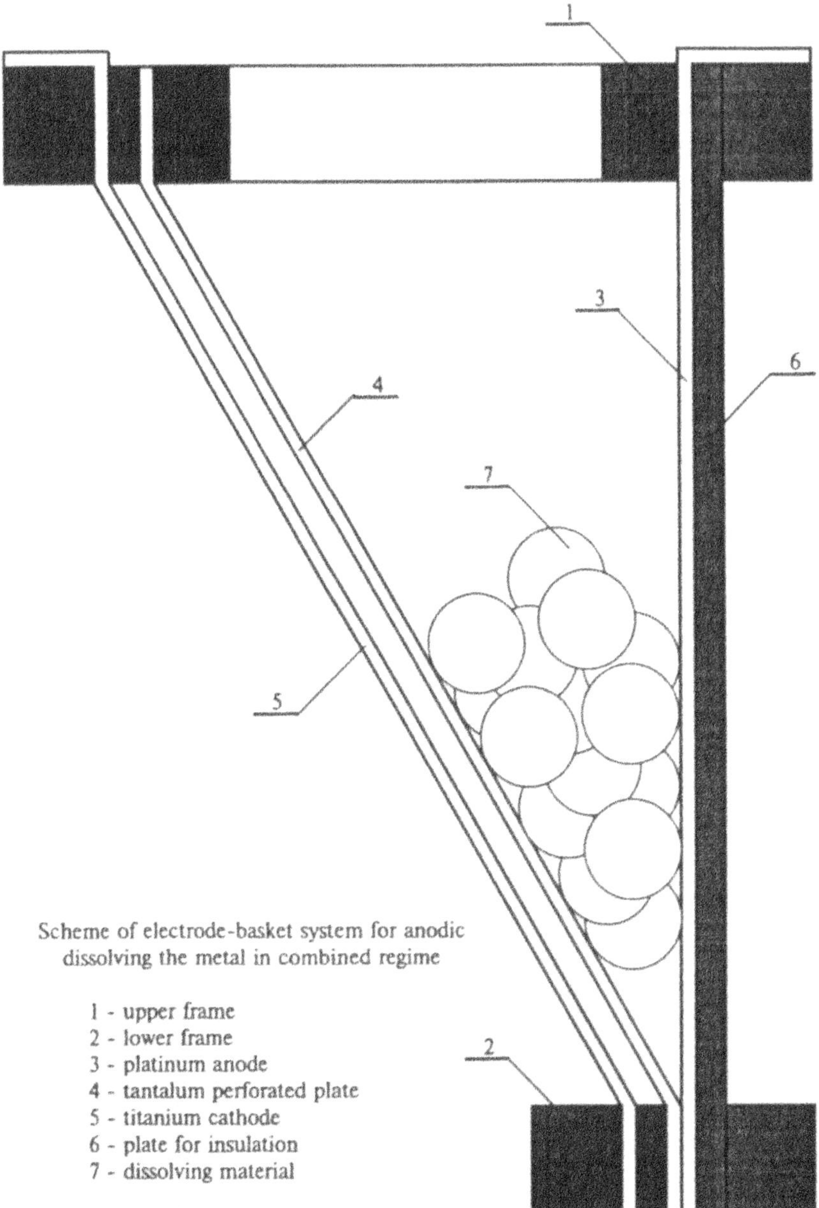

Scheme of electrode-basket system for anodic dissolving the metal in combined regime

1 - upper frame
2 - lower frame
3 - platinum anode
4 - tantalum perforated plate
5 - titanium cathode
6 - plate for insulation
7 - dissolving material

Fig.1

On one-time metal-plutonium chargings of 1 kg, current density 1÷2,5 A/cm^2, temperature 30÷65°C and [HNO_3]=4÷7,5 M dissolving rate in combined regime attains 200 g/h. Highest current efficiency is equal of 0,9÷1,0 g Pu/Ah. Dissolving rate significantly depends from the extent of basket charging. Dissolving rate in current field at the same apparatus geometry is half as many as dissolving rate in combined regime.

On anodic unpassivating the plutonium along with its anodic dissolving the chemical interaction of oxide-film uncoated plutonium with water is realized, with new solid phase (sludge) forming. This phase quickly settles and is well filtered. Mass fraction of sludge depends from nitric acid concentration and temperature, increasing with the augmentation of this parameters. For pure metal plutonium this fraction is small (5÷9,6% of initial raw material mass). The sludge dissolving is accomplished by electrochemical oxidizing technique. It consist in the sludge digestion with hot nitric acid (4 M), containing homogeneous catalyst ($6 \cdot 10^{-2}$ M Ce+certain additives in concentrations from $1 \cdot 10^{-2}$ M to $1 \cdot 10^{-7}$ M) at current density ~1 A/dm^2 and volume-mass relation of ~10:1. In this case for 2 hours 99% of plutonium, which is contained in sludge, go into solution in plutonyl-ion form. Plutonium stabilization in oxidation state-IV before the extraction is accomplished in the same apparatus-dissolver by current reverse without introducing in solution the otherwise substances (salt-free technique) [17].

Anodic plutonium dissolving in combined regime together with electrochemical oxidizing dissolution of sludge provides the going into solution more 99,9% of plutonium from initial material.

Advantage of using the complex homogenous catalyst containing cerium and the certain additives is that it "works" even under high radiation fields. So, in solutions with high radioactivity level Ag(II) (reoxidized by the ozone or electrochemically) is quickly reduced with radiolysis products, and in this case plutonium dioxide practically is not dissolved. However cerium(IV) with catalyst-additives effectively transfers into solution even high-calcined dioxide of plutonium-238.

In addition it is possible to use the alternate dissolving technique of materials with metal plutonium (in part, cutting). This technique involves Pu-oxidation to plutonium dioxide (by calcining or firing) with following dissolving the plutonium dioxide by electrochemical oxidizing technique. In this case at volume-mass relation of solvent/PuO_2=20/1 for 4 h it is possible to transfer into solution up to 97,1÷99,9% plutonium, which is contained in initial material. In this conditions substandard plutonium dioxide is dissolved for 1,5÷2 h.

Electrochemical oxidizing dissolution technique effectively transfers the plutonium into solution from diverse wastes containing oxidized plutonium forms (furnace sweepings, sublimates from chamber, ash from firing the paper, rags and so on).

Comparison of electrochemical oxidizing PuO_2-dissolution with its dissolving in mixtures nitric and hydrofluoric acids and certain melting techniques demonstrates indisputable advantages of the former. As to dissolving rate and degree of plutonium drawing out the electrochemical technique not only is not inferior to the dissolving in mixture of nitric and hydrofluoric acids, but often it exceeds the last-mentioned process. However, the negative features of fluoride-nitric dissolving are not inherent to electrochemical oxidizing technique; among them: multi-phase process, high fluoride concentration in finished solutions, the result of which is the corrosive and ecological hazards of this solutions and gas-exhausts and also the difficulty of reprocessing such solutions. The presence of high-temperature operation, multi-stage process and high salt-content of solutions are intrinsic for the melting techniques, also.

Contrary to considered methods the electrochemical oxidizing dissolving consist of one operation conducted at 90÷98°C. Finished nitric acid solutions ($[HNO_3]$=2.5÷4.0 M, [Pu]≤100÷120 g/l) are not contaminated with the corrosive product of apparatus and contain as the impurities no more than 7÷8.5 g/l cerium and very minor amounts (10^{-7}÷10^{-2} M) of catalytic additives, without interference for the following electrochemical stabilization of plutonium in the oxidizing state (IV) and the extraction or sorption reprocessing this solutions. Dual-cycle sorption reprocessing that solutions provides the purification of plutonium from cerium by the factor of 10^6, at extraction reprocessing the coefficient of Pu-purification from cerium exceeds 10^6 in one cycle.

The technique of electrochemical oxidizing dissolving is provided with construction materials. The apparatus must be maked up from the constructional titanium WT1-0. The platinized titanium is suited for anodes. In conditions of oxidizing dissolving the titanium corrosion does not exceed 0.02 mm/year. At used current density the corrosion of platinized anodes is no more than 1 μm/year. Investigation of the composition and release rate of electrolytic gases shows that the proposed technique is explosion-proof. The using 5 M HNO_3 as catholyte rules out the possibility of dangerously explosive hydrogen-concentration formation.

So, the proposed electrochemical oxidizing dissolving is the most progressive technique for plutonium transfer into solution from the sludge of anodic dissolving the metal plutonium, reversed products and wastes containing Pu-metal (if they are preferentially calcined (or fired)), and also from substandard plutonium dioxide and other materials containing oxidized forms of plutonium.

REFERENCES

1. Wheelwright, E.J., Fox, R.D. (1977) Development of Electrolytic Dissolver for Plutonium Metal, *Ind. Eng. Chem., Process Des. Dev.* **16**, 297-302.
2. Berry, J.W., Miner, F.J. (1968), Anodic Dissolution of Plutonium, USAEC Paper RFP-1136.
3. Tallent, O.K. (1976) Dissolution of Plutonium Dioxide. Pat.-3 976 775 (USA). Filed May 06, 1975.
4. Mills, A.L., Weatherlay, L.R., and Croft, J. A., (UKAEA) (1981) Method for Dissolution Acceleration and Solubility Increasing the Plutonium Dioxide, Mixed U/Pu Oxides and Their Wastes in Nitric Acid, Pat. 2 455 565 (France). Patented Apr. 29, 1980, *Bulletin Officiel de la Propriete Industrielle,* №1,2 p.44.
5. Deaton, R.L., Silver, G.L.(1972) Dissolution of Plutonium Dioxide, *Radiochem. Radioanal. Lett.* **10**, 277-283.
6. Amal, Th., Hoyraud, J., Trestini, J.Cl. (1981) Method for Dissolving Plutonium Dioxide Which is Contained in Wastes of Nuclear Fuel, CEA-Pat 2456545 (France). Patented May 14, 1979. *Izobreteniya v SSSR i Zarubejom (in Russian)* №10, p.30.
7. Ledebrink, F.R., Rosenkranz, W., Stoll W.(FRG) (1982) Method for Dissolving Hard-to-Dissolve Thorium and/or Plutonium Oxides, Pat 3038807 (FRG). Patented Oct.14, 1980, *Izobreteniya v SSSR i Zarubejom (in Russian)* №10, p.15.
8. Ledebrink, F.R., Rosenkranz, W., Stoll W.(FRG) (1986) Method for Dissolving Hard-to-Dissolve Thorium and/or Plutonium Oxides, Pat. 4 528 130 (USA), Patented July 09, 1985, *Izobreteniya Stran Mira (in Russian)* №6, p. 22.
9. Ryan, J.L., Bray, L.Al., Boldt, Al.L. (1983) Dissolution of PuO_2 or NpO_2 Using Electrolytically Regenerated Reagents, USAEA-Pat 00 89 185 A2. Priority March 11, 1982. Published Sept. 21, 1983, *Bulletin* 83/38.
10. Komarov E.V., Nikitina G.P., Pushlenkov M.F., Usachev V.A., Shumkov V.G. (1978) Oxidizing Dissolution Method of Transuranium Elements Dioxides Calcined to 800-1000°C, Inventor's Certificate. №440879 (Soviet Union). Filed. Jan. 04,1967, *Byuleten Izobretenii (in Russian),* №41.
11. Koehly, G., Bourges, J., Lecomte, M., Madic, Ch. (1984) Method and Equpment for Dissolving the Oxides of Plutonium and/or Neptunium, CEA-request 2 561 942 (France). Filed March 27, 1984, *Bulletin Officiel de la Propriete Industrielle* (1984) №40.
12. Akihiko Inoue, Takeshi Tsujino (1985) Dissolving Method of Plutonium Dioxide and/or Mixed U/Pu Oxides, Pat. 2 553560 (France). Patented Oct. 12, 1983, *Izobreteniya Stran Mira (in Russian)* №9, p.15.
13. Bourges, J., Madic, C., Koehly, G., Lecomte, M. (1986) Dissolution du Bioxide de Plutonium en Milieu Nitrique par L'Argente (II) Electrogenere (Dissolution of Plutonium Dioxide in Nitric Acid Media with Electrogenerated Silver (II)), *J. Less Common Metals,* **123**, 303-311.
14. Ryan, J.L., Bray, L.A., Boldt, A. L. (1988) Dissolution of PuO_2 or NpO_2 Using Electrolytically Regenerated Reagents, Pat. 4 686 019. Filed Sep. 26, 1984. Patented Aug.11, 1987.
15. Bernard, C., Chenevier F.(1988) Reprocessing of Thermal Oxide Fuel at La Hague: Overview of the UP-3 Reprocessing Plant, *Proceed. Int. Symp. on Reprocessing and Waste Management., Aug. 1988, Denver (USA).*
16. Gue', J.P., Philippe, M., Farrugia, J.M., Decobert, V. (1993) French Experience in MOX Fuel Dissolution, *Proceed. Int. Conf. and Technology Exposition on Future Nuclear Systems. Emerging Fuel Cycle and Waste Disposal Options. Global'93, Sept. 12-17, 1993, Seattle, Washington,* Published by the American Nuclear Society, Inc. La Grange Park, Illinois (USA).
17. Nikitina, G.P., Egorova, V.P., Jukova, I.N. (1995) The Reduction of Plutonium (VI) with Electrolytically Generated Nitrous Acid, *Radiokhimiya (in Russian),* **37**, №3 (in press).

Crystalline Ceramics:

Waste Forms for the Disposal of Weapons Plutonium

R.C. Ewing, Department of Earth & Planetary Sciences
University of New Mexico
Albuquerque, NM 87131 USA

W.J. Weber, Materials Science Department
Pacific Northwest Laboratory
Richland, WA 99352 USA

Werner Lutze, Center for Radioactive Waste Management
University of New Mexico
Albuquerque, NM 87131 USA

1. Abstract

At present, there are three seriously considered options for the disposition of excess weapons plutonium: *i)* incorporation, partial burn-up and direct disposal of MOX-fuel; *ii.)* vitrification with defense waste and disposal as glass "logs"; *iii)* deep borehole disposal [1]. The first two options provide a safeguard due to the high activity of fission products in the irradiated fuel and the defense waste. The latter option has only been examined in a preliminary manner, and the exact form of the plutonium has not been identified. In this paper, we review the potential for the immobilization of plutonium in highly durable crystalline ceramics apatite, pyrochlore, zirconolite, monazite and zircon. Based on available data, we propose zircon as the preferred crystalline ceramic for the permanent disposition of excess weapons plutonium.

2. Introduction

During the period from 1977 to 1982, there was a tremendous diversity in the types of nuclear waste forms under development. In the United States, much of this work ended with the decision to use borosilicate glass as the waste form for defense waste at Savannah River. Synroc, a ceramic waste form, was selected as the alternative waste form, but further development in the United States ended in the absence of funded programs. Major research and development programs for the development of Synroc continued in Australia culminating in the construction of a "cold" Synroc pilot-scale processing plant. Basic research on the properties of Synroc has been continued at the Australian National University and at the Australian Nuclear Science and Technology

E.R. Merz and C.E. Walter (eds.), Disposal of Weapon Plutonium, 65-83.

Organization with collaborative work at the Japan Atomic Energy Research Institute and AERE Harwell in the United Kingdom. Synroc remains perhaps the most thoroughly studied ceramic alternative to borosilicate glass. Lutze and Ewing [2] have provided a summary of work on waste forms through early 1987, as well as one of the first efforts to compare the performance of borosilicate glass with that of a crystalline ceramic waste form, Synroc [3].

Investigations into the properties and performance of other ceramic waste forms have revived during the past five years for application to special waste stream compositions. At Lawrence Livermore National Laboratory, a Mixed Waste Management Facility is being developed in order to demonstrate an alternative to incineration. The waste form is a derivative of a Synroc composition originally developed for the immobilization of reprocessed residues at Savannah River, and typical phases include zirconolite, perovskite, spinel, nepheline and rutile. Although the radioactivity is low, this does illustrate the ubiquity of a rather limited set of crystalline phases. Additionally, the zirconolite phase in Synroc has received increased attention as a host phase for actinides, particularly excess weapons plutonium [4]. At the Idaho National Engineering Laboratory, an iron-enriched basalt waste form has been under development, and the addition of $ZrO2$ and $TiO2$ has produced zirconolite crystals as an actinide host in a silicate ceramic. A "glassy" slag and a "vitreous" ceramic are being developed at Pacific Northwest Laboratory, both of which are composed of crystals which include zirconolite, perovskite and zircon (up to 80 volume percent) in a silicate glass matrix. At the Khlopin Radium Institute in St. Petersburg, Russia, a program has been initiated for the development of crystalline hosts for actinide wastes, with an emphasis on zircon, yttrium-aluminum garnet and cubic zirconia [5]. Most recently, spent fuel - a metal clad, ceramic oxide - has received important consideration as a waste form because its use eliminates the need for reprocessing, and it is highly stable under reducing conditions [6], and the MOX-fuel option for weapons Pu is a simple extension of the concept of direct disposal of spent nuclear fuel. Thus, there are considerable data and experience in the design, evaluation and selection of waste forms [7-9]. We draw on this experience in evaluating crystalline ceramics as waste forms for the immobilization and disposition of weapons plutonium [10].

3. Importance of the Waste Form

Despite the great challenge of handling chemically complex wastes, which are highly radioactive and of great volume, the greatest challenge still lies in the development and evaluation of the *long-term durability* of waste forms. Materials scientists will have to design materials to performance standards that are not measured in decades, but rather are measured in 10^4 to 10^6 years, particularly in the case of long-lived radionuclides such as ^{239}Pu. The issue of long-term durability is unusual in materials science and requires interdisciplinary research programs with rather unusual combinations of subdisciplines -- processing/synthesis technologies, materials properties, mineralogy and geochemistry. Thus, even after issues of technological feasibility and cost are considered and settled, the most difficult scientific question remains: *What is the long-term durability of the waste form?* and *What is the effect of improved durability on repository performance assessments and the calculated dose to humans?*

Thus, the development of durable waste forms can play a critical role in reducing the uncertainty in performance assessments. The reduced uncertainty should lead to greater public acceptance of proposed disposition and disposal concepts. The specific value of the waste form has been previously discussed [7], but it is useful to recall the requirements for waste form development and acceptance:

i.) careful consideration of synthesis and processing technologies;
ii.) a detailed characterization of the wastes and the waste form after immobilization;
iii.) an extensive data base of corrosion/alteration experiments over a wide range of conditions, and in some cases, for extended periods (both repository-relevant and special experiments designed to elucidate the corrosion mechanism);
iv.) kinetic models of the corrosion/alteration process and thermodynamic models that predict the formation and stability of phases which will control solution compositions;
v.) knowledge of the effects of other materials, e.g., canisters and bentonite overpack, on waste form stability;
vi.) studies of relevant natural phases or systems to confirm experimental and extrapolated results.

The last requirement is particularly important in the selection of waste forms for the disposition of weapons plutonium, as a twenty half-life durability criterion leads to disposal periods of a 500,000 years. This would reduce the Pu inventory of 100 metric tons to well below a critical mass for Pu; however, the 100 metric tons of Pu is transmuted to fissile ^{235}U. Using the same twenty half-life criteria, the disposal period of concern for the ^{235}U daughter product is 14 billion years. Thus, the discussion in this paper of crystalline phases is limited to those phases which occur naturally and for which some assessment of long-term durability is possible. Additionally, for ^{235}U, the decay product of ^{239}Pu, the waste form must provide a means of reducing the probability of criticality events.

4. Principles of Nuclide Isolation in Ceramic Waste Forms

In contrast to glass waste forms in which the radionuclides are in principle homogeneously distributed throughout the waste solid, ceramic waste forms incorporate radionuclides in two ways:

(1) Radionuclides may occupy specific atomic positions in the periodic structures of constituent crystalline phases, that is as dilute solid solutions. The coordination polyhedra in each phase impose specific size, charge, and bonding constraints on the nuclides that can be incorporated into the structure. This means that ideal waste form phases usually have relatively complex structure types with a number of different coordination polyhedra of various sizes and shapes and with multiple substitutional schemes to allow for charge balance with radionuclide substitutions. Extensive nuclide substitution can result in cation and anion vacancies, interstitial defects, and finally changes in structure type. One expects, and finds, the formation of polytypes and twinning on a fine scale. The point defects can themselves become sites for the radionuclides. Except in unusual situations (e.g., monazite, CeP04), the complexity of the waste composition results in the formation of a polyphase assemblage (e.g., Synroc consists of phases such as zirconolite, CaZrTi207; perovskite, CaTi03; and "hollandite", BaAl2Ti6016), with unequal partitioning of radionuclides between the phases, and some elements which are not accommodated at all (e.g., Mo and Tc) and

may exist in the reduced metallic state.. In Synroc, actinides partition preferentially into the zirconolite phase [11]. The polyphase assemblages are sensitive to waste stream compositions, and minor phases form, including glass, segregated along grain boundaries. Ideally, all waste stream elements, radioactive and non-radioactive, are important components in the phases formed. In some rare cases, a single phase (e.g., monazite or sodium zirconium phosphate, NZP) can incorporate nearly all of the radionuclides into a single structure.

(2) Radioactive phases, perhaps resulting from simply drying the waste sludges, can be encapsulated in non-radioactive phases. The most common approach has been to encapsulate individual grains of radioactive phases in TiO_2, Al_2O_3, or ZrO_2, mainly because of their extremely low solubilities. This usually requires major modifications to the waste stream composition and special processing considerations to maintain temperatures that are low enough to avoid volatilization of radionuclides. A similar approach may be taken with low temperature assemblages (e.g., mixing with concrete), but in this case there is the possibility of reaction between the encapsulating phase and the radioactive phases. Some phases, such as zircon, can both incorporate a radionuclide at the atomic scale (e.g., Pu substituting for Zr in the structure of zircon) and encapsulate unreacted materials from the synthesis (e.g., PuO_2 as inclusions in zircon crystals). The combined incorporation and encapsulation can lead to a very durable waste form.

Finally, the complexity of ceramic waste forms is mostly the result of the compositionally complex waste streams that must be solidified. In the case of weapons plutonium, the disposal of a pure nuclide is an ideal situation for the use of highly durable ceramics which have specific atomic sites for the incorporation of Pu and its decay product, U.

5. Advantages and Disadvantages of Ceramic Waste Forms

The main advantage of ceramic waste forms lies in the fact that they hold the potential for engineering a phase assemblage which provides unique structural hosts for specific radionuclides. Ideally these hosts should be thermodynamically stable, but for most repository environments this is unlikely (a notable exception is titanite, $CaTiSiO_5$, in the ground waters of the Canadian shield); but one can already demonstrate greater stability for some of the ceramic phases than for the metastable borosilicate glass. The development of ceramic waste forms which are stable at high temperatures has even more important implications: 1) higher thermal stability provides the possibility of higher waste loadings, and thus a reduction in the amount of material to be handled and space required for disposal; 2) higher thermal stability means that disposal can occur in rock units at greater depth or in canister arrays with closer spacings. The decreasing stability of borosilicate glass with increasing temperature is well known, and in fact, disposal concepts for borosilicate glass are shaped by the limitations imposed by its thermal instability. Even without the higher thermal stability, higher waste amounts are incorporated into ceramics because of their higher density (a factor of two over borosilicate glass).

A final consideration is that many of these ceramic phases occur in nature (e.g., zirconolite, pyrochlore, perovskite, zircon, monazite, uraninite, and coffinite). This provides the possibility of evaluating the long-term durability of these phases in the presence of aqueous solutions and at α–decay doses that reach values comparable to those which the waste form will experience in the first 1,000 to 10,000 years after disposal. The ability to validate projected long-term behavior is a critical part of the performance assessment of the nuclear waste containment strategies.

The disadvantages of ceramic waste forms are an inherent part of their complex microstructures. First, for polyphase ceramics, the corrosion process may be inherently more complicated than the corrosion process of an essentially single phase glass (although most glasses contain a proportion of crystalline phases). Despite a considerable amount of experimental data, there is no confirmed model for the long-term corrosion of Synroc. Second, the atomic periodicity of crystalline materials is disrupted by α–decay damage. Thus, there is the possibility of a radiation-induced transformation from the periodic-to-aperiodic state (amorphization or metamictization). This is a well known process in minerals which contain uranium and thorium and is observed in actinide-doped phases (e.g. Pu-doped zirconolite and zircon). The process is mitigated by natural annealing (which increases with higher temperatures), and some phases anneal at low enough temperatures that they are only found in the crystalline state (e.g., uraninite and monazite). Still, this transformation can result in decreased chemical durability, and the volume expansion associated with the transformation can cause microfracturing with an increase in the surface area exposed to corroding fluids. For any selected ceramic phase, there must be a through understanding of its behavior at high radiation doses.

6. Waste Forms for Plutonium

There has been no concerted research program to develop ceramic waste forms specifically for the disposal of excess weapons plutonium; however, there is extensive enough experience with crystalline phases to review at least five possible phases: apatite, pyrochlore, zirconolite, monazite and zircon. We restrict ourselves to these five phases, as they all occur in nature and may often contain uranium and thorium; thus, all five of these phases may be evaluated in terms of their long term durability in their natural geologic occurrences.

Additionally, one must note the special features of excess weapons plutonium which distinguish it from typical radioactive waste:

i.) The plutonium is almost isotopically pure ^{239}Pu; therefore, it is ideal for the synthesis of specific phases of high durability.

ii.) ^{239}Pu has a long half-life (24,500 years), as compared to most fission products; therefore, extended, confirmable durability is a highly desired attribute.

iii.) ^{239}Pu is fissile; hence, one must always be concerned about criticality events; furthermore, ^{239}Pu decays to ^{235}U which has a half-life of 700 million years and is also fissile. Thus, the durability of the waste form is not only essential to the retention of Pu and U, but is also required in order to maintain the presence of any added neutron poisons (e.g., B, Li, Gd, or Hf).

iv.) The multiple oxidation states of Pu and U make the processes of dissolution, oxidation, transport and reconcentration (= supergene enrichment) over geologic periods a possible event in the development of scenarios for performance assessments, particularly in oxidizing environments (e.g., the dissolution of volcanic glass, where U is in ppm concentrations, with the subsequent release and concentration of uranium to form high concentrations of uranium in roll-front ore deposits).

6.1. APATITE

Rare-earth silicates with the apatite structure have been observed or proposed as actinide host phases in a borosilicate nuclear waste glass [12], a multiphase ceramic waste form [13], a glass ceramic waste form [14], and a cement waste form [15]. These apatites generally have the compositions: Ca4-xREE6+x(SiO4)6-y(PO4)y(F,OH,O)2 (where REE = La, Ce, Pr, Nd, Pm, Sm, Eu, and Gd) and are isostructural with the natural mineral apatite, Ca10(PO4)6(F,OH)2. Plutonium and other actinides readily substitute for the rare earths in this hexagonal crystal structure. At least 6 wt. % actinides (^{244}Cm and ^{240}Pu) have been incorporated into an apatite phase in a devitrified waste glass [12], and a phase-pure apatite containing 1.8 wt % actinides (^{244}Cm and ^{240}Pu) has been prepared [19]. Much higher weapons Pu-loadings are feasible in the rare-earth apatites by substitution for the rare earths. In addition, rare earths, such as Gd, can be included as neutron poisons.

Natural apatites with appreciable Th- and U-contents provide analogues for the performance and properties of actinide-containing apatites over geologic periods. For example, natural apatites with significant amounts of rare earths and Th are reported to be partially metamict as a result of self-irradiation damage from the α–decay of Th [16]. Fission track analysis in U-containing natural apatites is used for geologic age dating, and annealing studies provide data on fission track recovery kinetics [17,18]. Of more relevance is the recent discovery of natural apatites formed near the Oklo natural reactors that indicate significant amounts of ^{239}Pu and U were incorporated into the apatites, leading to ^{235}U enrichment that has been retained for 2 billion years [17].

Natural apatites [16] and rare-earth silicate apatites containing actinides [12,13,19] do undergo an α–decay-induced crystalline-to-amorphous transformation. The mechanism [20] and temperature dependence [21] of this transformation is well understood and modeled. The macroscopic swelling associated with the amorphization process is 9.5 percent and has also been modeled [20]. Other property changes associated with this α–decay-induced transformation have been summarized in a recent review by Ewing et al. [22]. Fission track annealing studies of natural apatites [17] and ion-beam irradiation studies of natural and synthetic apatites [21,23] suggest that simultaneous thermal recovery processes will minimize or even prevent amorphization in Pu-containing apatites under deep borehole conditions (temperature > 150-200°C). The chemical durability of natural apatites is well demonstrated by the 2 billion year old, ^{235}U-enriched Oklo apatites [17]. In a study of a rare-earth silicate apatite containing ^{244}Cm and ^{240}Pu [24], the dissolution in deionized water at 90°C occurred congruently, and the measured Pu dissolution rate for the undamaged crystalline phase was 0.035 g/m^{2}d (this dissolution rate does not account for the 22 percent porosity of this material). Alpha-decay-induced amorphization increased the average Pu dissolution rate in this apatite by a factor of 12.

6.2. PYROCHLORE and ZIRCONOLITE

Pyrochlore (*Fd3m*, Z = 8), $^{VIII}A_2{}^{VI}B_2{}^{IV}X_6{}^{IV}Y$ is a derivative of the fluorite structure type in which the A-site contains large cations (Na, Ca, U, Th, Y and lanthanides) and the B-site consists of smaller, higher valence cations (Nb,Ta, Ti, Zr, Fe^{3+}). The essential feature of the structure is sheets of corner-sharing, BX_6 octahedra parallel to the (111) plane which are arranged into three- and six-membered rings (the hexagonal tungsten bronze structure). Monoclinic zirconolite (the principal actinide bearing phase in Synroc) is a derivative of the cubic pyrochlore structure. Actinides may be accommodated in the A-site, and charge balance is maintained by cation deficiencies in the A-site and substitutions on the B-site. Rare-earth titanates with the pyrochlore structure have been observed as actinide-host phases in nuclear waste glasses, in titanate ceramic waste forms, and in a glass ceramic waste form. A full summary of work on pyrochlore and zirconolite can be found in references [2,4,22].

The effect of α–decay on the pyrochlore phase Gd2Ti2O7 containing ^{244}Cm and ^{240}Pu has been extensively investigated by x-ray diffraction analysis and transmission electron microscopy [25,26]. At a dose of 3.1 x 10^{18} α–decay events/g, the material became fully amorphous with no evidence of any residual crystallinity. Wald and Weber [24] studied the radiation-induced changes in dissolution kinetics of Cm-doped Gd2Ti2O7 by testing fully-damaged (amorphized specimens and a second set of specimens that had been fully recrystallized to the original structure by a 12 hour anneal at 1,100°C). Specimens were immersed in distilled, deionized water at 90°C for 14 days. The measured ^{240}Pu release rate was 0.009 g/m^{2}d in the crystalline material and increased by a factor of 50 due to amorphization. The concentration of Ti in solution was below the detection limit, suggesting some dissolution resistance of the titania network. Analysis for non-network Gd was not performed. These results suggest that the dissolution may occur incongruently, selectively leaching the non-network ions. Similar behavior is observed for zirconolite [24]. The increase in dissolution rate due to amorphization is probably due to the increase in the size of the network tunnels as the structure disorders and expands.

Systematic studies [27-30] have been completed of α–decay event damage of natural, isometric pyrochlore structure types. Actinides may be accommodated in the A-site, and in natural pyrochlores the substitution of uranium can be extensive (up to 30 wt. %). Charge balance is maintained by cation deficiencies in the A-site and substitutions on the B-site (typically, Nb, Ta and Ti). Depending on the age of the sample, the total calculated displacements per atom, dpa, may be as high as 80. Careful analysis of samples from the same deposit, e.g., the Harding Pegmatite in New Mexico which is 1300 million years old, have allowed detailed studies of radiation damage over a range of dpa (0.2 to 49) as function of the uranium concentration (0.05 to 8.6 wt. % UO2). Members of the pyrochlore group are quite susceptible to hydrothermal and lower-temperature surface alteration [32-34]. In general the alteration involves the loss (or ion exchange) of large cations (Na and Ca) and the loss of Pb due to long-term diffusion. The alteration effects are enhanced by amorphization and fracturing due to the α–decay damage.

6.2.1. *Zirconolite* (monoclinic CaZrTi2O7) is one of the three principal phases of Synroc, and zirconolite is one of the most extensively studied waste form phases, as it is the primary actinide host. Monoclinic zirconolite is a fluorite-derivative structure closely related to pyrochlore. The (001) layers of the corner-sharing arrangement of TiO6 octahedra are similar to the (111) layers in pyrochlore and the (001) layers in the hexagonal tungsten bronzes. The A-site cations (Ca, Zr and actinides) are "sandwiched" between the (001) sheets of B-site cations (corner-sharing Ti06 octahedra). There can be several polytypes depending on the stacking arrangements of the sheets. In this section, we discuss the damage response as investigated by study of synthetic compositions doped with the short half-life actinides, ^{238}Pu (87.7 y, α–decay to ^{234}U) and ^{244}Cm (18.1 y, α–decay to ^{240}Pu), and natural zirconolites damaged by α–decay of uranium and thorium.

6.2.1.1. ^{238}Pu-substituted zirconolite can be synthesized with the substitution of PuO_2 for ZrO_2. X-ray absorption spectroscopy demonstrated that the ^{238}Pu was incorporated into the A-site [35]. The effect of this chemical substitution is to convert the zirconolite composition to an isometric, pyrochlore derivative structure [36,37]. A major change in $CaPuTi_2O_7$ with increasing dose is the large decrease in density. Storage at ambient temperature (approximately 350 K, due to self-heating) resulted in macroscopic swelling, up to a saturation value of approximately 5.5 vol. % in material that had reached an α–decay dose of 2-3 x 10^{25} α–decay events/m^3. Storage at 575 K resulted in less swelling, which required a longer time (300 days) to reach saturation. When this material was held at 875 K, swelling was minimal.

Transmission electron microscopy conducted after various storage times and temperatures [36,37] showed that with sufficient α–decay damage, Pu-zirconolite at 350 and 575 K converts from the crystalline to the aperiodic, metamict state.

EXAFS/XANES studies completed on ^{238}Pu-substituted zirconolite held at 350 K for at least 500 days helped to identify details of the damage state [35]. For comparison, measurements were made on essentially undamaged material (that is, substituted with ^{239}Pu with a half-life of 24,500 years, as compared to that of ^{238}Pu of 87.7 years). In the heavily-damaged material, the Pu/O nearest-neighbor bond length contracted by 0.003 nm, while expansion was observed for the more distant coordinating atoms, corresponding to a 6 to 7% overall volume expansion. This observation is close to the measured volume swelling of 5.5 %.

6.2.1.2. ^{244}Cm-doped zirconolite has been synthesized and self-damage studies of zirconolite have also been carried out by substituting ^{244}Cm, an α–active isotope with a half-life of 18.1 y, and ^{240}Pu for Zr [26]. Changes in swelling, microstructure, hardness, fracture toughness, leachability, and recrystallization behavior were assessed. The Cm-doped zirconolite retained the monoclinic structure; whereas, the ^{238}Pu-doped zirconolite was isometric (although up to concentrations of 5 mole % PuO_2, the monoclinic, zirconolite structure was retained [36]). More importantly, the composition of the Cm-doped zirconolite was not greatly changed by the substitution; whereas, the Pu-substituted structure was significantly changed by replacement of ZrO_2 by PuO_2.

Density measurements at 350 K indicate a saturation swelling value of 6.0 vol.% at a dose of 2 x 10^{25} α–decays events/m^3. These results are in good agreement with those for the Pu-substituted material, which exhibited saturation swelling of 5.5 vol.% at nearly the same damage level [36,37]. TEM images and diffraction patterns confirmed that Cm-doped zirconolite amorphized by the accumulation of damage tracks produced by the ^{240}Pu recoil ions that are emitted during α–decay of ^{244}Cm.

The dissolution (leaching) characteristics of the Cm-doped zirconolite were determined [26] by testing fully-damaged (amorphized) specimens and a second set of specimens that had been fully recrystallized to the original structure by a 12 h anneal at 1,100°C. Testing was performed in distilled, deionized water at 90°C for 14 d. The measured ^{240}Pu release rate was 0.002 g/m^{2}d and increased by a factor of 15 due to amorphization. The high final pH is believed to limit the solubility of the Cm. The low concentrations of Ti, as compared to that of Ca, suggests that the dissolution is incongruent, with the Ti ions (which are part of the TiO_6 octahedra comprising the network structure) being more resistant to leaching than the interlayer cations (e.g., Pu and Ca). The mechanism for the enhanced leaching of the radiation-induced amorphous state may be associated with the broken bonds and the increased free volume (a more open framework) of the amorphous state.

6.2.1.3. Natural zirconolite occurrences are rare, but selected specimens have been the subject of extensive studies [38-43]. Samples from Sri Lanka (550 million years old) typically can have concentrations of UO_2 of 2 weight percent and of ThO_2 of 20 weight percent, thus reaching doses of 10^{16} α–decay events/mg (=10^{26} α–decay events/m^3). This is a dose equivalent to nearly two displacements per atom.

X-ray diffraction analysis of fully-metamict zirconolite shows no diffraction maxima, confirming that zirconolite is X-ray diffraction amorphous. Because of the paucity of samples which span a range of α–decay event doses, it is not possible to follow the change of diffraction maxima intensity and position with increasing α–decay event dose. A zirconolite annealed at 1130°C did give a diffraction pattern with sharp peaks which could be refined on the basis of a monoclinic unit cell [44].

High resolution transmission electron microscopy showed no evidence of crystallinity [44], and in approximately ten percent of the grains examined, there were subspherical microvoids ranging in diameter from 100 to 400 nm [39]. The microvoids are attributed to He-gas accumulation which is a result of the radioactive decay. Annealed samples (1,100°C) recrystallized producing crystals up to 2,000 nm in length. The crystallites were often twinned at the unit cell scale as a result of different stacking sequences of the TiO_6 sheets, thus resulting in laminar intergrowths of the various zirconolite polytypes. The same texture of fine-scale twinning has been observed in synthetic zirconolite in Synroc C [45] and synthetic zirconolite of $Ca_{0.6}Sm_{0.4}ZrNb_{0.4}Mg_{0.4}Ti_{1.2}O_7$ composition [46]. The defects along the twin boundaries may become sites at which actinide elements are incorporated. In some instances [44], there was evidence for the formation of crystals (5-100 nm) with *d* -spacings similar to those of the fluorite structure type. Lumpkin et al. [44] suggested that metamict zirconolite may recrystallize initially with a disordered, fluorite-type structure. Continued heating to higher temperatures (1,000° to 1,200°C) appears to favor the highly twinned, monoclinic zirconolite structure. Differential thermal analysis (30° to 1,200°C) showed exothermic recrystallization at 780°C with a heat of recrystallization of 50 J/g for a zirconolite with a total dose of > 10^{26} α–decay events/m^3. This is similar to the value obtained by Foltyn et al. [47] for synthetic $CaPuTi_2O_7$ at nearly the same α–decay event dose of 10^{26} α–decay events/m^3 after 1,200 days. Although the measured energy release for the natural zirconolites and Pu-doped zirconolites of equivalent dose are nearly identical, both of these values are distinctly lower than the value (84 J/g) measured for Pu-doped zirconolite that had experienced a smaller dose (approximately 10^{25} α–decay events/m^3) [47] or for Cm-doped zirconolite (126 J/g) and $(Gd,Cm)_2Ti_2O_7$ at doses of 2.0 X 10^{25} α–decay events/m^3 [26,48]. Foltyn et al. [47] have suggested that the decrease in released energy with increasing dose is due to a continuing damage process than results in a reordering of atoms in the metamict state.

Leaching studies of natural zirconolites (which have accumulated significant α–decay event damage) have leach rates in the range of 1 to .001 g/m^2d at temperatures in the range of 95 to 200°C (based on Ca release) at times between 10 to 100 days [49]. The chemical durability of natural zirconolite has been confirmed by microprobe analyses which reveal only minor alteration (comparison of core to rim areas in the natural samples) [44]. The low release of actinides has been confirmed in laboratory experiments with Synroc-C formulations. Measured leach rates are in the range of 10^{-4} to 10^{-8} at 70°C in experiments that have run up to 1200 days [50].

6.3. MONAZITE

Monazite has been proposed as a single-phase ceramic to incorportate a wide variety of nuclear wastes, particularly those rich in actinides [51,52]. The mineral monazite is a mixed lanthanide orthophosphate, LnPO4 (Ln = La, Ce, Nd, Gd, etc.), that often contains significant amounts of Th and U (up to 27 wt. % combined). As a result, many natural monazites have been subjected to significant α–decay event damage over geologic time (some samples are over two billion years of age). In spite of the large radiation doses received by natural monazites, Ewing [53,54] has noted that monazite is always found in the crystalline state. The apparent resistance of natural monazites to radiation-induced amorphization was an important factor in the initial proposal of monazite ceramics as potential candidates for the immobilization of nuclear wastes. Further, the durability of monazite has led to its increasing use in U/Pb dating of minerals in both igneous and metamorphic rocks [55]. One should note that the closure temperature for Pb-loss for monazite (approximately 725°C) is lower than that of zircon.

Karioris et al. [56] have established that natural monazite (Ce,La,Y)PO4 can be readily transformed to an amorphous state by irradiation with 3 MeV Ar^+ ions to moderate doses. In natural monazite that was irradiated to a partially amorphous state (diffraction intensities were reduced by one-third), complete recovery of the fully crystalline state occurred after annealing at 300°C for 20 hours. The apparent stability of natural monazite is attributed to this relatively low temperature of recovery of the radiation damage. Differential scanning calorimetry (DSC) measurements of natural [57] and synthetic [57,58] monazites irradiated with 3 MeV Ar^+ ions have been performed. In natural monazite, the stored energy released during recrystallization was 33 J/g, the peak in the release rate occurred at 450°C, and the activation energy associated with the recovery process was estimated to be 2.77 eV. For synthetic monazite, CePO4, the stored energy release was 30.5 $\pm$ 1.3 J/g, the temperature for maximum release was slightly lower at 370°C, and the activation energy was estimated to be 2.7 $\pm$ 0.3 eV.

Leaching studies on synthetic monazite containing 20 wt. % simulated Savannah River waste (MCC-1 leach test, 28 days at 90°C) show release rates of uranium to be on the order of .001 gm/m^2d [59]. The leach rate of the host matrix of a synthetic monazite, LaPO4, containing simulated waste remains low even after the material has been transformed to an amorphous state by irradiation with 250 keV Bi^+ ions. However, work by Eyal and Kaufman [60] on natural monazite indicate that there is a preferential dissolution of the radionuclide daughter-products, ^{234}U, ^{230}Th, and ^{228}Th, by factors of between 1.1 and 10 relative to the structurally incorporated parent isotopes ^{238}U and ^{232}Th. This isotopic fractionation is attributed to radiation damage in the tracks of the recoil nuclei emitted during α–decay of the parent isotopes. While there have been some concerns regarding these results [52], the increases in dissolution rates are similar to those observed in other actinide-host phases. The same phenomenon has been identified in Th-doped borosilicate glass [61].

6.4. ZIRCON

Recently, zircon, ZrSiO4, (*I4₁/amd*; Z = 4) has been proposed as the preferred crystalline ceramic for the disposal of excess weapons plutonium [5 and 62, note extensive references therein]. Zircon occurs in nature with uranium and thorium concentrations up to 5,000 ppm. Zircon is an extremely durable mineral, often found as a heavy mineral in stream sediments that after transport over great distances shows limited chemical alteration or physical abrasion. Indeed, zircon has been extensively studied, because it is a phase commonly used in the U/Pb dating of samples that may be as old as four billion years . Numerous studies have analyzed the conditions under which the U/Pb systematics are disturbed as a result of radiation damage, thermal events, or alteration. Additionally, zircon has been identified as an actinide-bearing phase in polyphase ceramic waste forms.

The zircon structure is well known, consisting of triangular dodecahedral $ZrO8$ groups which form edge-sharing chains parallel to the *a*-axis, and $SiO4$ tetrahedral monomers which form edge-sharing chains with alternating $ZrO8$ groups parallel to the *c*-axis. In natural zircons, U and Th replace the Zr in low concentrations; however, compositions of $ASiO4$, in which A^{4+} = Zr, Hf, Th, Pa, U, Np, Pu and Am, have been synthesized. The regular increase in the unit cell volume with the increasing ionic radius of the A-site cation confirms the homologous topologies of these structures [63]. Four of these compositions, hafnon ($HfSiO4$), zircon, coffinite ($USiO4$) and thorite ($ThSiO4$), occur naturally. Present structure refinements and structural analyses suggest complete miscibility between $ZrSiO4$ - $HfSiO4$, but there are miscibility gaps on the $ZrSiO4$ - $USiO4$ - $ThSiO4$ joins [64]. Zircon with 9.2 atom percent plutonium (8.1% Pu^{238}; 1.1% Pu^{239}) substituting for Zr has been synthesized. This is equal to a waste loading of 10 wt.% Pu [65], but the maximum extent of the solubility of Pu in zircon has not been determined. The fact that a pure, endmember composition, $PuSiO4$, has been synthesized suggests extensive substitution of Pu for Zr is possible.

6.4.1. *Radiation damage* studies of zircon have a long history. Zircon undergoes the radiation-induced transformation from the periodic-to-aperiodic state (metamict state) at doses over the range of 10^{18} to 10^{19} α–decay events/g (= 0.2 to 0.6 dpa) with a density decrease and a corresponding volume expansion of 18% [66-71]. These studies include analysis of natural zircons which have accumulated α–decay event damage up to doses of 0.7 dpa over 550 million years, ^{238}Pu-doped zircons (^{238}Pu has a half-life of 87.7 years), up to doses of 0.7 dpa in 6.5 years and heavy ion beam irraidiations using 2 Mev He^+, 0.8 MeV Ne^+, 1.5 MeV Ar^+, 1.5 MeV Kr^+, 0.7 MeV Kr^+ and 1.5 MeV Xe^+ to doses of 0.2 to 2.3 dpa in times of less than an hour. All three types of damage experiments include detailed studies of annealing kinetics [72]. These experiments cover a dose rate range of $>10^8$ and provide a firm basis for predicting the microstructure of the radiation-damaged zircon as a function of dose, temperature, and type of radiation. There are few crystalline ceramics for which such a wide variety of data are available. Based on these data, for a waste loading of 10 wt.% of ^{239}Pu, the zircon will reach the saturation value of damage (1.2 x 10^{19} α–decay events/g or 0.8 dpa) in approximately 1,700 years; thus, the properties of zircon must be considered in light of its radiation-damaged, aperiodic state. By comparison a borosilicate waste glass with a similar waste loading will attain an apparent saturation damage at a dose of 3 x 10^{18} α–decay events/g in approximately 400 years and will have a partially decomposed network containing a significant concentration of helium/oxygen bubbles. Studies of borosilicate glasses to higher doses have never been performed; consequently, the effects at high doses are not predictable. Finally, transmutation effects (^{239}Pu decays to ^{235}U) are important in crystalline materials, as they may lead to phase instability. In the case of zircon, the solubility of U in zircon is only known approximately (4 $\pm$ 2 mole percent); thus, for higher concentrations of U one may expect the formation of USiO4 which also has the zircon structure.

6.4.2. *Stored energy* values have been carefully determined by transposed temperature drop calorimetry over the range of the periodic-to-aperiodic transition on a suite of zicons from Sri Lanka (550 million years old) [73]. The energy released during annealing varies sigmoidally as a function of α–decay event dose reaching a saturation value of 322 ± 16 J/g at doses greater than 5 X 10^{18} α–decay events/g. This is greater than values typical of nuclear waste glasses, which are generally less than 150 J/g and saturate at dose on the order of 10^{18} α–decay events/g; however, sudden release of this energy is not anticipated to cause a signficant rise in temperature for either the glass or zircon. The magnitude of the enthalpy of annealing suggests that the damage is pervasive on the scale of fractions of nanometers, perhaps leading to the formation of micro-domains of amorphous SiO2-rich and ZrO2-rich regions in the metamict state. This suggestion is consistent with observations made by secondary ion mass spectrometry (SIMS) and high resolution transmission electron microscopy (HRTEM) and extended x-ray absorption fine structure spectroscopy (EXAFS) of annealed zircons [69,70,74].

6.4.3. *Alteration* of natural zircons has important implications to the use of U/Th/Pb techniques in geochronology; thus, there is an extensive literature that describes the alteration [75-79]. Physical degradation can occur due to microfracturing that is the result of the volume expansion caused by the defect accumulation with increasing α–decay dose and results in an increase of surface area [68]. Chemical alteration can involve disturbed U/Pb systematics due to diffusion of Pb and an increase of leach rate. Under extreme geologic conditions, e.g., zircons subjected to deformation in shear zones and altered by hydrothermal solutions, disturbed U/Pb systematics are clearly documented. Differential etching experiments (using 48% HF) have shown by scanning electron microscopy the removal of radiation damaged zones with improved concordance of U/Pb systematics in the remaining, unaltered material. Despite this evidence of alteration in samples that may be hundreds of millions to several billions of years old, one must state that the alteration is minor and the disturbed U/Pb systematics are of elements that are present in concentrations of hundreds to thousands of ppm. The minor alteration of zircon over long periods of time and under rather extreme conditions (e.g., meteorite impact [75]) stands in contrast to the observations of natural glasses which are readily altered and rarely have ages in excess of tens of millions of years. Natural glasses have been studied extensively as natural analogues for the long-term behavior of nuclear waste borosilicate glasses and are certainly less stable than zircon [84].

6.4.4. *Leaching studies* of natural zircons under extreme laboratory conditions have confirmed the loss of U, Th and Pb under hydrothermal conditions (e.g., 2 molal NaCl solution at 500°C and 1,000 bars pressure). The lead loss can be the result of grain boundary or volume diffusion (in which there is no dissolution of the zircon), or with time the dissolution of zircon [77]. However, at lower temperatures (< 80°C) and near neutral pH values, i.e., conditions more pertinent to nuclear waste disposal, zircon is extremely insoluble [78]. There are much less data in the literature than necessary for a full evaluation of zircon as a waste form. Crystalline zircon is stable to such an extent that the equilibrium concentrations of Zr and Si are in the order of 10^{-9} moles/L (0.1 ppb) at 25°C. Dissolution of amorphous zircon (cyrtolite) followed a first order reaction, based on Si concentrations. Zr concentrations remained below 0.05 ppm, the instrument detection limit, due to precipitation of $ZrO2$ and $ZrSiO4$. The dissolution of natural zircons at 87°C as a function of α–decay event damage causes an increase in weight loss (10^{-3} to 10^{-1} weight percent for the transition from the crystalline ($\leq 10^{16}$ α/g) to the amorphous state ($\geq 10^{19}$ α/g)). To detect zirconium in solution, these experiments had to be performed in KHCO3 solution. Comparing the dissolution rate for cyrtolite , R = 10^{-7} moles/m^2/d at 80°C, pH = 5, with the long-term rate of a nuclear waste borosilicate glass, R = 10^{-5} moles/m^2/d (= 10^{-3}g/m^2/d) in water at 98°C shows that the dissolution rate of amorphous zircon is still considerably lower than that of glass in stagnant, silica-saturated solutions. In an open system (e.g., moving ground water), the leach rate for zircon used in this comparison does not increase; however, the leach rate of borosilicate glass may increase by three orders of magnitude until reaching the forward rate of reaction. Thus, one of the main advantages of zircon may be its high durability where large volumes of water have access to the waste form, as this allows considerably greater flexibility in repository design.

6.4.5. *Physical properties* of zircon have been determined as a function of increasing α–decay event dose, 0.15 to 0.65 dpa [68]. The radiation-induced softening leads to a decrease in hardness of 40 percent and of the bulk elastic modulus of up to 70 percent, but there is an increase in fracture toughness probably caused by crack-tip blunting by the aperiodic domains.

6.4.6. *Criticality* must be a concern for any Pu-waste form, both during processing as well as in the final waste form configuration. This concern can be mitigated by adjusting the waste loading in the zircon, and also by including neutron "poisons" in the formulation. Natural zircons can contain up to several thousand parts per million rare earths, including up to 500 ppm of Gd (or up to 5,000 ppm at Oklo [17]). Thus, neutron absorbing nuclides may be incorporated into the zircon structure as dilute solid solutions or, at higher concentrations, used as a binding material intimately mixed with the zircon. There is no evidence of selective loss of gadolinium from zircons, but laboratory experiments with natural and synthetic material should confirm this observation.

6.4.7. *Production technologies* are always an important consideration in the adoption of a waste form, as one should not underestimate the difficulties of processing large volumes of radioactive material. In the case of zircon, phase relations in the system ZrO2 - SiO2 were investigated as early as 1967 and the system ZrO2 - SiO2 - ThO2 - UO2 has been investigated by Mumpton and Roy [64] in which they report the solubility limits of Th (< 4 mole %) and U (4 ± 2 mole %) in zircon. As already discussed, Pu-doped zircons have been synthesized with 9.1 mole percent Pu. The apparent increase of solubility of Pu over U and Th in zircon is expected as the ionic radius of the A^{4+}-site cation approaches that of Zr^{4+}. Komarneni and Roy [80] have reviewed the well established hydrothermal and solid-state synthesis of zircon as either single crystals or powders, as well as sol-gel synthesis. Most recently, Mori and others [81] have synthesized zircon by sintering (1,200 to 1,300°C) and have obtained yields of nearly 90 percent in less than one day. Although larger-scale pilot demonstrations are required, as well as detailed studies of the effect of Pu (and any potential impurity) on zircon synthesis conditions, much is already known about the requirements for synthesis. In the case of weapons plutonium, one can take advantage of the essentially reagent grade of the waste in engineering the process and the waste form. The final processing technology can draw on the experience of producing mixed-oxide fuels from powders. Some modifications of the industrial MOX process are necessary to make zircon. Safeguard and criticality aspects of the MOX process have been reviewed by the IAEA and need not be reestablished if zircon is produced by this process. Theft and recovery of Pu from zircon may be a concern prior to disposal in deep borehole; however, this will be difficult, as zircon is a highly refractory substance (melting point > 1,500°C).

In summary, zircon is an extremely durable phase which may be used as a Pu-host for the disposal of weapons plutonium. The long-term durability is confirmed from natural occurrences in diverse and extreme geologic environments over geologic time. Leaching experiments confirm the extremely low solubility. The very low solubility, as compared to borosilicate glass, will insure that Pu is not concentrated by later cycles of geochemical alteration to values that might lead to criticality. Several processing options, sol-gel or sintering, have been demonstrated on a laboratory scale. Zircon does experience a radiation-induced crystalline-to-amorphous transition, but even the aperiodic (metamict) material is extremely durable. Finally, the lower volume (due to higher waste loading and the higher density of zircon as compared to glass) and greater durability, particularly at elevated temperatures, expand the range of geologic disposal possibilities, e.g, deep boreholes, a third, less studied option suggested by the U.S. National Academy of Sciences. In fact, retention of the most mobile element, Pb, in zircons to depths of over 4,000 meters (>300°C) has already been demonstrated [82,83].

7. Conclusions

There is every reason to expect that waste form performance can be much improved over what is now accepted for borosilicate glass. Ceramic phases, such as zircon or zirconolite, have already demonstrated this improved performance under certain conditions (e.g., hydrothermal, up to 300°C). Prudence requires that further research and development of these ceramic waste forms be conducted for the following reasons:

i.) Any strategy of isolation should emphasize the near-field containment of the radionuclides. This is primarily a function of waste form or "waste package" performance. Strategies that rely solely on long travel times, dispersal or dilution, implicitly presume release and movement of radionuclides. In the case of the disposal of weapons plutonium, this leads inevitably to concerns over criticality.

ii.) The long-term performance assessment of the success of radionuclide containment requires the development of deterministic models of the future physical and chemical behavior of each part of the barrier system. Although difficult, it is certainly easier to model the chemistry and physics of corrosion and alteration of waste forms, with the subsequent release or retention of radionuclides over some range of conditions, than it is to develop coupled hydrologic, geochemical and geophysical models of the movement of radionuclides through the far-field of a geologic repository or a deep borehole. The extrapolation of corrosion behavior over long periods rests on a firmer scientific foundation for a waste form as chemically simple as zircon than the extrapolated behavior of, as an example, hydrologic systems that are site specific and highly dependent on idealized boundary conditions (e.g., climate and hydrologic recharge). This is a particularly important consideration in proposals for deep borehole disposal where the geophysical, geochemical and hydrologic regimes are more uncertain and less amenable to engineered barriers.

iii.) Natural phases, minerals, provide a means of "confirming" the hypothesized long-term behavior of waste form phases in specific geochemical environments to very high α–decay event doses. Indeed, the study of natural phases provides an essential confirmation of performance assessments in which the corrosion or alteration behavior of waste form phases over long periods is an essential component.

iv.) Finally, the availability of a well established process (e.g., the modified MOX-process for the production of zircon) for the handling of large quantities of fissile material is certainly an important criterion in the selection of a waste form for plutonium.

Considering each of the above criteria and based on the present state-of-knowledge, we propose that zircon is the preferred crystalline, ceramic waste form for the disposal of excess weapons plutonium.

8. Acknowledgements

We thank Professor Erich Merz for the vision and fairness of including a discussion of crystalline ceramics in this workshop. Major portions of this work were supported by the Office of Basic Energy Sciences, United States Department of Energy under Contract DE-AC06-76RLO (PNL) and Grant DE-FG03-93ER45498 (UNM). Finally, R.C. Ewing thanks his colleagues in the Department of Earth Sciences of Aarhus University for their support where much of the writing of this paper was completed.

9. References

1. National Academy of Sciences (1994) Management and Disposition of Excess Weapons Plutonium. (National Academy Press, Washington, D.C.) 31 pages.
2. Lutze, W. and Ewing, R.C. (1988) *Radioactive Waste Forms for the Future,* North-Holland, Amsterdam, 778 pages.
3. Lutze, W. and Ewing, R.C. (1988) Summary and evaluation of nuclear waste forms in *Radioactive Waste Forms for the Future,* W. Lutze and R.C. Ewing, editors, North-Holland, Amsterdam, 699-740.
4. Vance, E.R. (1994) Synroc: A suitable waste form for actinides. *Materials Research Society Bulletin* **XIX**, 28-32.
5. Anderson, E.B., Burakov, B.E. and Vailiev, V.G. (1993) Creation of a crystalline matrix for actinide waste at the Khlopin Radium Institute, Proceedings of SAFEWASTE '93, **2**, 29-33; Anderson, E.B., Burakov, B.E. and Pazukhin, E.M. (1993) High-uranium zircon from "Chernobyl Lavas", *Radiochimica Acta* **60**, 149-151.
6. Johnson, L.H. and Werme, L.O. (1994) Materials characteristics and dissolution behavior of spent nuclear fuel. *Materials Research Society Bulletin* **XIX**, 24-27.
7. Ewing, R.C. (1992) Preface. *Journal of Nuclear Materials* **190**, vii - x; Ewing, R.C., editor (1992) Special thematic issue on Nuclear Waste Forms. *Journal of Nuclear Materials* **190**, 347 pages.
8. Ewing, R.C. and Lutze, W. (1991) High-level nuclear waste immobilization with ceramics, *Ceramics International* **17**, 287.

9. Ewing, R.C. and Lutze, W. (1994) Materials science of radioactive waste forms. *Materials Research Society Bulletin* **XIX**, no. 12, 16-18.
10. Lutze, W. and R.C. Ewing (in press) International Symposium on "Environmental Issues and Waste Management Technologies in Ceramic and Nuclear Industry" R.A. Palmer and V. Jain, editors, Proceedings of the American Ceramic Society meeting, May 1-4, 1994.
11. Lumpkin, G.R., Smith, K.L., and Blackford, M.G. (1995) Partitioning of uranium and rare earth elements in Synroc: Effect of impurities, metal additive, and waste loading, *Journal of Nuclear Materials* **224**, 31-42.
12. Weber, W. J., Turcotte, R. P., Bunnell, L. R., Roberts, F. P. and Westsik, J. H. (1979) Radiation effects in vitreous and devitrified simulated waste glass. In: *Ceramics in Nuclear Waste Management*,, T. D. Chikalla and J. E. Mendel, Eds. CONF-790420 (National Technical Information Service, Springfield, Virginia) 294-299.
13. Turcotte, R. P., Wald, J. W., Roberts, F. P., Rusin, J. M. and Lutze, W. (1982) Radiation damage in nuclear waste ceramics. *J. Amer. Ceram. Soc.* **65**, 589-593.
14. Dé, A. K., Luckscheiter, B., Lutze, W., Malow, G. and Schiewer, E. (1976) Development of glass ceramics for the incorporation of fission products. *Ceramic Bulletin* **55**, 500-503.
15. Jantzen, C. M. and Glasser, F. P. (1979) Stabilization of nuclear waste constituents in portland cement. *Ceramic Bulletin* **58**, 459-466.
16. Lindberg, M.L. and Ingram, B. (1964) Rare-earth silicate apatite from the Adirondack Mountains, New York. *Professional Paper 501-B*, U.S. Geological Survey, 64-65.
17. Koul, S.L. (1979) On the fission track dating and annealing behaviour of accessory minerals of eastern ghats (Andhra Pradesh, India). *Radiation Effects* **40**, 187-192; J. Carpéna (in press) From the natural to the synthetic analogue: Interest of the Oklo programme for devising high retention-potential matrices. Proceedings of the final meeting of the Oklo Natural Analogue Working Group, June, 1995, H. von Maravic and P.-L. Blanc,editors, CEC EUR series.
18. Bertagnolli, E., Keil, R. and Pahl, M. (1983) Thermal history and length distribution of fission tracks in apatite: Part 1, *Nuclear Tracks* **7**, 163-177.
19. Weber, W.J. (1983) Radiation-induced swelling and amorphization $Ca_2Nd_8(SiO_4)_6O_2$. *Radiation Effects* **77**, 295-308.
20. Weber, W.J. (1993) Alpha-decay-induced amorphization in complex silicate structures. *Journal of the American Ceramic Society* **76**, 1729-1738.
21. Weber, W.J. and Wang, L.M. (1994) Effect of temperature and recoil-energy spectra on irradiation-induced amorphization in $Ca_2La_8(SiO_4)_6O_2$. *Nuclear Instruments and Methods in Physics Research* **B91**, 22-29.
22. Ewing, R.C., Weber, W.J., and Clinard, F.W. Jr. (1995) Radiation effects in nuclear waste forms for high-level radioactive waste, *Progress in Nuclear Energy* **29(2)**, 63-1127.
23. Wang, L.M., Cameron, M., Weber, W.J., Crowley, K.D. and Ewing, R.C. (1994) In situ TEM observation of radiation induced amorphization of crystals with apatite structure. *Proceedings of the Materials Research Society* (CRC Press) 243-249.
24. Wald, J. W. and Weber, W. J. (1984) Effects of self-radiation damage on the leachability of actinide-host phases. In: *Advances in Ceramics, Vol. 8, Nuclear Waste Management*,. G. G. Wicks and W. A. Ross, Eds. (The American Ceramic Society, Columbus, Ohio) 71-75.
25. Weber, W. J., Wald, J. W. and Matzke, Hj. (1985) Self-radiation damage in $Gd_2Ti_2O_7$. *Materials Letters* **3**, 173-180.
26. Weber, W.J., Wald, J.W. and Matzke, Hj. (1986) Effects of self-irradiation damage in Cm-doped $Gd_2Ti_2O_7$ and $CaZrTi_2O_7$. *Journal of Nuclear Materials* **138**, 196-209; Weber, W. J. and Matzke, Hj. (1986) Radiation effects in actinide host phases. *Radiation Effects* **98**, 93-99.
27. Lumpkin G.R. and Ewing R.C. (1985) Natural pyrochlores: Analogues for actinide host phases in radioactive waste forms. In *Scientific Basis for Nuclear Waste Management VIII*, C.M. Jantzen, J.A. Stone and R.C. Ewing, Eds., Materials Research Society Proceedings, vol. 44, (Materials Research Society, Pittsburgh) 647-654.
28. Lumpkin G.R. and Ewing R.C. (1986) High-resolution transmission electron microscopy of microlite from the Harding pegmatite, Taos County, New Mexico. In:

Microbeam Analysis 1986, Romig A.D. Jr and Chambers W.F., Eds. (San Francisco Press, San Francisco)145-147.

29. Lumpkin, G.R., Foltyn, E.M. and Ewing, R.C. (1986) Thermal recrystallization of alpha-recoil damaged minerals of the pyrochlore structure type. *Journal of Nuclear Materials* **129**, 113-120.

30. Lumpkin, G. R. and Ewing, R. C. (1988) Alpha-decay damage in minerals of the pyrochlore group. *Physics and Chemistry of Minerals* **16**, 2-20.

31. Lumpkin, G.R., Hart, K.P., McGlinn, P.J., Payne, T.E., Gieré, R. and Williams, C.T. (1994) Retention of actinides in natural pyrochlores and zirconolites. *Radiochimica Acta* **66/67**, 595-600.

32. Lumpkin, G.R. and Ewing, R.C. (1992) Geochemical alteration of pyrochlore group minerals: Microlite subgroup. *American Mineralogist* **77**, 179-188.

33. Lumpkin, G.R. and Ewing, R.C. (in press) Geochemical alteration of the pyrochlore group minerals: Pyrochlore subgroup. *American Mineralogist.*

34. Lumpkin, G.R. and Ewing, R.C. (submitted) Geochemical alteration of the pyrochlore group minerals: Betafite subgroup. *American Mineralogist.*

35. Greegor, R.B., Lytle, F.W., Livak, R.J. and Clinard, J.W., Jr. (1988) X-ray spectroscopic investigation of Pu-substituted zirconolite. *Journal of Nuclear Materials* **152**, 270-277.

36. Clinard, F.W.,Jr., Peterson, D.E. and Rohr, D.L. (1984) Self-irradiation effects in ^{238}Pu-substituted zirconolite. I. Temperature dependence of damage. *Journal of Nuclear Materials* **126**, 245-254.

37. Clinard, F.W., Jr., Rohr, D.L. and Roof, R.B. (1984) Structural damage in a self-irradiated zirconolite-based ceramic. *Nuclear Instruments and Methods in Physics Research* **B1**, 581-586; Clinard, F.W., Jr., (1986) Review of self-irradiation effects in Pu-substituted zirconolite. *Ceramic Bulletin* **65(8)**, 1181-1187.

38. Ewing, R.C., Haaker, R.F.,Headley, T.J. and Hlava, P (1982) Zirconolites from Sri Lanka, South Africa and Brazil. In: *Scientific Basis for Nuclear Waste Management,IV*, Materials Research Society Proceedings, vol. 6, S. V. Topp, Ed. (North-Holland, New York) 249-256.

39. Ewing, R.C. and Headley, T.J. (1983) Alpha-recoil damage in natural zirconolite ($CaZrTi_2O_7$). *Journal of Nuclear Materials* **119**, 102-109.

40. Ewing, R.C., Chakoumakos, B.C., Lumpkin, G.R., Murakami, T., Greegor, R.B. and Lytle, F.W. (1988) Metamict minerals: Natural analogues for radiation damage effects in ceramic nuclear waste forms. *Nuclear Instruments and Methods in Physics Research* **B32**, 487-497.

41. Oversby, V.M. and A.E. Ringwood (1981) Lead isotopic studies of zirconolite and perovskite and their implications for long range Synroc stability. *Radioactive Waste Management* **1**, 289-308.

42. Sinclair, W. and Ringwood, A. E. (1981) Alpha-recoil damage in natural zirconolite and perovskite. *Geochemical Journal* **15**, 229-243.

43. Sinclair, W. and Eggleton, R.A. (1982) Structure refinement of zirkelite from Kaiserstuhl, West Germany. *American Mineralogist* **67**, 615-620.

44. Lumpkin, G.R., Ewing, R.C., Chakoumakos B.C., Greegor R.B., Lytle F.W., Foltyn E.M., Clinard F.W. Jr., Boatner L.A., Abraham M.M. (1986b) Alpha-recoil damage in zirconolite ($CaZrTi_2O_7$). *Journal of Materials Research* **1**, 564 - 576.

45. White, T.J., Segall, R.O., Hutchison, J.L. and Barry, J.C. (1984) Polytypic behaviour of zirconolite. *Proceedings of the Royal Society of London* Ser. **A392**, 343-358.

46. White, T.J. (1984) The microstructure and microchemistry of synthetic zirconolite, zirkelite and related phases. *American Mineralogist* **69**, 1156-1172.

47. Foltyn, E.M., Clinard, F.W., Jr., Rankin, J. and Peterson, D.E. (1985) Self-irradiation effects in ^{238}Pu-substituted zirconolite: II. Effect of damage microstructure on recovery. *Journal of Nuclear Materials* **136**, 97-103.

48. Weber, W.J., Wald, J.W. and Matzke, Hj. (1985) Self-radiation damage in actinide host phases of nuclear waste forms. In: *Scientific Basis for Nuclear Waste Management VIII*, vol.

8, C.M. Jantzen, J.A. Stone and R.C. Ewing, Eds (Materials Research Society, Pittsburgh) 679-686.
49. Ringwood, A.E., Kesson, S.E., Reeve, K.D., Levins, D.M. and Ramm, E.J. (1988) Synroc. In: *Radioactive Waste Forms for the Future*, W. Lutze and R.C. Ewing, Eds. (North-Holland, Amsterdam) 233-334.
50. Jostons, A. (1994) Development of Synroc for HLW management, Proceedings of the 9th Pacific Basin Conference, Sydney, Australia on "Nuclear Power & Industry" (National Conference Publication no. 94/6) 1.
51. Boatner, L.A., Beall, G.W., Abraham, M.M., Finch, C.B., Huray, P.G. and Rappaz, M. (1980) Monazite and other lanthanide orthophosphates as alternative actinide waste forms. In: *Scientific Basis for Nuclear Waste Manegement*, vol. 2, C.J.M. Northrup, Jr., Ed. (Plenum Press, New York) 289-296.
52. Boatner, L.A. and Sales, B.C.(1988) Monazite. In: *Radioactive Waste Forms for the Future*. Lutze, W. and Ewing, R.C., Eds., (North-Holland, Amsterdam) 495-464.
53. Ewing, R.C. (1975) The crystal chemistry of complex niobium and tantalum oxides. IV. The metamict state. *American Mineralogist* **60**, 728-733.
54. Ewing, R.C. and Haaker, R.F. (1980) The metamict state: Implications for radiation damage in crystalline waste forms. *Nuclear and Chemical Waste Management* **1**, 51-57.
55. Parrish, R.R. (1990) U-Pb dating of monazite and its application to geological problems. *Canadian Journal of Earth Science* **27**, 1431-1449.
56. Karioris, F.G., Gowda, K.A. and Cartz, L. (1981) Heavy ion bombardment of monoclinic $ThSiO_4$, ThO_2, and monazite. *Radiation Effects Letters* **58**, 1-3.
57. Gowda, K.A. (1982) Heavy ion bombardment of zircon, monazite and other crystal structures. Ph.D. thesis, Department of Mechanical Engineering, Marquette University, Milwaukee, Wisconsin.
58. Ehlert, T.C., Gowda, K.A., Karioris, F.G. and Cartz, L. (1983) Differential scanning calorimetry of heavy ion bombarded synthetic monazite. *Radiation Effects* **70**, 173-181.
59. Sales, B.C., White, C.W. and Boatner, L.A. (1983) A comparison of the corrosion characteristics of synthetic monazite and borosilicate glass containing simulated nuclear waste glass. *Nuclear and Chemical Waste Management* **4**, 281-289.
60. Eyal, Y. and Kaufman, A. (1982) Alpha-recoil damage in monazite: Preferential dissolution of radiogenic actinide isotopes. *Nuclear Technology* **58**, 77-83.
61. Eyal, Y. and Ewing, R.C. (1993) Impact of alpha-recoil damage on dissolution of thoriated glass. In: Proceedings of the International Conference on Nuclear Waste Management and Environmental Remediation, Vol. 1, Low and Intermediate Level Radioactive Waste Management, Eds., D. Alexandre, R. Baker, R. Kohout and J. Marek, sponsored by the American Society of Mechanical Engineers, Prague, Czech Republic, September 5 - 11, 191 - 196.
62. Ewing, R.C., Lutze, W., and Weber, W.J. (1995) Zircon: A host-phase for the disposal of weapons plutonium. *Journal of Materials Research*, **10(2)**, 243-246.
63. Keller, C. (1963) Untersuchungen über die germanate und silikate des typs ABO_4 der vierwertigen elemente thorium bis americium, *Nukleonik* **5(2)**, 41-47.
64. Mumpton, F.A. and Roy, R. (1961) Hydrothermal stability studies of the zircon-thorite group. *Geochimica et Cosmochimica Acta* **21**, 217-238.
65. Exarhos, G.R. (1984) Induced swelling radiation damaged $ZrSiO_4$. *Nuclear Instruments and Methods in Physics Research* **B1**, 538-541.
66. Murakami, T., Chakoumakos, B.C., and Ewing, R.C. (1986) X-ray powder diffraction analysis of alpha-event radiation damage in zircon (ZrSiO4). In: *Advances in Ceramics: Nuclear Waste Management II*, vol. 20, D.E. Clark, W.B. White and J. Machiels, Eds., (American Ceramic Society, Columbus, Ohio) 745-753.
67. Murakami, T., Chakoumakos, B.C., Ewing, R.C., Lumpkin, G.R. and Weber W.J. (1991) Alpha-decay event damage in zircon. *American Mineralogist* 76, 1510-1532.
68. Chakoumakos, B.C., Oliver, W.C., Lumpkin, G.R., and Ewing, R.C. (1991) Hardness and elastic modulus of zircon as a function of heavy-particle irradiation dose: I. In situ a-decay event damage. *Radiation Effects and Defects in Solids* **118**, 393-403; Chakoumakos,

B.C., Murakami, T., Lumpkin, G.R., and Ewing, R.C. (1987) Alpha-decay-induced fracturing in zircon: The transition from the crystalline to the metamict state, *Science* **236**, 1556-1559.

69. Farges, F. and Calas, G. (1991) Structural analysis of radiation damage in zircon and thorite: An X-ray absorption spectroscopic study. *American Mineralogist* **76**, 60-73.

70. Farges, F. (1994) The structure of metamict zircon: A temperature-dependent EXAFS study. *Physics and Chemistry of Minerals* **20**, 504-514.

71. Holland, H.D. and Gottfried, D. (1955) The efect of nuclear radiation on the structure of zircon. *Acta Crystallographica* **8**, 291-300.

72. Weber, W.J., Ewing, R.C. and Wang, Lu-Min (1994) The radiation-induced crystalline-to-amorphous transition in zircon. *Journal of Materials Research* **9(3)**, 688-698.

73. Ellsworth, S., Navrotsky, A. and Ewing, R.C. (1994) Energetics of radiation damage in natural zircon ($ZrSiO_4$). *Physics and Chemistry of Minerals* **21**, 140 - 149.

74. McLaren, A.C., Fitzgerald, J.D. and Williams, I.S. (1994) The microstructure of zircon and its influence on the age determination from Pb/U isotopic ratios measured by ion microscope. *Geochimica Cosmochimica Acta* **54**, 993-1005.

75. Kamo, S.L. and Krogh, Th.E. (1995) Chicxulub crater source for shocked zircon crystal from the Cretaceous-Tertiary boundary layer, Saskatchewan: Evidence from new U-Pb data. *Geology* **23**, 281-284.

76. Ludwig, K.R., Zartman, R.E., and Goldich, S.S. (1984) Lead retention in zircons. *Science* **223**, 834.

77. Pidgeon, R.T., O'Neil, J.R., and Silver, L.T. (1986) Uranium and lead isotopic stability in a metamict zircon under experimental hydrothermal conditions. *Science* **154**, 1538-1540.

78. Tole, M.P. (1985) The kinetics of dissolution of zircon ($ZrSiO_4$), *Geochimica et Cosmochimica Acta* **49**, 453-458; Sinha, A.K., Wayne, D.M. and Hewitt, D.A. (1992) The hydrothermal stability of zircon: Preliminary experimental and isotopic studies. *Geochimica et Cosmochimica Acta* **56**, 3551-3560.

79. Ewing, R.C., Haaker, R.F., and Lutze, W. (1982) Leachability of zircon as a function of alpha-dose. In: Scientific Basis for Radioactive Waste Management - V, W. Lutze, Ed. (North-Holland, New York) 389-397.

80. Komarneni, Sridhar and Roy, R. (1988) Synthesis of zircon in *Zircon, Science and Technology*, S. Somiya, editor (Uchidarokakuho, Japan) 289-298.

81. Mori, T.,Yamamura, H., Kobayashi, H., Mitamura, T. (1993) Formation mechanism of ZrSiO4 powders. *Journal of Materials Science* **28**, 4970 (1993).

82. Gentry, R.V., Sworski, T.J., McKown, H.S., Smith, D.H., Eby, R.E., and Christie, W.H. (1982) Differential lead retention in zircons: Implications for nuclear waste containment. *Science* **216**, 296-297.

83. Gentry, R.V. (1984) Lead retention in zircons. *Science* **223**, 835.

84.Ewing, R.C. and Jercinovic, M.J. (1987) Natural analogues: Their application to the prediction of the long-term behavior of nuclear waste forms. *Scientific Basis for Nuclear Waste Management X*, J.K. Bates and W.B. Seefeldt, Eds., Materials Research Society Proceedings, vol. 84, 67-86; Jercinovic, M.J. and R.C. Ewing (1988) Basaltic glasses from Iceland and the deep sea: Natural analogues to borosilicate nuclear waste form glass. JSS Project Technical Report 88-01 of the Swedish Nuclear Fueld and Waste Management Co., Stockholm, Sweden, 221 pages.

THE CRYSTALLINE HOST-PHASES FOR IMMOBILIZATION OF WEAPONS PLUTONIUM AND WASTE ACTINIDES

B. E. BURAKOV,
Mineralogical Group,
V. G. Khlopin Radium Institute,
28, 2-nd Murinsky Ave., St. Petersburg, 194021, Russia

E. B. ANDERSON,
Deputy Director General,
V. G. Khlopin Radium Institute,
28, 2-nd Murinsky Ave., St. Petersburg, 194021, Russia

B. Ya. GALKIN, V. A. STARCHENKO, V. G. VASILIEV,
Department of radioactive materials technology,
V. G. Khlopin Radium Institute,
28, 2-nd Murinsky Ave., St. Petersburg, 194021, Russia

Abstract

The management of weapons plutonium released as a result of conversion, is considered inseparably with the general problem of the actinide fraction of long-lived high-radioactive wastes. It is proposed for different versions of plutonium and actinides management to use polymineral ceramics based on crystalline host-phases: zircon $ZrSiO_4$ and zirconium dioxide ZrO_2. It is shown that plutonium and actinides can form with these phases broad areas of substitutional solutions. A synthesis of polymineral ceramics based on $ZrSiO_4$ and ZrO_2 by the hot pressing method has been conducted on the laboratory level. An original method, developed by the authors as a result of studying high-uranium zircon $(Zr,U)SiO_4$ from Chernobyl "lavas", was applied for the incorporation of plutonium simulators into the $ZrSiO_4$ and ZrO_2 structure in the course of synthesis.

Introduction

Weapons plutonium as well as transplutonium elements (actinides) of high-level wastes: Am, Np, Cm, – are exceptionally toxic long-lived radionuclides. The removal of these materials from the human activity sphere is considered at present within the slope of the two main concepts:

E.R. Merz and C.E. Walter (eds.), Disposal of Weapon Plutonium, 85-89.

1. Plutonium and radioactive actinides of nuclear wastes must be transferred into thermodynamically stable crystalline compounds similar to most durable natural minerals. The final isolation of solidified wastes is into deep geological formations or even to outer cosmic space.
2. Plutonium and actinides may be "burned", undergoing transmutation as components of nuclear fuel. It is supposed that the use of mixed uranium-plutonium fuel UO_2-PuO_2 is more justified economically than its isolation as radioactive wastes. It should be noted that such an approach does not exclude an application of especially stable crystalline host-phases for plutonium, while for actinides transmutation they, theoretically, are necessary.

In the present paper we tried to reflect briefly investigations by specialists of the V. G. Khlopin Radium Institute (St. Petersburg, Russia) in the field of synthesis of crystalline host-phases capable of incorporating considerable amounts of plutonium and actinides into their lattice. We consider possible not to oppose to each other the two main concepts, but to regard them with account of technologies existing in Russia.

The choice of mineral phases

The possibility to apply these or those minerals as host-phases for the disposal of plutonium and actinides is determined first of all by their physico-chemical stability and the isomorphous capacity of their crystalline lattice.

If it is supposed to dispose crystalline matrices with plutonium and actinides into deep geological formations, then one more requirement is demanded of the host-phases. It is the existence of natural analogues, which proves these minerals' exclusive stability to the impact of a real geological medium during long time ($>10^4...10^5$ years). In the case of using plutonium as nuclear fuel, as well as in elimination of actinides by the transmutation method, there are restrictions to the chemical composition of host-phases. These minerals must not consist of element-absorbers of neutrons.

Finally, there are requirements of economical expediency of applying these or those crystalline materials as host-phases. And these requirements are determined not only by the cost of starting materials, but also by the extent of their knownness, the existence of a synthesis technology, etc.

Thus, having considered all the restrictions to the choice of mineral phases for the immobilization of plutonium and actinides, we drew a conclusion that the most promising materials are zircon $ZrSiO_4$ and cubic zirconium dioxide ZrO_2.

Note the main characteristics of these minerals:

- zirconium dioxide ZrO_2 forms broad areas of solid solutions in the PuO_2–ZrO_2 system [1], as well as with oxides of rare-earth elements – simulators of radioactive actinides [2];
- zircon $ZrSiO_4$ can incorporate into its lattice considerable amounts of uranium [3, 4, 5], plutonium [6] and rare-earth elements [7];
- zircon and cubic zirconium dioxide ZrO_2 are exclusively durable minerals, they have natural analogues. These phases are widely discussed as promising materials for the immobilization of plutonium and actinides [5, 6, 8, 9, 10];
- irradiated nuclear fuel reprocessing in Russia is connected with release of great amounts of zirconium, the main component of fuel claddings. The plasmo-chemical calcination technology existing in Russia, allows to use this source of zirconium for production of ZrO_2 immediately from nitrate solutions [10].

An experimental synthesis of ceramics on the basis of zircon $ZrSiO_4$ and zirconium dioxide ZrO_2 by the hot pressing method

In developing an industrial technology for the synthesis of ceramics for plutonium and actinides immobilization there are certain restrictions in Russia:

- the synthesis of crystalline matrices must be done by the hot pressing method;
- in the course of synthesis the whole volume of plutonium and actinides must be isomorphously incorporated into the lattice of the host-phases.

These restrictions cause a problem of observing strict stoichiometry in the fabrication of the starting oxide mixture.

Another difficulty is in the fact that crystalline zirconium dioxide ZrO_2 is a chemically inert and thermally stable material. To release an interaction between crystalline oxides of zirconium, silicon, plutonium and actinides in the course of hot pressing, requires a considerable temperature of pressure, or time expenses. This may put to doubt the economical expediency of the given technology.

At the V. G. Khlopin Radium Institute (St. Petersburg, Russia) the technology of crystalline matrix synthesis is developed on the following basis:

- the final product of synthesis is prefixed polymineral ceramics based simultaneously on two host-phases: zircon $ZrSiO_4$ and cubic zirconium dioxide ZrO_2;
- the starting materials for the synthesis are ultra-fine powders of oxides of plutonium, silicon, zirconium and actinides, obtained by the plasmo-chemical method. Plasmochemical powders are distinguished by partial or

full amorphousness, i. e. they are more chemically reactionable. In some cases the simultaneous plasmochemical denitration of a mixture of radioactive and nonradioactive solutions allows to obtain phases of solid solutions, for example, $(Zr,U)O_2$ [10];

☞ a zirconium metal powder is added to the starting mixture of oxides; this admixture secures electrical conductivity of the charge during induction heating. In also promotes incorporation of plutonium and actinides into the host-phase lattice by the scheme deciphered as a result of investigation of high-uranium zircon from Chernobyl "lavas" [5]:

$Zr + MeO_X \rightarrow Zr\text{-}Me\text{-}O$, where *Me* – U, Pu, Am, Np, Cm, ...;

$Zr\text{-}Me\text{-}O + SiO_2 \rightarrow (Zr,Me)SiO_4$ – zircon;

$Zr\text{-}Me\text{-}O + O_2 \rightarrow (Zr,Me)O_2$ – zirconium dioxide.

The excess of zirconium metal in the course of synthesis is oxidized to ZrO_2. This allows not to observe a strict stoichiometry of the starting charge's chemical composition. Real experiments on the synthesis of polymineral ceramics based on zircon $ZrSiO_4$ and zirconium dioxide ZrO_2 by the hot pressing method, were conducted at the V. G. Khlopin Radium Institute in a broad range of conditions. The maximum temperature of synthesis did not exceed 1500 °C, the maximum pressure was $4.5 \cdot 10^9$ Pa, the synthesis time from 1 minute to 30 minutes. A plasmochemical powder of cerium oxide CeO_2 was used as a simulator of plutonium and actinides.

All the experiments were successful. The maximum yield of zircon $(Zr,Ce)SiO_4$ was 38%*mass*, but the main phases in all the samples were mixtures of cubic, tetragonal and monoclinic zirconium oxides $(Zr,Ce)O_2$. The developed method of zircon ZrSiO4 synthesis doped with isomorphous admixtures has been patented [11] and experimental work is going on at present.

Conclusion

The work on the synthesis of crystalline matrices is not exhausted by the problem of weapons plutonium and waste actinides. Results of these investigations can be used in other fields of human activity, too. We may expect the creation of new types of especially strong ceramics, sources of ionizing radiation for work under conditions of high temperature and aggressive medium, etc.

References

1. Carrol, D. F. (1963) The system PuO_2-ZrO_2. *Journal of American Ceramic Society – Discussion and Notes* **46**, p. 194.
2. Rouanet, A. (1971) Contribution a l'etude des system zircon – oxides, des lanthanides au voisinage de la fusion. *Rev. Int. Hautes Temper et Refract* **8**, 161–180.
3. Mumpton, F. A., Roy, R. (1961) Hydrothermal stability studies of the zircon–thorite group. *Cosmochimica Acta* **21**, 217.
4. Anderson, E. B., Burakov, B. E., Pazukhin, E. M. (1961) High uranium zircon (Zr,U)SiO_4 from "Chernobyl lavas". *Radiochimica Acta* **60**, 149–151.
5. Burakov, B. E. (1993) A study of high-uranium technogenous zircon (Zr,U)SiO_4 from Chernobyl "lavas" in connection with the problem of creating a crystalline matrix for high-level waste disposal. *Proceedings of international conference SAFEWASTE'93 13–18 /06/ 1993, Avignon, (France)* **2**, 243–246.
6. Ewing, R. C., Lutze, W., Weber, W. J. (1995) Zircon: A host-phase for the disposal of weapons plutonium. *Journal of Material Research* **10**, 243–246.
7. O'Donoghue, M. A. (1976) Resent developments in the synthesis of possible gem materials. *Journal of Gemology* **XV**, 119–124.
8. Heimann, R. B., Vandergraaf, T. T. (1988) Cubic zirconia as a candidate waste form for actinides: dissolution studies. *Journal of Material Science* **7**, 583–586.
9. Anderson, E. B., Burakov, B. E., Pazukhin, E. M. (1992) Materials testing aspects of the problem of the Chernobyl NPP 4-th Unit's high-level radioactive products burial. *High Level Radioactive Waste Management. Proceedings of the Third International Conference, Las Vegas, Nevada, April 12–16, 1992* **2**, 29–33.
10. Anderson, E. B., Burakov, B. E., Vasiliev, V. G. (1993) A creation of crystalline matrix for actinide waste in Khlopin Radium Institute. *Proceedings of international conference SAFEWASTE'93 13–18 /06/ 1993, Avignon, (France)* **2**, 29–33.
11. Burakov, B. E., Anderson, E. B., Pazukhin, E. M., Mikheeva, E. E. (1991) A method for obtaining crystalline zircon doped with isomorphous admixtures. Patent of Russia, by the claim No. 5020357/26 of 12.11.91.

ON THE LONG-TERM STORAGE OF WEAPONS PLUTONIUM IN FORM OF CRITICALITY-FREE CERAMICS CONTAINING LITHIUM

J.A. TRUTNEV, B.A. NADYKTO, N.S.PRUDOVA,
J.M. KHIRNY
RFNC-VNIIEF
Arzamas-16

Nuclear stockpile reductions have caused safe disposal problems for nuclear materials recovered from dismantled weapons. Of these materials, the most environmentally and radiologically hazardous are the nuclear active plutonium and uranium enriched in U-235 isotope, both being of finite critical masses.

For long-term storage it is suggested to transfer nuclear active materials into states such that neither reactions with oxygen nor critical mass would occur, thus avoiding any inadvertent or intended nuclear accident or fire, and, accordingly, large-scale dispersal and scattering of these materials at fire. This method can make it impossible for terrorist or saboteurs to reproduce a nuclear weapon without recycling the stolen materials, which would take much time, efforts, risks and expenses. Clearly, this recycling would require special equipment and know-how.

For long-term storage purpose, plutonium and uranium are initially transformed into powder oxides (PuO_2 and UO_2), and then mixed with the oxides of other elements which can efficiently absorb neutrons over a wide energy range from thermal to megaelectronvolte level. These materials are lithium-6 and boron-10 isotopes, with the contents being 7.4% and 18.7% for natural lithium and boron, respectively.

Mixed powders are inconvenient to store, moreover, their large density difference can bad to inadmissible separation of heavy and light fractions at storage. Therefore, the mixtures are to be compacted and sintered in furnace at temperatures allowing to produce ceramic or vitrified granules. These granules represent rather tough formations offering high thermal stability, and remain unchanged under temperatures as high as 1000°° C. To facilitate production of granules and impart them suitable properties, other element oxides can be added to plutonium (uranium), lithium and boron mixtures.

E.R. Merz and C.E. Walter (eds.), Disposal of Weapon Plutonium, 91-92.

As calculations show, critical mass of PuO_2 mixed with Li_2O 30% in mass enriched with Li-6 to 90% will be 10 tons. The critical mass for PuO_2 mixed with B_2O_3 accounting for 34% by mass, 90%-enriched in B-10, is 8 tons. For 50 percent weight fractions of Li_2O or B_2O_3, their mixtures with PuO_2 will be of no critical mass.

There may be certain difficulties in storing plutonium arising from its radioactive decay that would create heating, gamma and neutron sources. Intensity of the heat source is 2 W/kg, while neutron sourses intensity depends on the isotopic composition of plutonium. In addition to spontaneous fission, there is another neutron source in plutonium mixed with lithium or boron, which is (α, n) reaction involving Pu-239 decay α-particles. Lithium-7 has (α, n) reaction threshold 4.38 Mev, and lithium-6 - 6.62 Mev. Therefore, (α, n) reaction involving plutonium decay alphas will not go (E=5.2 Mev) for Li-6 isotope. The neutron yield can be estimated by scaling with respect to the experimental neutron yield from Pu-Be source. For Li-6 case (90%), the estimated neutron yield is $3.5\ 10^5$ n/s per kilogram of Pu. The actual yield may be of lower value due to oxygen and Li-6, that can be found experimentally.

Further investigations are to be pursued with emphasis upon developing a technology for ceramic material based on plutonium oxide mixed with neutroncapture materials. The basic possibility to produce such material was tested using a uranium oxide of natural isotopic composition as an example. The uranium oxide was mixed, compacted and then sintered with addition of Li_2CO_3, i.e. similarly to lithium niobate production. Addition of a small percentage of Al_2O_3 is also useful like for other ceramics materials. This technique can be used to obtain ceramic material containing 35% of Li_2O and 65% UO_2. With sintering, the granule diameter reduces by 10%. The sintered granule represents a rather tough form with its properties remaining unchangeable after 3 days' stay in water at room temperature.

With this technique for material reprocessing, the storage of large amounts of plutonium can be made safer and less expensive. One of the approaches to storing this critmassless plutonium is to use a specifically drilled hole about 300 m in depth, 1 or 2 m diameter. About 10 or 15 tons of Pu can be placed into this hole. It is essential that plutonium storage should be subject to continuous monitoring and inspections. This could be provided with leak tight Pu containment vessels being mounted on hoisting rods inside a borehole. The borehole can be filled up with water to cool the containers, if necessary.

INCORPORATION OF Pu AND OTHER ACTINIDES IN BOROSILICATE GLASS AND IN WASTE CERAMICS

Hj. Matzke and J. van Geel

European Commission, Joint Research Centre, Institute for Transuranium Elements, Postfach 2340, D-76125 Karlsruhe, Federal Republic of Germany

Abstract

A large series of actinide containing borosilicate waste glasses with different compositions and of different potential waste ceramics (including SYNROC) was produced in the European Institute for Transuranium Elements some years ago. The actinides were Pu, Np, and also Am and Cm. These latter two short-lived actinides were used to observe effects of radiation damage and of He-accumulation in a short time scale, thus simulating the long-time behavior of glasses or ceramics containing high level waste or Pu only. A number of interesting properties such as solubility of actinides in the matrix, its hardness, fracture toughness, Youngs modulus, and leach rates in water at different temperatures, were measured in the as-produced products and subsequently again as a function of α-decay damage. Volume changes (swelling) and - for the ceramics - phase changes from the crystalline to the amorphous phase were also detected and damage annealing was investigated. This data base, though mainly measured on Cm-containing glasses and ceramics, is used to provide predictions on the performance of Pu-matrices.

1. INTRODUCTION

Work on incorporation of actinides into borosilicate waste glasses has been performed at the European Institute for Transuranium Elements since 1975 [1]. In addition, different potential waste ceramics containing actinides were characterized and analysed [e.g. 2, 3]. Most emphasis was placed on Cm in order to study the effects of α-decay damage in an acceptably short time, but waste matrices containing Th, U, Np, Pu and Am were also produced and studied. Much of this work was done in close collaboration with scientists and institutes with expertise in glasses and ceramics to solidify high level waste (HLW) and actinides.

E.R. Merz and C.E. Walter (eds.), Disposal of Weapon Plutonium, 93-105.

These scientists and their institutions are mentioned in the acknowledgements and in the list of references.

In the present article, relevant results on the solubility of Pu in borosilicate glasses will be summarized, the main results of the radiation damage experiments will be given, new data on the release of He after long storage time will be reported and two new methods will be presented to detect gas bubbles in glasses and to determine boron release by leaching in water.

2. MATERIALS

The glasses investigated were the German glasses VG 98/12 (base glass, no high level waste added) and GP 98/12 (containing HLW) and some precursors of this glass, the French product SON 68.18.17 L1C2A2 Z1 (or R7T7) and the US product MCC 76-68. The compositions are given in Table I. All glasses are borosilicate glasses. The HLW is simulated and inactive. The actinides were incorporated by adding their oxides to the glass frit and melting the mixture.

The ceramics used are summarized in Table II.

3. SOLUBILITY AND SEGREGATION PROCESSES OF ACTINIDES

3.1 Borosilicate glasses

The actinide content of HLW from reprocessing of spent nuclear fuel is small (< 1 wt.% for all actinides, i.e. the sum of UO_2, NpO_2, PuO_2, AmO_2 and CmO_2) and can safely be assumed to be in solution.

To determine the solubility and/or to observe possible segregation of PuO_2 at higher Pu-contents, the German borosilicate waste glass was produced with up to 10 wt.% PuO_2 by mixing the proper quantity of the glass frit with or without denitrated waste solution and with Pu (or also Am). Melting was performed in Pt-crucibles. The resultant glass was pulverized and molten again in graphite crucibles at 1150 °C. Discs were cut from the glass cylinders produced in this way. Alpha-autoradiographs showed no visible segregation for the highest AmO_2 content used, i.e. 4.9 wt.% AmO_2, whereas very obvious segregation (spots very rich in PuO_2) were seen when 10 wt.% PuO_2 were added [1]. Additionally, a base glass without HLW and the glass with high HLW loadings, i.e. either 20 wt.% HLW, or 20 % HLW with additionally 10 wt.% Gd_2O_3, were produced [4] with 5 wt.% UO_2 or 5 wt.% PuO_2, respectively. The actual aim of these early experiments was to study the feasibility of vitrifying actinide-rich waste, e.g. from sludges or incinerator ashes. 80 mm long cylinders were kept molten (1175 °C) in Pt-crucibles for 25 h. Subsequently, discs were cut from the bottom, the middle and the top section of these rods. The discs were coated with an 150 Å Al conducting layer and analyzed by electron probe microanalysis, epma. These analyses showed the absence of sedimentation of actinides. In contrast, the noble metal particles were enriched in the bottom section [4].

Table I: Composition of the (simulated) nuclear waste glasses (wt.%)

	SON 681817 L1C2A2Z1+	GP 98/12	Precursor GP*	MCC 76-68
SiO_2	45.5	48.3	40.2	46.3
B_2O_3	14.0	10.7	10.8	10.6
Al_2O_3	4.9	2.3	1.1	0.6
Na_2O	9.9	14.9	21.9	14.8
MgO	-	1.8	-	-
CaO	4.0	3.5	2.2	2.7
TiO_2	-	4.0	3.3	3.4
Fe_2O_3	2.9	0.3	1.9	4.4
ZnO	2.5	-	-	4.0
Li_2O	2.0	-	-	-
HLW[a]	14.3	14.2	18.6	13.2

[a] contains fission products, corrosion products and actinides (0.3 wt.% U in GP 98/12, 0.85 wt.% U+Th in SON). Note that ~ 15 wt.% HLW corresponds to ~ 6 mol% because of the high molecular weight of actinide and fission product oxides. The added fission product oxides vary slightly between different glasses and are, typically: BaO, CeO_2, Cs_2O, Eu_2O_3, Gd_2O_3, La_2O_3, Nd_2O_3, MnO_2 (for Tc), MoO_2, PdO, Pr_2O_3, Rb_2O, Rh_2O_3, RuO_2, Sm_2O_3, SrO, TeO_2, Y_2O_3 and ZrO_2

+Also nominated "R7T7" LWR

* Used for solubility and segregation studies (Section 3)

Table II: Waste ceramics investigated in TUI for actinide incorporation and radiation damage (Cm-doping)

SYNROC	a **syn**thetic **roc**k composed of (predominantly) 3-4 phases simulating minerals: perovskite $CaTiO_3$, Ba-hollandite $BaAl_2Si_2O_8$, zirconolite $CaZrTi_2O_7$ and rutile TiO_2, together with e.g. 10 wt% of added calcined HLW
$CaZrTi_2O_7$	zirconolite, monoclinic
$Gd_2Ti_2O_7$	pyrochlore structure, cubic
$Ca_2Nd_8(SiO4)_6O_2$	apatite structure, hexagonal
$ZrSiO_4$	zircon, tetragonal
and	
Celsian glass ceramic -	mainly celsian $BaAl_2Si_2O_8$ and a rare earth titanate $(RE)_2\ Ti_2O_7$ with some scheelite (Ca,Ba,Sr) MoO_4 and spinel $(Ni, Fe, Cr)_3O_4$, and residual glass.

Glasses doped with 5 wt.% AmO_2 or 3 at% UO_2 were still vitreous with no crystalline phase being present. In contrast, the glasses containing 5 wt.% PuO_2 consisted all of two phases, the major vitreous phase and a minor crystalline phase. In detail

- the base glass contained 4.6 wt.% PuO_2 in solution. The second phase was PuO_2
- the glass with 20 % HLW contained less PuO_2 in solution, i.e. 4.4 wt.%, the crystalline phase being again PuO_2
- the glass with 20 % HLW and 10 % of the neutron poison Gd_2O_3 showed a very small solubility of only 1.2 wt.% PuO_2 and contained segregated crystalline phases of an (average) composition of $(Pu_{0.68}Ce_{0.2}Gd_{0.12})O_2$.

These solubility values are large compared with the actinide contents in HLW solutions, but they may be of concern for the fabrication of Pu-containing glasses, in particular, if fission products (in particular rare earths) are added to these glasses.

3.2 Ceramics

Most materials of Table II (zirconolite, pyrochlore, apatite, celsian glass ceramic) were doped with ~ 3 wt.% Cm. X-ray diffraction did not indicate any segregation of Cm. Epma measurements were not performed on these materials.

A very careful examination was made with the titanate waste ceramics (type SYNROC) using electron probe microanalysis (epma), autoradiography, X-ray diffraction (XRD), scanning electron microscope backscattered electron imaging (BSE), analytical electron microscopy (AEM), and selected area electron diffraction (SAD) [4,5]. 10 wt.% NpO_2 could be incorporated into SYNROC B, where it was shown to be partitioned between zirconolite (10.7 wt.% Np) and perovskite phases (9.5 wt.% Np) whereas the remaining two phases of hollandite and rutile contained $\leq$ 0.2 wt.% Np [5] . Other titanate-based, polyphase waste forms were synthesized with a loading of 10 wt.% HLW and with 0.5 wt.% UO_2, 2.0 wt.% NpO_2, 0.6 wt.% PuO_2 and 0.005 wt.% Am_2O_3 [4]. The greater portion of the actinides partitioned again between perovskite and zirconolite. However, autoradiography revealed clusters of Pu-atoms interpreted as unreacted PuO_2 or Pu_2O_3 though the Pu content was low. The conclusion is therefore that though transuranium elements can be successfully incorporated into polyphase titanate waste forms, some portion does not enter the expected phases but precipitates as oxides. The construction of experimentally well supported ternary and quaternary phase diagrams and more careful characterization of polyphase waste with different contents of e.g. Pu are needed before a tailor-made matrix can be suggested with a high solubility of Pu.

4. RADIATION DAMAGE EFFECTS

The effects of α-decay were studied on all Cm-doped glasses and ceramics (Tables I and II). Fig. 1 shows the relations between damage accumulation, expressed as number of α-decays per m^3, and the storage time for a borosilicate glass containing 1.5 wt.% Cm-244 (lower abscissa). On the upper abscissa, possible equivalent storage times are shown for a loading with a typical HLW from reprocessing of spent UO_2 fuel (33 GWd/tM burnup) and for a glass loaded with 7 wt.% Pu-239. The time scales depend, of course, on amounts and composition (elemental and isotopic) of the actinides loaded into the glass (or ceramic). The damage studies at ITU were usually terminated at about 15 to 25×10^{24} α-decays/m^3. These damage levels were achieved within a few years (typically 5 years) with the products containing Cm-244. A damage level of 15×10^{24} α-decays/m^3 will be reached with Pu-239 containing matrices with, say, a 7 wt.% loading in about 1000 years.

The damage caused by α-decay events is predominantly due to the atomic displacements created by the daughter atom of the decaying actinide, e.g. the U-235 atoms formed in the decay of Pu-239. Typically, the recoil energy of these heavy daughter atoms is 100 keV. About 90 % of this energy is used to produce displaced atoms creating a dense collision cascade along the range of ~ 25 nm of the recoil atom and producing about 1200 to 1500 atomic defects. In contrast, the α-particle with a typical energy of 5 MeV looses 99 % of its energy by ionization, it has a large range of ~ 12 μm in glasses and it produces only ~ 200 atomic defects (see e.g. ref. [6]).

At ITU, different ion implantation techniques are also used to simulate the α-decay damage. Fission damage is also simulated with high energy heavy ions. These experiments are not described here, but they serve as an additional data base for the conclusions reached from actinide doping studies.

The time needed to achieve a given damage level is very short in ion implantation studies, (typically minutes to hours) as compared with actinide doping (some years, see Fig. 1). Also, often U- or Th-containing minerals are available in nature which have accumulated damage over very long time scales (typically 10^8 years). A comparison of the results on these different materials gives information on dose-rate effects, i.e. it shows whether or not, or how much, damage effects vary if a given damage level is reached in very different time intervals. Dose rate effects are known to exist in some ceramics due to partial annealing of the damage at small dose rates, and they could be expected in glasses because of the low damage recovery temperatures (~ 200 to 500 °C). They were, in fact, observed in measurements of Na diffusion in the glass GP 98/12 [7]. The problem of dose rates in glasses was discussed in detail at a workshop on radiation damage in nuclear waste materials [6] with the conclusion that some damage recovery occurs in the time scale of Cm-doping, and under real storage conditions of active HLW glasses, but not in the short times of ion implantation. No significant differences in damage effects were observed between the three glasses GP 98/12, R7T7 and MCC 76-68 with 0.5 or 1.5 wt.% Cm, respectively, hence for a factor of 3 in dose rate. We conclude therefore, that the results obtained with Cm-doped glasses can be used to predict the behavior of Pu-glasses.

4.1 The properties measured on the Cm-doped materials

Immediately following fabrication, at different time intervals during damage ingrowth and at the end of the experiments at high damage levels, a number of properties of the Cm-doped glasses (g) or ceramics (c) were measured: These properties are

- density and volume, usually with immersion techniques (g+c)
- crystallinity, with X-ray or electron diffraction (c)
- leaching, i.e. corrosion in water, usually in autoclaves with a ratio of sample surface to water volume, S/V ~ 10 m^{-1} at 90 to 200 °C for various times (g+c)
- mechanical properties: hardness, fracture toughness, Youngs modulus with indentation techniques (g+c)
- damage recovery, recrystallization and stored energy (c)
- He-release (g).

4.2 Radiation stability of the glasses

Density and volume: very small changes in density ($\Delta\rho \leq 0.6$ %) were observed in the three glasses of Table I up to a damage level of ~ 5 x 10^{24} α-decays/m^3. The dose dependence which could be fitted to an ingrowth equation of the type

$$\Delta\rho = A\,(1\text{-}\exp\{\text{-}BD\})$$

where A is the change in density at saturation, B is the fraction of the glass totally damaged by each α-decay, and D is the dose. For instance, for the French glass, R7T7, A = 0.55 ± 0.05 % and B = (0.095 ± 0.005)x10^{-17} g. Saturation was observed at ~ 5x10^{24} α-decays/m^3 (or 2x10^{18} α/g) suggesting that an equilibrium state is reached between defect formation and instantaneous defect recovery at this damage level. Similar small density changes (range ± 0.5 %) were found for other glasses including UK glasses containing 2.5 wt.% Pu-238 [8, 9].

Leaching: Figs. 2 and 3 show typical results for leaching of the Cm-244 doped glasses GP 98-12 and R7T7 (static test, H_2O, 14d, 150 °C). The damage level achieved is of the order of 1 dpa (displacement per atom) and corresponds roughly to the level achieved in a real HLW glass after ~ 10^5 years, or in a Pu-glass after ~ 10^3 years (see Fig. 1). No significant effect of damage is seen. The leach rates actually tend to decrease rather than to increase with increasing α-decays. Fig. 2 shows the total normalized mass loss, and Fig. 3 shows the trend for 7 of the 30 elements analyzed. The trend for all 30 elements was that of the total weight loss: the leaching behavior is not significantly affected by radiation damage if a realistic damage production, i.e. actinide doping, is used.

Mechanical properties: Hardness H, fracture toughness K_{Ic}, and Youngs modulus E were measured with the same three Cm-doped glasses as a function of α-decays, using indentation techniques [10]. H decreased by ~ 30 %. This decrease occurred fast, at about 3x10^{24} α-decays/m^3, for all 3 glasses. Simultaneously, the probability

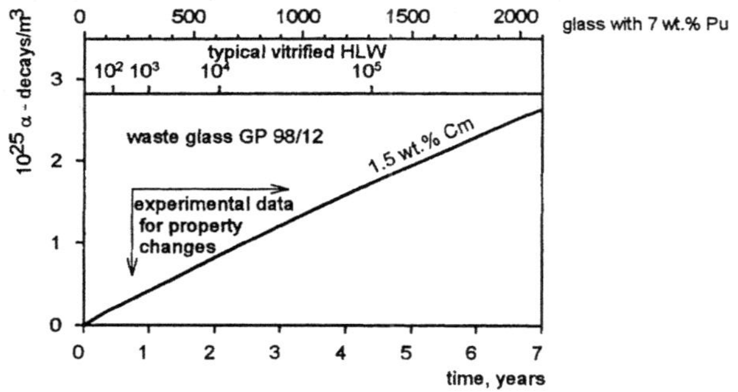

Fig.1: Relation between accumulating damage due to α-decays and storage time for the glass GP 98/12 containing either 1.5 wt.%Cm-244 or HLW from reprocessing,or 7 wt.%Pu-239

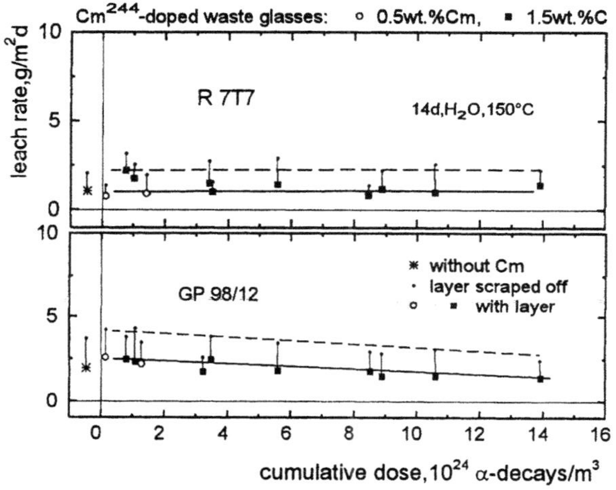

Fig.2: Leach rate of two borosilicate glasses doped with 0.5 or 1.5 wt.% Cm-244 as function of α-decay damage (static test,H_2O,150°C,14d)

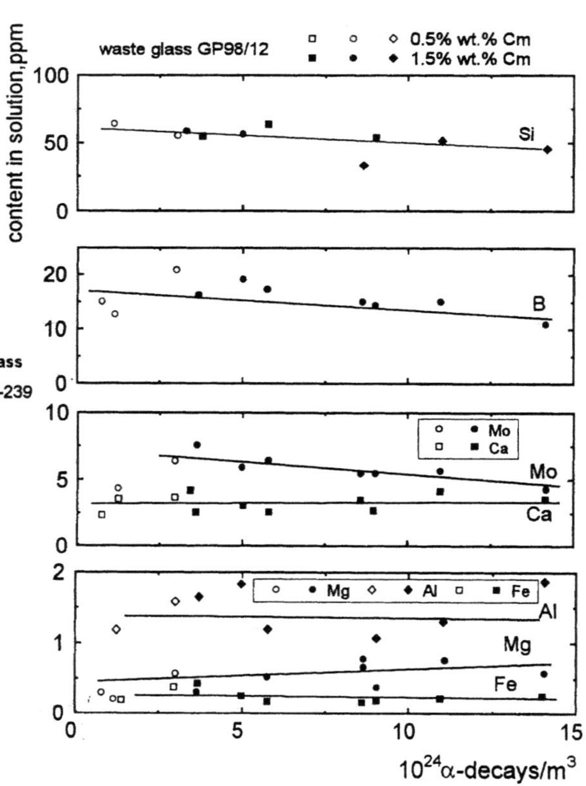

Fig. 3: Content in solution of Si,B,Mo,Ca,Al,Mg and Fe as a function of α-decay damage for the GP 98/12 glass of Fig. 2

of crack formation and the crack lengths at Vickers indentations decreased, corresponding to a significant increase in K_{Ic} by up to 100 %. This behavior is explained to be most probably due to quenched-in stresses along the path of the recoil atoms of the α-decay, possibly supported by the formation of He-bubbles. In the same dose range {(2-3)x10^{24} α-decays/m^3}, E decreased by ~ 30 %.

The net result of these changes are reduced brittleness and better resistance against crack propagation, i.e. technologically positive effects.

He-behavior and release: due to the large amount of α-decays, a high concentration of He is built-up during storage, since the highly energetic α-particles are forming He-atoms after coming to rest in the glass and after capturing two electrons. For instance, more than one unit volume of He under standard temperature and pressure conditions per unit volume of the glass, actually 1.3 cm^3 He (STP)/cm^3 glass, had been formed in pieces of the R7T7 glass with 1.5 wt.% Cm-244, when these pieces were heat-treated to observe swelling and He-release. An anneal for 10 h at 300 °C caused no swelling, the measured density change being 0.28 % only and thus within the scatter band of the density values for different glass pieces. On direct annealing to 500 °C, in contrast, the corresponding glass GP 98/12 "foamed" to about twice its starting volume.

The R7T7 glass pieces were further annealed in a shielded Knudsen cell [11]. He release compatible with single He atom diffusion with an activation enthalpy of 175 to 185 kJ/mol was observed between 300 and 420 °C, involving < 1 % of the total He. Between 420 and 550 °C, most gas was released with kinetics compatible with gas bubble diffusion. On further annealing up to 800 to 1000 °C, a few large release peaks were found similar to those observed in nuclear fuel when cracks are formed. The glass had swollen during this treatment.

The question of the onset of He-bubble formation is still open. In the present study, no swelling was seen at a very high He-content and at 300 °C. In contrast, in a Cm-doped Japanese glass, large He-bubbles (0.15 to 0.35 μm radius) were reported to be formed even without annealing, and these bubbles shrank during annealing.

A potential means of following He bubble formation has been recently developed; this is the nondestructive method of small-angle X-ray scattering, SAXS [13]. Thin foils of the glass GP 98/12, either as-produced or following ion implantation with He, Kr or Xe, were analysed with SAXS. Two distinct radiation effects were easily seen by SAXS and could be confirmed by transmission electron microscopy, TEM.

i) The He-implanted specimens with a high He-concentration (within the range of the implanted ions) of about 3x10^{20} He/cm^3, hence a concentration higher than that of the above Cm-doped glasses, showed increased SAXS intensity with little angular dependence. This suggests the formation of density fluctuations due to radiation-induced materials distortions having atomic to molecular sizes, similar to point defects. In agreement, the TEM examinations revealed a featureless material, hence He-bubbles were not present, in agreement with the SAXS results [13].

ii) In contrast, the foils implanted with Kr- or Xe-ions showed a very clear rise in SAXS intensity with decreasing angle at small angles, as would be expected for gas bubbles of ~ 18 nm size. Such bubbles were indeed seen in TEM [13].

4.3 Radiation stability of the investigated ceramics

All of the ceramics doped with Cm (see Table II) swelled significantly. Volume changes of up to 9 % were observed (e.g. refs. [2, 9], in contrast to the very small (~ 0.5 %) volume changes in borosilicate glasses. Also at - or slightly below - the α-decay level causing saturation in swelling, a phase change from crystallinity to a radiation-induced amorphous or metamict state was observed. This amorphization resulted in higher leach rates [2, 9], and the volume increase causes stress on the container and/or crack formation of the ceramics. As a technologically positive feature, the hardness H decreases and the fracture toughness K_{Ic} increases with increasing exposure to α-decays. Fig. 4 shows, as an example, the results for Cm-doped zirconolite $CaZrTi_2O_7$ [2, 14]: swelling, X-ray intensity, hardness, H/E-ratio and fracture toughness are plotted as a function of cumulative dose. Arrows in Fig. 4a show that slightly below the dose level causing saturation in swelling, metamictization occurs, i.e. the crystalline sample becomes amorphous. For the amorphized zirconolite, the leach rate is higher by about a factor of 10 as compared to that of the original crystalline material.

Annealing of damage was also studied. It occurs at higher temperatures than those needed for glasses. Complete recovery of swelling, for instance, is typically achieved at 700 - 800 °C. He-release has not been measured, but it is worth mentioning that the specimens did not fracture into pieces upon annealing, as some radiation-damaged minerals do.

Radiation effects in the above ceramics (see Table II) are thus much more important than radiation effects in glasses, at least if damage is accumulated at or near ambient temperature. Large volume changes, amorphization and increased leach rates are negative features, a certain improvement in mechanical properties (as in glasses) is a technologically positive consequence of α-decay damage.

5. DISCUSSION AND APPLICATION TO THE LONG-TIME BEHAVIOR OF PU-CONTAINING MATRICES

The present paper gives a summary of work performed at the European Institute for Transuranium Elements, Karlsruhe which is of relevance to the problem of incorporation of Pu into glass or ceramics. It is not the aim of the paper to summarize all of the available relevant literature. Much of this is presented in other contributions to the present workshop, or it can be found in the book on Radioactive Waste Forms for the Future, edited by W. Lutze and R.C. Ewing [15]. The data base available at ITU can be applied to predict certain properties of Pu-matrices, to define important open questions and to suggest helpful experiments.

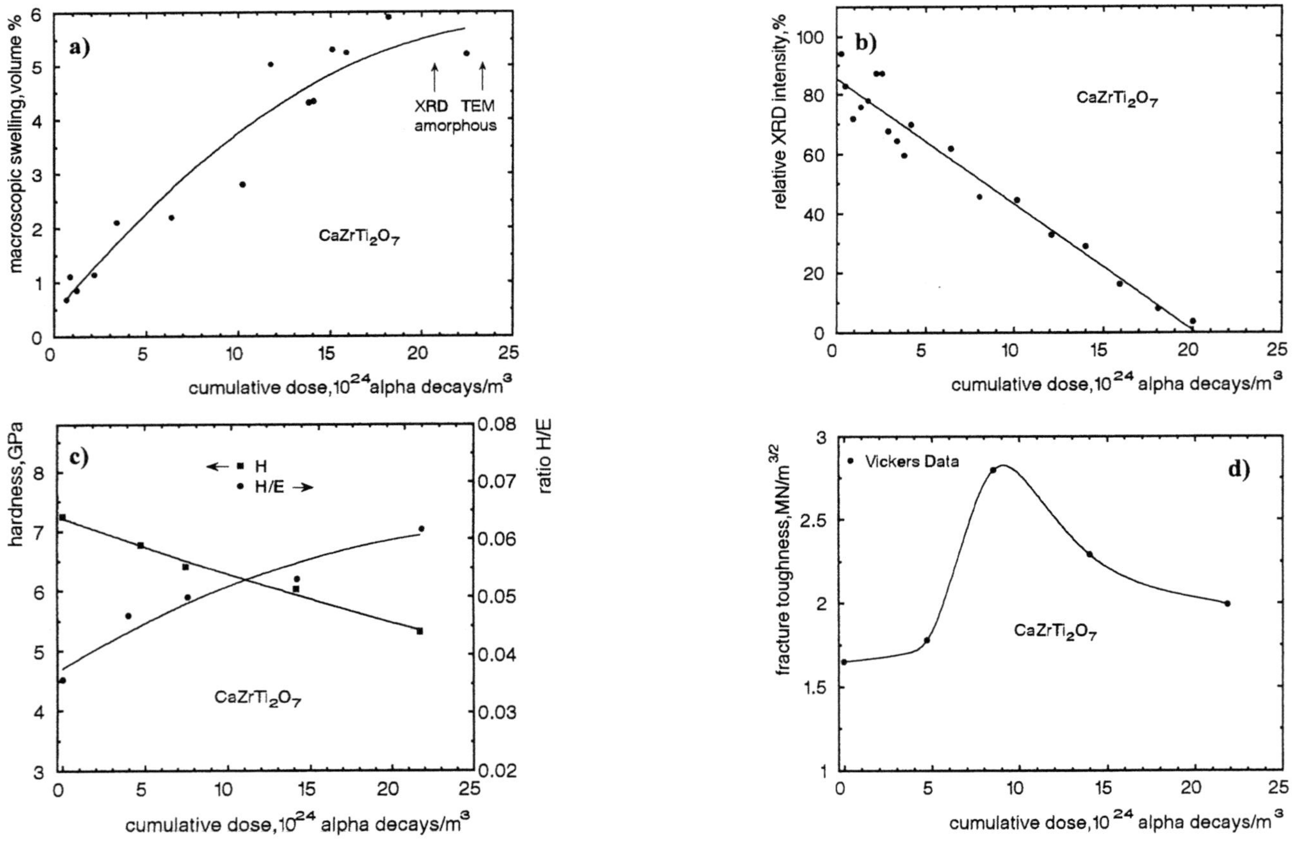

Fig. 4: Radiation damage effects in Cm-244 doped zirconolite $CaZrTi_2O_7$ as a function of cumulative dose.
a) macroscopic swelling b) total integrated X-ray intensity c) hardness H and ratio hardness to Youngs modulus,H/E,and d) fracture toughness K_{Ic}

Solubility of Pu: careful experimental work has indicated the existing problems. Solubilities of Pu in glasses vary with glass type and composition. In particular, a low Pu solubility was found in a glass containing a large concentration of rare earths. Also for ceramics, more work is needed before Pu can be safely assumed to be in solid solution. Extrapolating of actinide behavior from the known phase behavior of rare earth elements is not always possible.

Radiation damage effects: if proper damage sources are used, borosilicate glasses can be shown to be satisfactorily stable against α-decay damage. Density and volume changes are small ($\leq$ 0.5 %), the leach rate does not increase appreciably with increasing damage level and there is a technologically positive change in that hardness and brittleness are reduced and the resistance against crack propagation is increased.

In contrast, the Cm-doped ceramics swelled significantly (by up to 9 %), with the related risk of cracking. Also, metamictization was observed at high doses, with significantly increased leach rates for the radiation-damaged amorphous state. The mechanical properties were affected in a positive way, as for the glasses.

Whereas saturation in damage is achieved with the Cm-doped materials within 3-5 years, the same level will be achieved in a much longer time span in a matrix for Pu-239. This time span depends on the Pu-content used. For 7 wt% Pu, saturation will be reached in (typically) 4000 to 10000 years, depending on the matrix. We do not expect any significant differences in damage effects if active fission products are added.

An important property for long term final storage is the leach rate in the (unlikely) event of ground water having access to the repository. As is well known, there are other parameters affecting leaching as well, e.g. the type of container material and of backfill. For instance, the glass GP 98/12 (see Table I) showed very significantly increased leaching in the presence of a steel container [16]. For most silicate minerals, the dissolution rate is controlled by the activity of the silicic acid in the solution. As a consequence, the initial large corrosion rate (typical for dynamical leaching with a high flow rate) decreases soon to the much lower static leach rate [8]. In the presence of corroding steel, the formation of iron oxide precipitates and adsorption of Si on these precipitates decrease the activity of the silicic acid and cause the large initial leach rate of the glass to be maintained over much longer times. Careful selection of container and backfill materials is thus recommendable.

For the specific case of Pu-matrices, preferential leaching of boron from borosilicate glass may be a potential problem for criticality considerations. Leach rates for boron are known to be high, and B depletion can extend comparitively deep into the glass. This was recently shown with a nondestructive nuclear method using thermal neutrons to induce the reaction B-10(n,α)Li-7 and obtaining the B-profile by measuring the energy loss of the α-particles [17]. Of course, this method is only applicable to the base matrix before loading with Pu.

In summary, a large data base exists at ITU, and in the literature, to suggest matrices and loading levels with Pu. For a fully optimized, tailor-made matrix, some more research seems to be necessary.

Acknowledgments

A larger number of scientists of different organisations have contributed to accumulate the data summerized here. We like to thank, in particular, J.P. Hiernaut, C. Ronchi, E. Toscano and C.T. Walker (ITU), T.J. White (Univ. Queensland, Australia), W.J. Weber (PNL Battelle, USA), E. Vernaz (CEA, France), Y. Eyal (Technion Haifa, Israel) and D. Fink (HMI Berlin, Germany).

References

[1] K. Scheffler, U. Riege, K. Louwrier, Hj. Matzke, I. Ray and H. Thiele: "Long Term Leaching of Silicate Systems: Testing Procedure, Actinides Behaviour and Mechanism". German Report KfK 2456 and EUR-5509.e (1977)

[2] W.J. Weber and Hj. Matzke: "Radiation Effects in Actinide Host Phases", Radiation Effects 98 (1986) 93

[3] Hj. Matzke, B.W. Seatonberry, I.L.F. Ray, H. Thiele, H. Trisoglio, C.T. Walker and T.J. White: "The Incorporation of Transuranic Elements in Titanate Nuclear Waste Ceramics", J. Amer. Ceram. Soc. 73 (1990) 370

[4] K. Scheffler, U. Riege and C.T. Walker: "Chemical Compatibility of HLW Borosilicate Glasses with Actinides: Solubilities and Segregation Processes", German Report KfK 2552 and EUR-5750 e (1978)

[5] Hj. Matzke, E. Toscano, C.T. Walker, and A.G. Solomah: "Neptunium Doping of the Crystalline Ceramic Waste Form SYNROC B", Advanced Ceram. Mater. 3 (1988) 285

[6] Hj. Matzke, ed.: "Nuclear Waste Materials. A Workshop on Radiation Damage Effects in Nuclear Waste Materials". Special issue Nucl. Instrum. Methods in Phys. Research B32 (1988) 453-517
and therein Hj. Matzke: "Radiation Damage Effects in Nuclear Materials", p. 455

[7] M. Yamashita and Hj. Matzke: "Self-diffusion of Na in GP 98/12 Waste Glass under α-ray Irradiation" in Modifications Induced by Ion Irradiation in Glasses, Ed. P. Mazzoldi (Elsevier, Amsterdam, 1992) p. 53

[8] Hj. Matzke and E. Vernaz: "Thermal and Physicochemical Properties Important for the Long Term Behavior of Nuclear Waste Glasses", J. Nucl. Mater. 201 (1993) 293

[9] E. Vernaz, A. Loida, G. Malow, J.A.C. Marples and Hj. Matzke: "Long-Term Stability of High-Level Waste Forms" in Proc. 3rd EC Conf. on Radioactive Waste Management and Disposal, Luxembourg, Sept. 17-21, 1990. Ed. L. Cecille, Elsevier, Amsterdam (1991) p. 302

[10] Hj. Matzke and E. Toscano, eds.: "Indentation Techniques", Special Issue Europ. Appl. Res. Rep. 7, No. 9 (1990)

[11] J.P. Hiernaut and C. Ronchi: "Shielded Knudsen Cell for Fission Product Source Term Analysis" in Report EUR 15154EN (1992) p. 44

[12] Y. Inagaki, H. Furuya, K. Idemitsu, T. Banba, S. Matsumoto and S. Muraoka, "Microstructure of Simulated High-Level Waste Glass Doped with Short-Lived Actinides, Pu-238 and Cm-244"," Mater. Res. Soc. Symp. Proc. 257 (1992) 199

[13] R. Evron, Y. Cohen, O. Regev, Y. Eyal, Hj. Matzke and K. Tinschert: "Ion-Implantation Induced Microstructural Damage in a Nuclear Waste Glass", Proc. Meeting Nucl. Soc. Israel 18 (1994) VIII-2

[14] W.J. Weber, W.J. Wald and Hj. Matzke: "Effects of Self-Irradiation Damage in Cm-Doped $Gd_2Ti_2O_7$ and $CaZrTi_2O_7$", J. Nucl. Mater. 138 (1986) 196

[15] W. Lutze and R.C. Ewing, eds.: "Radioactive Waste Forms for the Future", North Holland, Amsterdam (1988) 778 pages

[16] Hj. Matzke: "Nuclear Waste Materials", in Ion Beam Modification of Insulators, eds. P. Mazzoldi and G. Arnold, Elsevier, Amsterdam (1987) p. 501

[17] J.P. Biersack, D. Fink, R. Henkelmann and K. Müller: "The Use of Neutron Induced Reactions for Light Element Profiling and Lattice Localization", Nucl. Instrum. Methods 149 (1978) 93

FEASIBILITY OF ROCK-LIKE FUEL AND GLASS WASTE FORM FOR DISPOSAL OF WEAPONS PLUTONIUM

H. FURUYA
Department of Nuclear Engineering, Kyushu University
Hakozaki, Fukuoka 812, JAPAN
S. MURAOKA and T. MUROMURA
Japan Atomic Energy Research Institute
Tokai-mura, Ibaraki 319-11, JAPAN

1. INTRODUCTION

The steering law on atomic energy in Japan strictly confines its research and development only in the field of peaceful use. Then, most of people, more or less, have interest in the dismantling of nuclear weapon. Recently, Japan Atomic Energy Research Institute (JAERI) started the special program, PROFIT (Plutonium Rock-like Fuel Integrated R&D), which is attempting to almost completely burn up the plutonium recovered from dismantled warheads in LWR and directly dispose its spent fuel (once-through burning process). The details of this project were presented in the previous Workshop held in October 1994, at the Institute of Physics and Power Engineering in Obninsk, Russia[(1)].

On the other hand, several studies on the borosilicate waste glass containing fission products and plutonium in the cooperative works between Kyushu University and JAERI, in the connection with the disposal of high-level radioactive waste originating from spent fuel reprocessing. Emphasis has been put on the safety of borosilicate waste glass such as volume change, helium release, microstructural change and so on. This knowledge is applicable and important to the disposal of plutonium recovered from dismantled warheads.

In the first part of this paper, a brief outline of this special project, PROFIT, and recent results after the previous Workshop are described. The latter section in this paper reviews the details of studies on the long term stability of borosilicate glass mixed with fission products and plutonium.

2. ROCK-LIKE FUEL AND ITS DIRECT DISPOSAL (ONCE-THROUGH BURNING OF PLUTONIUM)

JAERI has organized the special team which consists of over 20 researchers from the field of material science and reactor physics, and promoted the project, PROFIT. Now, fundamental R&D on the formation of fuels and spent fuels, reactor burn-up calculation and safety analysis are under progress as Phase I study.

2.1. OUTLINE OF ONCE-THROUGH BURNING OF PLUTONIUM

In this system, it is desirable that chemically stable fuel should be designed to satisfy the requirements of proliferation resistance, economy(cost efficiency) and environmental safety, as shown in Table 2.1[(2,3,4)]. From the comparison of chemical, thermodynamical and crystalline properties of several ceramic materials and minerals, two oxide systems have been chosen for the stable fuel that has rock-like structures and compositions: PuO_2-ThO_2-Al_2O_3-MgO and PuO_2-stabilized ZrO_2-Al_2O_3-MgO systems.

E.R. Merz and C.E. Walter (eds.), Disposal of Weapon Plutonium, 107-121.

They consist of fluorite, corundum and spinel type phases, and subsequently the actinide elements and fission products originating from the burning are fixed in the spent fuel matrix by forming the stable phases as shown in Table 2.2.

In reactor burn-up study, the survey calculations have been performed on the transmutation of plutonium, based on some simple cell models of LWR, FBR and HTGR. Figure 2.1 shows the transmutation of plutonium as a function of burn-up. It can be seen that LWR can transmutate more than 98 % of Pu-239 and 85 % of total plutonium, and therefore appears to be the most promising for the once-through plutonium burning, compared with other two reactors.

2.2. PRESENT STATUS

2.2.1. Fuel study

In the preliminary examination of the rock-like fuels, extensive investigations on the phase relations of fuels and spent fuels have been carried out by the reaction between inert additives, PuO_2 and simulated fission products. The phases evolved in the specimen heated around 1,400℃ were identified by X-ray diffraction analyses (XRD). The distribution of fission products was determined by scanning electron microscope and electron probe micro-analysis (EPMA). Figure 2.2(A) and (B) show tentative phase diagrams for the fuel and spent fuel, respectively.

It is found from the XRD and EPMA measurements as predicted from Table 2.1 that the actinide and lanthanide elements are mainly dissolved in such fluorite phases as thorianite (Th, An, Ln)O_2 and zirkelite (Zr, An, Ln)O_{2-x}, and the noble metals and Mo form Ru-based alloys, respectively. The alkaline-earth elements are fixed in the phase of hibonite (magnetoplumbite) type $SrO \cdot Al_2O_3$ which is found in meteorites containing Pu-244[5,6], with corundum Al_2O_3.

The leach rates of constituent elements from the simulated spent fuel spinel were measured in deionized water at 130℃. Figure 2.3 shows the typical results. Alkali metals, alkaline earth metals and Mo are most mobile and leachable in the spent fuel. Of the cations in the inert additives, Mg in spinel is most leachable. The leach rate of Al is in the order of U that will be the most leachable element in the actinide ones.

From the consideration of these results with neutron irradiation behaviors of ceramics, it appears that two candidates are favorable for the rock-like fuel: the fluorite and a two-phase fuel of fluorite + spinel. A three-phase fuel of fluorite + spinel +small amount of corundum would also be applicable.

2.2.2. Reactor burn-up study

In the reactor burn-up and safety analyses, the core burn-up calculation was carried out to accurately estimate the Doppler reactivity coefficients of zirconia fuel (PuO_2-stabilized ZrO_2-Al_2O_3) in PWR. Figure 2.4 shows the plutonium transmutation characteristics obtained by the 2-dimensional core calculation. More than 80 % of total plutonium and almost all of Pu-239 are transmuted, and the quality of plutonium becomes very poor in the spent fuels. A total of 0.9 tonne of plutonium would be denatured every year on the assumption of a 1 GWe PWR operating at 80 % availability.

As shown in Table 2.3, the void and Doppler reactivities for PuO_2-ZrO_2(Y,Gd)-Al_2O_3 fuel are negative both at the beginning and at the end of burnup cycle. The reactivities and β_{eff} are, however, very small at the beginning of cycle. To confirm the effect of these small reactivity coefficients on reactor safety, reactivity accident event of zirconia fueled PWR was studied. The result of the study showed that the Doppler reactivity of the fuel should be 3 - 5 times larger than the value in Table 2.3, or the maximum fuel enthalpy exceeds the limited value in conventional PWR.

For the improvement of the Doppler reactivity of the zirconia type fuel PWR, additives such as ThO_2, UO_2 and W were found to be effective. For example, 10mol% of depleted UO_2 can increase the reactivity about 4 times, as shown in Fig.2.5. Modified fuel systems are now examined such as

Table 2.1. Requirements and Candidate Systems of the Rock-like Fuels

REQUIREMENTS	MEANS	POSSIBLE CHEMICAL FORMS	CANDIDATE SYSTEMS
(1)Proliferation Resistance	Chemical stability	Oxide	Stable ceramics, Oxide
(2)Economy	Technical background of fuel production	Oxide, Carbide, Nitride, Alloys	Stable ceramics, Oxide
(3)Environmental Safety	Chemical stability	Oxide	Stable minerals, Oxide

Table 2.2. Phase Equilibria of Fuel Matrix and Spent Fuels

	MATRIX PHASE(S) OF FUEL (distribution of Pu)	PHASE(S) OF SPENT FUEL
SINGLE PHASE MATRIX FUEL	(1)Fluorite(Pu)	Fluorite + (Perovskite) + Alloys
	(2)Spinel (+ PuO_2)	Spinel + Fluorite
MULTI- PHASE MATRIX FUEL	(3)Fluorite(Pu) + Corundum	Fluorite + Corundum + Hibonite + Alloys
	(4)Fluorite(Pu) + Corundum + Spinel	Fluorite + Corundum + Spinel + Hibonite + Alloys
	(5)Fluorite(Pu) + Spinel	Fluorite + Spinel + Alloys
	(6)Spinel + Corundum (+ PuO_2)	Spinel + Corundum + Hibonite + Alloys

Table 2.3. Void and Doppler Reactivities and the Effective Delayed Neutron Fraction(β_{eff}) Estimated for PuO_2-ZrO_2(Y,Gd)-Al_2O_3 Fueled PWR Based on 2-Dimentional Core Burnup Calculation

	BOEC	EOEC
Void reactivity (% $\Delta k / k$)		
0% to 40% void	- 0.322	- 9.74
0% to 95% void	- 17.7	- 92.6
Doppler reactivity (% $\Delta k / k$)		
900 to 1200K	- 0.0982	- 0.201
600 to 900K	0.132	0.234
β_{eff}	2.87×10^{-3}	3.50×10^{-3}

BOEC : Beginning of the Equilibrium Cycle, EOEC : End of the Equilibrium Cycle

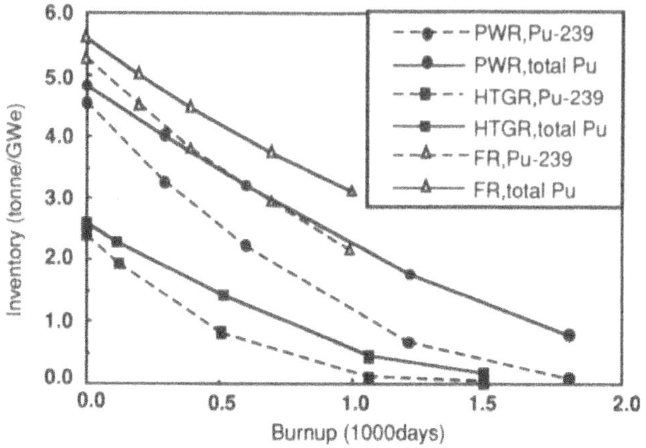

Fig. 2.1. Comparison of the burning plutonium amounts estimated by cell calculation in the PuO_2-ThO_2-Al_2O_3 fueled PWR, HTGR and FR.

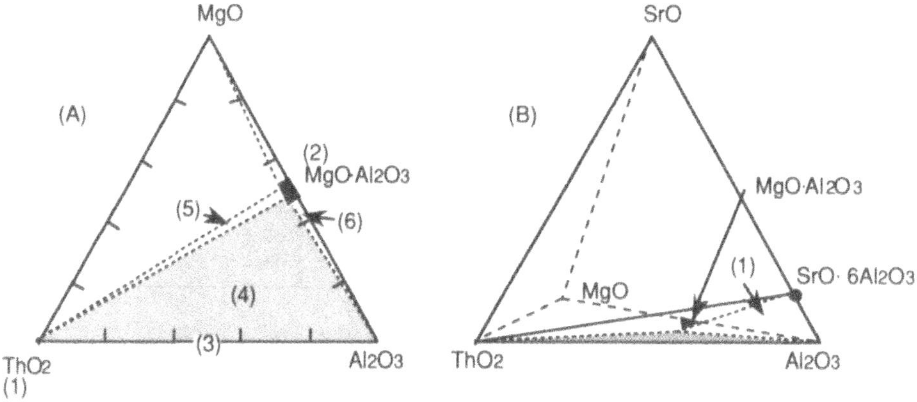

Fig. 2.2. Phase diagrams of fuel systems.

(A) ThO_2-Al_2O_3-MgO system at around 1400 ℃, (1) ThO_2(or stabilized ZrO_2) with fluorite structure, (2) MgO · Al_2O_3(spinel), (3) two phase region of ThO_2+Al_2O_3, (4) three-phase region of ThO_2+MgO · Al_2O_3+Al_2O_3, (5) two-phase region of ThO_2+MgO · Al_2O_3, (6) two-phase region of MgO · Al_2O_3+Al_2O_3, PuO2 is dissolved in ThO2 by formation of solid solutions.

(B) SrO-ThO_2-Al_2O_3-MgO system at around 1400℃, (1): four-phase region of ThO_2+MgO · Al_2O_3 +Al_2O_3+SrO · 6Al_2O_3.

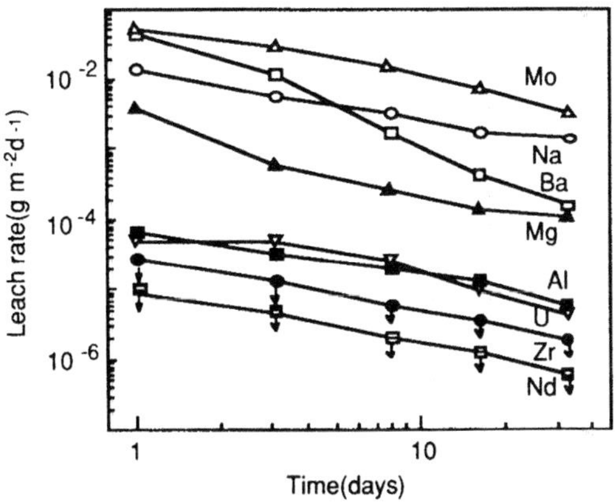

Fig. 2.3. Leachability of the simulated spent fuel.

Composition of spent fuel(mol%): 11simulated FPs(corresponding to PuO_2) + $17ZrO_2$ (Y,Gd) + $62Al_2O_3$ + 10MgO, Composition of simulated FPs(mol%):10.3Zr + 15.5Mo + 11.6Ru + 1.4Rh + 11.8Pd + 8.8Na + 5.9Ba + 15.5Ce + 19.3Nd, Composition of ZrO_2(Y,Gd)(mol%): $88.8ZrO_2$ + $11.0Y_2O_3$ + $0.2Gd_2O_3$, Condition: 130℃ in deionized water.

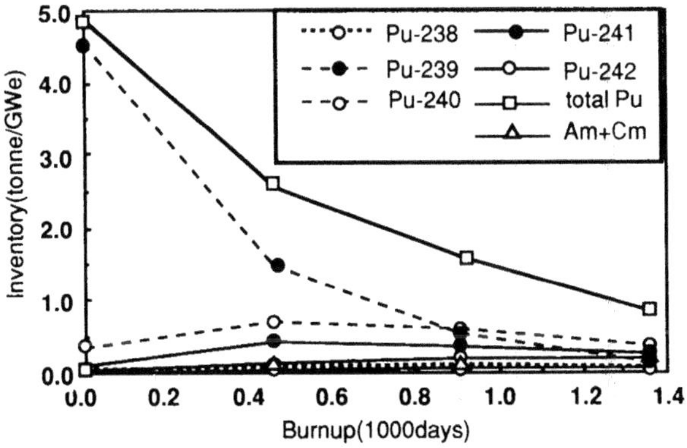

Fig. 2.4. Burnup dependence of plutonium composition obtained by 2-dimensional core calculation in PuO2-Zr(Y,Gd)-Al2O3 fueled PWR.

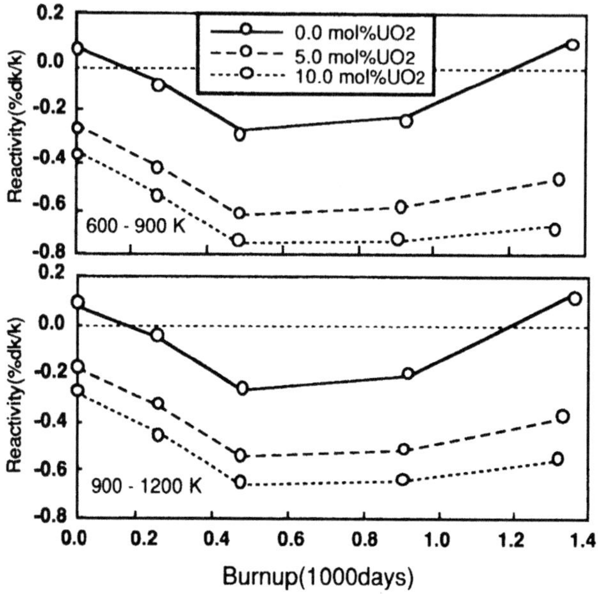

Fig. 2.5. Doppler reactivity of PuO_2-UO_2-ZrO_2-Al_2O_3 fuel PWR cell.

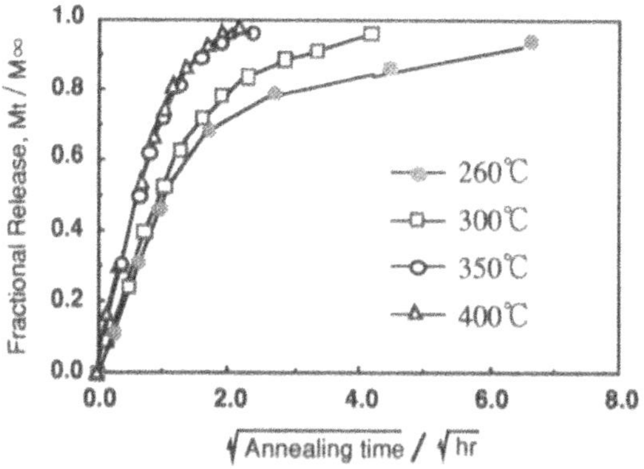

Fig.3.1. Fractional release of helium from irradiated glass.

PuO_2-(U or Th, Zr)O_2-Al_2O_3-MgO and PuO_2-ZrO_2-Al_2O_3-MgO-W metal. It will be also examined a heterogeneous core, where these fuels are partially charged in the conventional UO_2 PWR, to improve the reactivity.

3. LONG-TERM STABILITY OF BOROSILICATE GLASS WASTE FORM

To date, borosilicate glass has been accepted as the most suitable form for solidification of high-level nuclear waste. However, detailed and adequate studies on the radiation effects and leaching behavior are required in order to reveal the long-term stability. In this chapter, the details in the radiation effects from α-decay in the waste glass were investigated. A simulated waste glass was self-irradiated with doping of short-lived actinoids(^{238}Pu and ^{244}Cm), and changes in the glass properties were measured and analyzed. It was observed that the irradiation results in formation of helium bubbles in the glass, which leads to the changes in the volume and mechanical properties. However, it is expected that the radiation effects in waste glass are sufficiently small compared with other waste forms. In the study on the leaching behavior, Pu leaching from a simulated waste glass was investigated.

3.1. HELIUM RELEASE FROM IRRADIATED GLASS

Behavior of helium built up in the irradiated waste glass was investigated. A simulated nuclear waste glass of borosilicate type(similar composition to R7T7 glass) was self-irradiated from α-decay with doping of short-lived actinoids(^{238}Pu and ^{244}Cm). The composition is given in Table 3.1, and the size of glass was ϕ3mm x 1mm. The cumulative number of α-decay in the glass was 1.7×10^{25} α-decay/m^3, which corresponds to the irradiation dose at a few hundred thousand years after solidification of HLW from the spent fuel irradiated up to 33,000 MWD/MTU(7). The irradiated glass was annealed, and helium built up in the irradiated glass was released at fixed temperatures. The helium release rate was measured with a quadrupole mass spectrometer as a function of annealing time.

The fractional release(M_t/M_∞) of helium as a function of the square root of annealing time is shown in Fig.3.1, as symbols. It can be seen that the behavior of helium release from the irradiated glass is not controlled by only a simple diffusion process. The behavior of helium release was analyzed on the basis of the helium diffusion model with two chemical processes, i.e. trapping at bubbles and re-solution from bubbles into glass matrix(8,9). In the model, helium atoms in the glass were analyzed with the situation when diffusion occurs in a medium containing a stable array of trapping sites(helium bubbles). Schematic representative of the situation is shown in Fig.3.2, in which D is the diffusion coefficient of helium in glass matrix, g the trapping rate constant per unit time and b the re-solution rate constant per unit time. Diffusion equation of this process in a cylindrical glass specimen is described by,

$$\frac{\partial C}{\partial t} = D\left(\frac{\partial^2 C}{\partial r^2} + \frac{1}{r}\frac{\partial C}{\partial r} + \frac{\partial^2 C}{\partial z^2}\right) - gC + bM, \quad (1)$$

where C is the concentration of helium in glass matrix, t is the annealing time and M is the number of helium atoms trapped in bubbles per unit glass volume. The balance for the trapped helium is described by,

$$\frac{\partial M}{\partial t} = gC - bM. \quad (2)$$

Table 3.1. Composition of a Simulated Waste Glass.

Component	wt %				
Additives		MnO_2	0.26	SeO_2	0.02
SiO_2	45.15	Ag_2O	0.03	RuO_2	0.80
B_2O_3	13.90	CdO	0.03	Fe_2O_3	2.90
Al_2O_3	4.89	SnO_2	0.02	NiO	0.40
Li_2O	2.00	Sb_2O_3	0.01	Cr_2O_3	0.50
Na_2O	9.78	TeO_2	0.23	P_2O_5	0.30
CaO	4.00	Cs_2O	0.97	Ru	0.12
ZnO	2.47	BaO	0.62	Rh	0.15
		La_2O_3	0.14	Pd	0.43
Wastes		CeO_2	0.28	*Cm Oxides	3.04
Rb_2O	0.12	Pr_6O_{11}	0.14	**Pu Oxides	0.96
SrO	0.34	Nd_2O_3	0.45		
Y_2O_3	0.06	Sm_2O_3	0.09	Total	100.00
ZrO_2	2.64	Eu_2O_3	0.02		
MoO_3	1.73	Gd_2O_3	0.01		

* including ^{244}Cm, ** including ^{238}Pu

Table 3.2. Values of parameters obtained in helium release from irradiated glass.

$$D = 2.0 \times 10^{-7} \exp(-12.7\ \text{kcal/RT})\ [\text{m}^2/\text{sec}]$$

$$b = 1.1 \times 10^{5} \exp(-15.7\ \text{kcal/RT})\ [\text{sec}^{-1}]$$

$$g = 0.1\ [\text{sec}^{-1}]$$

Table 3.3. Best fit values of parameters, $\alpha vR/(vR+k\tau)$ and $vR+k\tau$.

T[K]	$\alpha vR/(vR+kT)$ [-]			$vR+kT$ [sec^{-1}]		
	H	E	KIC	H	E	KIC
573	-0.06	-0.10	0.30	8.0×10^{-4}	5.0×10^{-5}	1.0×10^{-4}
673	-0.001	-0.001	0.20	2.0×10^{-4}	1.0×10^{-4}	1.5×10^{-4}
723	-0.0005	-0.0005	0.30	3.5×10^{-4}	2.5×10^{-4}	1.0×10^{-4}

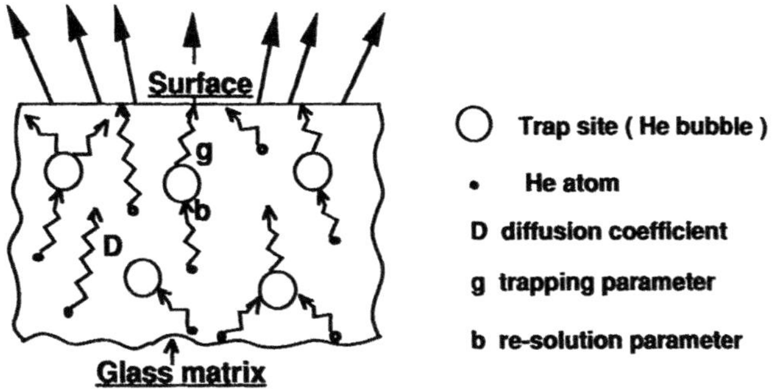

Fig.3.2. Schematic representative of helium diffusion model with two chemical processes.

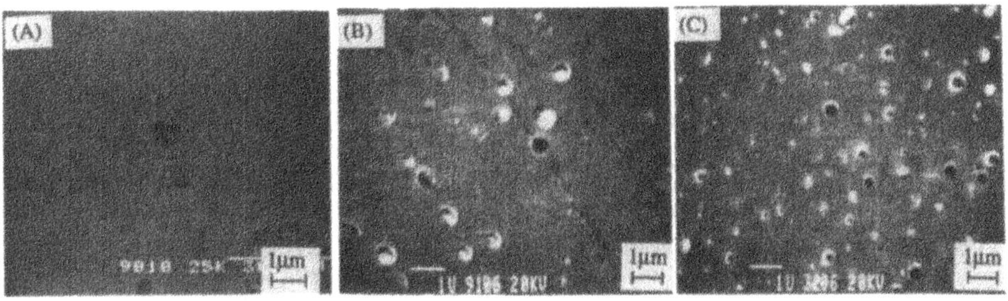

Fig.3.3. SEM micrographs for the simulated waste glass. (A) Unirradiated glass. (B) Glass irradiated at a dose of 2.75 x 10^{25} α-decay/m^3. (C) Glass annealed at 673K for 3 hours after irradiation.

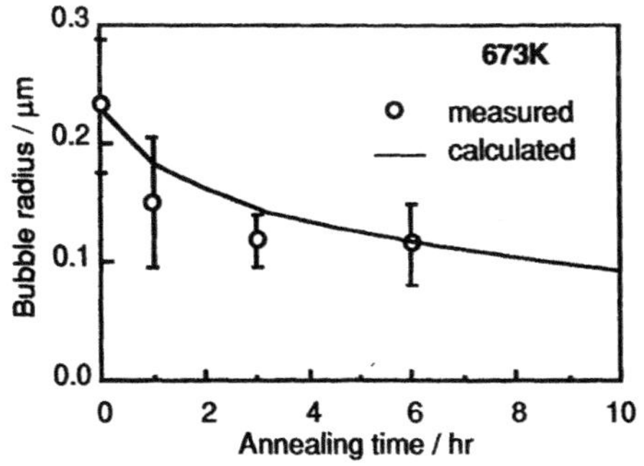

Fig.3.4. The average bubble radius during annealing at 673K as a function of annealing time.

Before annealing helium atoms are assumed to be partitioned between glass matrix and trap sites by the initial trapped fraction y. Defining C_0 as the total amount of helium in unit glass volume before annealing, the initial and boundary conditions are,

$$C(r, z, t) = (1 - y) C_0, \quad 0 \le r < a, 0 \le z < L/2, t = 0, \tag{3}$$
$$M(r, z, t) = y C_0, \quad 0 \le r < a, 0 \le z < L/2, t = 0. \tag{4}$$
$$C(a, z, t) = 0, \quad 0 \le z < L/2, t \ge 0, \tag{5}$$
$$C(r, L/2, t) = 0, \quad 0 \le r < a, t \ge 0, \tag{6}$$

where a and L represent a radius and a height of the glass specimen, respectively. Equation(1) and Eq.(2) can be solved by a numerical method, and the fractional release can be described as a function of annealing time,

$$\frac{M_t}{M_\infty} = 1 - \frac{8}{\pi a^2 L C_0} \int_0^{\pi/2} \int_0^{L/2} \int_0^{a} \{C(r,z,t) + M(r,z,t)\} dr dz d\theta. \tag{7}$$

The calculated fractional release at each temperature is shown in Fig.3.1 as a solid line. The good agreement between the calculated and measured values was obtained. The values of parameters obtained from the calculation are given in Table 3.2.

3.2. CHANGES IN MICROSTRUCTURE

Changes in microstructure of the glass by α-irradiation were observed using the doped simulated waste glass described in the previous section[(10)]. The cumulative number of α-decay in the glass was 2.75 x 10^{25} α-decay/m^3, and the size of the glass was ϕ3mm x 5.4mm. The irradiated glass was annealed at 673 K for the period up to 6 hours. Microstructures of a series of glasses were observed by use of a preshadowed carbon replica technique in combination with scanning electron microscopy(SEM)[(11)]. The series included the unirradiated glass, the irradiated glass and the glass annealed after irradiation.

Fig.3.3(A) and (B) show the microstructure of the unirradiated and irradiated glass, respectively. In the unirradiated glass only some flaws and segregate phases of metals were observed, while in the irradiated glass bubbles with a radius from 0.15 μm to 0.35 μm were observed. The average bubble radius and bubble density were 0.23 μm and 1.0x10^{17} bubbles/m^3, respectively. The microstructure of the glass annealed at 673 K for 3 hours after irradiation is shown in Fig.3(C), in which bubbles with the average radius of 0.12 μm were observed. The average bubble radius during annealing at 673 K as a function of annealing time is shown in Fig.3.4 as white circles. The bubble radius was observed to decrease with annealing time. The bubble behavior during annealing was analyzed on the basis of the helium diffusion model with two chemical processes described in the previous section. Equation(1) and Eq.(2) were solved by a numerical method, and the number of helium atoms in bubbles per unit glass volume as a function of annealing time, M(t), was obtained. The equilibrium condition for a gas bubble with a radius R embedded in a glass is described as,

$$R = \frac{2\gamma}{p}, \tag{8}$$

where γ and p represent the surface tension of the glass and the gas pressure in a bubble, respectively. Supposing ideal gas behavior, and assuming that the bubble density C_B during annealing is constant and all trap site makes a bubble with uniform size, the bubble radius R can be calculated as a function of annealing time,

$$R(t) = \left(\frac{3\, k_B\, T}{8\, \pi\, \gamma\, C_B}\ M(t) \right)^{0.5}, \qquad (9)$$

where T and k_B are the absolute temperature and Boltzmann's constant, respectively. The values of parameters, D, g and b, listed in Table 3.2 were applied to the calculation. The value of y was estimated to be 0.081 from the results of microstructural observation. Because the value of γ is unknown, the following relation, which had been obtained by the best fit for the data, was applied,

$$\frac{2\gamma}{k_B\, T} = 1.0 \times 10^{20}\ [m^{-2}]\,. \qquad (10)$$

The calculated bubble radius at 673K as a function of annealing time is shown in Fig.3.4 as a solid line. The agreement between the calculated and observed bubble radius was relatively good, which suggests that the bubble behavior is controlled by helium diffusion with trapping and re-solution. It is expected that the bubble behavior has a great influence especially on the volume and mechanical properties of the glass.

3.3. CHANGES IN VOLUME

Changes in volume of the glass by α-irradiation were measured using the doped simulated waste glass described in the previous sections. The cumulative number of α-decay in the glass was 1.7×10^{25} α-decay/m^3, and the size of glass was ϕ3mm x 1mm. The relative volume change by α-irradiation($\Delta V/V_0$) was measured to be + 0.4 % to + 0.6 %, which magnitude is small compared with other waste forms(12). The irradiated glass was annealed, and the recovery of volume change during annealing was measured. The volume change at 673K as a function of annealing time is shown in Fig.3.5 as open circles. It can be seen that the volume tends to recover to the original value with annealing time.

Changes in microstructure of the glass can lead to the volume change of the glass. As a result of the microstructural observation, it was found that bubbles were formed in the irradiated glass. Using the average bubble radius(0.23 μm) and bubble density(1.0×10^{17} bubbles/m^3) observed in the microstructural observation, the volume change of the glass resulting from bubble formation can be estimated to be,

$$\frac{\Delta V}{V_0} \cong \frac{4}{3} \pi R^3 C_B = +\,0.51\ \%\,. \qquad (11)$$

The estimated volume change resulting from bubble formation, +0.51%, is close to the measured volume change by α-irradiation(+ 0.4 % to + 0.6 %). The volume change resulting from the behavior of helium bubble can be calculated as a function of annealing time using the helium diffusion model. The calculated curve is shown in Fig.3.5 as a solid line. The calculated curve is in good agreement with the measured volume change, which suggests a large portion of the volume change by α-irradiation results from bubble formation.

3.4. CHANGES IN MECHANICAL PROPERTIES

Changes in mechanical properties of the glass by α-irradiation were measured using the doped simulated waste glass described in the previous sections(13). The cumulative number of α-decay in

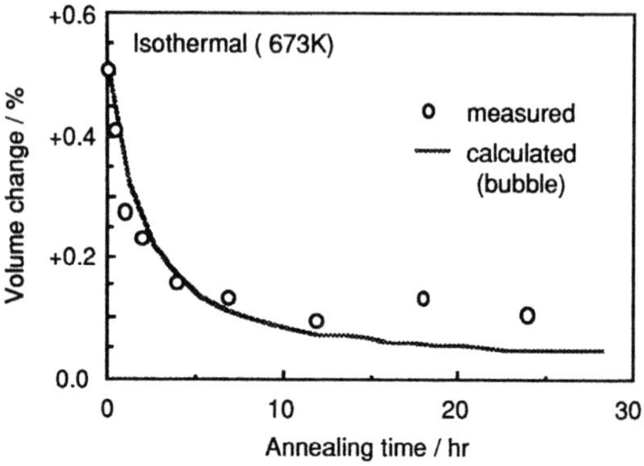

Fig.3.5. Volume change of the glass during postirradiation annealing at 673K.

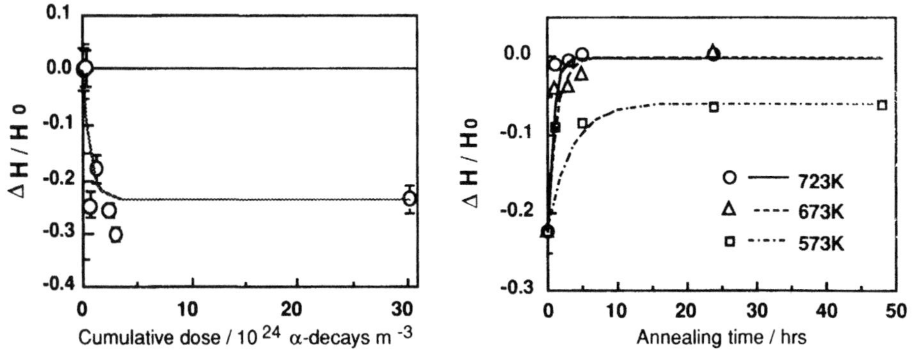

Fig.3.6. Changes in hardness of the waste glass, (A) as a function of cumulative dose, (B) as a function of annealing time.

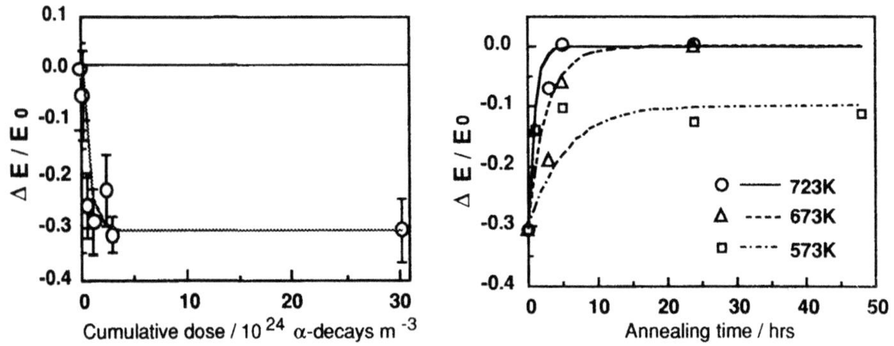

Fig.3.7. Changes in Young's modulus of the waste glass, (A) as a function of cumulative dose, (B) as a function of annealing time.

the glass was 3.0 x 10^{25} α-decay/m^3, and the size of glass was ϕ3mm x 1.5mm. The changes in the mechanical properties, hardness(H), Young's modulus(E) and fracture toughness(K_{IC}), were measured with a microhardness testing machine. The hardness was measured using the Vickers indentation, and the Young's modulus and fracture toughness were measured using the indentation technique with Vickers and Knoop indentation[14,15]. The relative changes in mechanical properties, $\Delta H/H_0$, $\Delta E/E_0$ and $\Delta K_{IC}/K_{IC0}$ as a function of the cumulative dose are shown in Fig.3.6(A) to Fig.3.8(A) as open circles. It can be seen that the hardness and Young's modulus decrease with the dose and approach saturation at the dose of 1.0 x 10^{24} α-decay/m^3, while the fracture toughness increases with the dose. The irradiated glass was annealed, and the recovery of the changes in mechanical properties was measured. The changes in mechanical properties as a function of annealing time are shown in Fig.3.6(B) to Fig.3.8(B) as symbols. It can be seen that the hardness and Young's modulus recover to the original values at the temperature above 673K within 10 hours, while the recovery of the fracture toughness is small in the region of temperature and time. The results of changes in the mechanical properties were analyzed by use of Marples' model[16], in which the changes in the glass properties are proportional to the volume fraction of damaged zone, F, and the recovery of F is first order. The changes in glass properties are described as following equation,

$$\Delta P/P_0 = \alpha F , \qquad (12)$$

where $\Delta P/P_0$ represents $\Delta H/H_0$, $\Delta E/E_0$ or $\Delta K_{IC}/K_{IC0}$, and α is a proportionality constant. While, the volume fraction of damaged zone, F, is described by the following equation,

$$\frac{dF}{dt} = v R (1 - F) - k_T F , \qquad (13)$$

where v is the volume damaged by each α-decay, R is the rate of α-decay, k_T is the recovery constant at temperature T and t is the irradiation time. Equation (13) can be integrated with the condition of irradiation at room temperature, to give:

$$F = \frac{vR}{vR + k_{298K}} [1 - \exp\{ - (vR + k_{298K}) t \}] . \qquad (14)$$

On the other hand, it can be integrated with the condition of annealing after irradiation, to give:

$$F = \frac{vR}{vR + k_{298K}} \exp\{ - (vR + k_T) t \} + \frac{vR}{vR + k_T} [1 - \exp\{ - (vR + k_T) t \}] . \qquad (15)$$

The changes in mechanical properties calculated by Marples' model are shown in Fig.3.6 to Fig.3.8 as solid lines, and the values of parameters obtained from the calculation are listed in Table 3.3. Assuming the value of vR equals 1 x $10^{-7}sec^{-1}$[1-8], the values of k_T for each temperature are shown on an Arrhenius plot in Fig.3.9. There is some scatter in the data, but in general they lie on reasonably straight lines for the data of H and E, which suggests that the changes in both hardness and Young's modulus are controlled by radiation defects with either a single broad annealing stage of Arrhenius type or the superposition of a large number of overlapping stages. On the other hand, the fracture toughness is considered to be strongly related to the behavior of bubbles, in addition the radiation defects. It must be revealed how the bubble behavior is responsible for the change in K_{IC}.

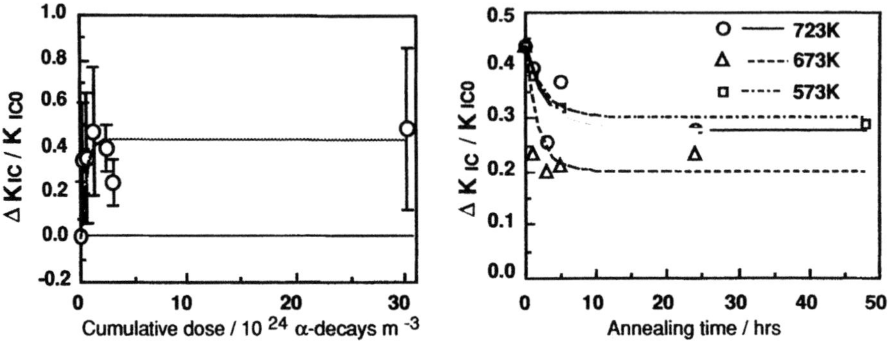

Fig.3.8. Changes in fracture toughness of the waste glass, (A) as a function of cumulative dose, (B) as a function of annealing time.

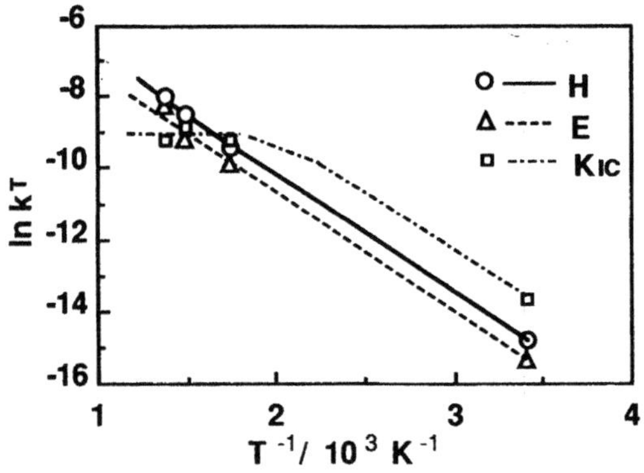

Fig.3.9. Arrhenius plots of the recovery constants kT.

4. CONCLUSION

The fundamental researches on the structural properties and burn-up characteristics of rock-like fuel have been carried out as scheduled in the R & D of once-through plutonium burning process. In this system, the similar MOX fuel technology can be applicable to the fabrication of rock-like fuels which have the chemical and geological stability. The plutonium in this type of fuels can be almost completely burned in LWR. Therefore, this system has such benefits as (1) proliferation resistance, (2) environmental safety, and (3) economical advantage by shorter R & D time.

Long term stability of borosilicate glass containing plutonium has been well studied mainly on the radiation damage and leachability. From these results, the vitrification of plutonium into borosilicate glass and its disposal are also one candidate of disposition route for plutonium stemmed from dismantled warheads.

ACKNOWLEDGEMENTS

The authors would like to thank their colleagues, members of PROFIT project team and in particular Dr.T.Banba(JAERI) and Dr.Y.Inagaki(Kyusyu Univ.) for their assistance in the preparation of this paper.

REFERENCES

(1) T.Muromura et al., Proc."Mixed Oxide Fuel(MOX) Exploitation and Destruction in Power Reactor", held in Obninsk, Russia, Oct.1994.
(2) H.Akie, T.Muromura, H.Takano and S.Matsuura, "A New Concept of Once-through Burning of Nuclear Warheads Plutonium", Seventh International Conference Emerging Nuclear Systems (ICENES '93), (Ed.) H.Yasuda, World Scientific, Singapore (1994)298-302.
(3) H.Akie, H.Takano, T.Muromura and N.Nitani, "A New Idea of Excess Plutonium Once-through Burning in Light Water Reactor", Progress Nucl. Energy, (1995), (to be published).
(4) H.Akie, T.Muromura, H.Takano and S.Matsuura, "A New Fuel Material for Once-Through Weapons Plutonium Burning", Nucl. Technol., 107(1994)182-192.
(5) R.S.Rajan and A.S.Tamhane, Earth and Planetary Science Letters, 58(1982)129-135.
(6) A.J.Fahey, J.N.Goswami, K.D.McKeegan and E.Zinner, Geochimica et Cosmochimica Acta, 51(1987)329-350.
(7) W.J.Weber and F.P.Roberts, Nucl. Technol. 60, 178 (1983).
(8) D.R.Olander, Fundamental Aspects of Nuclear Reactor Fuel Elements, (ERDA, Springfield Virginia, 1976) p.307.
(9) R.P.Turcotte, "Radiation Effects in Solidified High-level Waste Part 2-Helium Behavior", BNWL-2051, Battelle Pacific Northwest Laboratories (1976).
(10) Y.Inagaki, H.Furuya, K.Idemitsu, T.Banba, S.Matsumoto and S.Muraoka, Mat.Res.Soc. Proc., Vol.257(1992)199.
(11) S.P.Faile and D.M.Roy, J.Am.Ceram.Soc., Vol.49, No.12(1966)638.
(12) K.D.Reeve and J.L.Woolfrey, J.Aust.Ceram.Soc.,16(1980)10.
(13) Y.Inagaki, H.Furuya, Y.Ono, K.Idemitsu, T.Banba, S.Matsumoto and S.Muraoka, Mat.Res. Soc.Proc., Vol.294(1993)191.
(14) D.B. Marshall, T. Noma, and A.G. Evans, J. Amer. Ceram. Soc., 65, C175(1982).
(15) D.B. Marshall and A.G. Evans, J. Amer. Ceram. Soc., 64, C182(1982).
(16) J.A.C. Marples, Nucl. Instr. and Meth. B32, 480(1988).

POSSIBLE WAYS OF THE WEAPON PLUTONIUM PROCESSING TO PRODUCE THE STEADY GLASS MATERIALS SUITABLE FOR FURTHER SAFE DISPOSAL.

MATYUNIN Yu.I., KUSHNIKOV V.V., DEMIN A.V.
State Scientific Centre of Russian Federation, VNIINM,
All-Russia Scientific and Research Institute of Inorganic Materials,
Rogov St., 5, Moscow, 123060, Russia

1. INTRODUCTION

The questions about safe management, salvaging and disposal of weapon plutonium are appeared at present time in connection with significant reduction of military programs and removal with armament of large number of nuclear war-head.

The storage of nuclear materials, extracted from war-head after of their dismantles, is connected with necessity of reliable control of their warehousing, which guaranteed the impossibility of their unattended using (terrorism activity, nuclear blackmail and others).

The preliminary processing of military plutonium (transmutation on nuclear reactors, using as a fuel in nuclear power stations, localization in steady solid compound with purpose of their further disposal) excluding of it military uses is more preferable and safe from the point of view of listed above factors.

This paper is connected with consideration of probable ways of weapon plutonium processing with receiption of glass compound satisfying to equirements of their subsequent disposal in deep geological formations.

The technologies of processing and localization of high level wastes to produce glass-like (phosphate and borosilicate glasses), glass-ceramic and mineral-like materials, enabling reliably to fix the significant quantity of radio-toxic elements in matrices, are widely investigated in the countries, using nuclear power. Russia and some other states already use these technologies at present time.

It is possible to consider the existing technologies from the point of view of weapon plutonium vitrification. The data of plutonium localization researches in glass-like materials (phosphate and borosilicate glasses), indicating on opportunity of inclusion and high level of this radionuclide fixing in received (α-isotope leaching speeds from vitrification products by water, definite on standard technique, were $7 \cdot 10^{-8}$ g $\cdot$ sm$^{-2}\cdot$ day through some weeks) are presented in number works.

In these papers the plutonium dioxide and the solution of plutonium nitrate were used as initial solidified products. The presented results of plutonium solubility in matrixes and it distribution on volumes of molten compositions indicated more preferable using of the phosphate glasses as the matrix materials (high level of plutonium inclusion in matrix , not resulting on dispersed plutonium dioxide particle formations in the glass melts and more suitable technological processes of fluxing and dosage of solidified products).

E.R. Merz and C.E. Walter (eds.), Disposal of Weapon Plutonium, 123-128.

The various technological receptions and methods of vitrification of high level liquid wastes (HLLW) are presented the opportunity of 3-5 wt.% inclusion of plutonium dioxide alongside with fission product in phosphate compound.

The presented literary data, as well as preliminary experimental studies indicate on opportunity of receiption purely plutonium-phosphate glasses with fixing to 20 wt.% of plutonium dioxide. And in dependences on qualities and structure of initial solidified products (metal, dioxide or nitric acid solution of plutonium) the various technological approaches of execution of localization process can be offerred.

2. JOINT HLLW AND PU-CONTENT LIQUID PRODUCTS VITRIFICATION

Investigated with using of various methods of analysis the phosphate materials were received by calcining and vitrification of fluxed solutions.The compositions of vitrified solutions and solid materials are presented in tab. 1.

TABLE 1. Settlement compositions of received materials

Type of materials	Settlement composition of received materials (% mas.)				
	Na_2O	Al_2O_3	P_2O_5	PuO_2	$(FP+CP)_xO_y$
PGP-19	25.9	18.9	51.8	3.4	-
PGP-21	25.9	20.9	49.8	3.4	-
PGP-23	25.9	22.9	47.8	3.4	-
GP	26.0	23.2	50.0	0.8	-
GPM	24.0	19.0	51.7	0.8	3.90

Research of received materials have shown, that glasses of series "PGP" with holding time in melt of 8 hours were opaque and devitrified (PGP-19 (8 h) - grey, PGP-21 (8 h) and PGP-23 (8 h) -yellow). The materials, with holding time 100 and 200 hours, in volume were transparent and had deep-brown colours. The bottom layers of various hight in these materials were painted in yellow colour and opaque.

Adduced on fig.1 γ-scanning results of received materials indicate distinction in plutonium distribution on hight of glass block in dependence on time of melts endurance and material compositions.The data of γ-scanning show that the distribution of plutonium in materials, with melt holding time of 8 hours, is practically uniform.

With holding time of melts increase till 200 hours, the redistribution of plutonium on hight of glass block are observed that resulting in to formations of bottom layers, considerably enriched by plutonium.

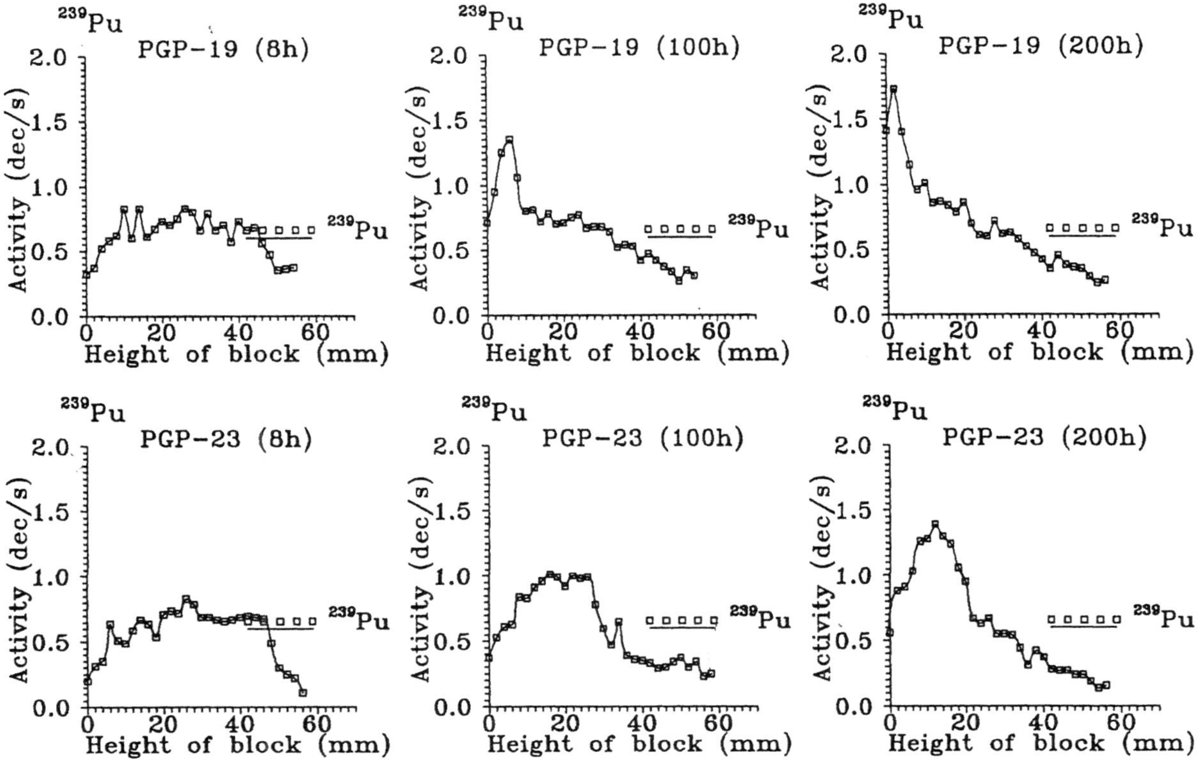

Figure 1. Distribution of plutonium in phosphate glass blocks (γ-scanning data)

X-ray phase analysis of samples from top and bottom layers of glass blocks PGP-23 (8 h) and PGP-23 (200 h) indicate redistribution of plutonium dioxide phase in melts with increase of their endurance time.

Table 2 submits the results of analyses of the glasses before and after of their chemical opening.

Researches of phosphate glasses have shown, that the content and distribution of plutonium in volume of solidified material is defined by composition of final product and time of endurance of glass melt. The distribution of plutonium dioxide in phosphate glass, received by direct vitrification of fluxed solutions at 1000 ºC (8 hours), is practically uniform, if it content is 3.4 wt.% or less. For matrix composition with lowered contents of Al_2O_3 (PGP-19) in these conditions some enrichment of bottom layers of molten glass by plutonium is already observed, that can be connected with greater speed of PuO_2 heterogeneous particles sedimentation in melt having low significance of viscosity.

However, it is follows to note the increasing solubility of plutonium in glasses at short time of melt endurance(~ 8 h) at reduction of aluminium oxide contents in matrix materials.

TABLE 2. The contents of plutonium in the different zones of glasses

Type of materials	Holding time (hours)	Plutonium contents (wt.% PuO_2)					
		Solid materials		Top		Bottom	
		top	bottom	solut.	sedim.	solut.	sedim.
PGP-19	8	2.98	4.98	2.10	0.88	1.12	3.86
PGP-19	100	1.61	17.60	1.20	0.41	1.72	15.88
PGP-19	200	0.86	28.60	0.77	0.09	0.94	27.66
PGP-21	8	3.36	4.17	0.42	2.94	0.61	3.56
PGP-21	100	1.27	9.94	0.96	0.31	0.28	9.66
PGP-21	200	0.89	27.70	0.71	0.18	0.30	27.40
PGP-23	8	3.21	3.32	0.31	2.90	0.20	3.02
PGP-23	100	0.66	8.15	0.58	0.08	0.42	7.73
PGP-23	200	0.59	8.75	0.37	0.22	0.71	8.04

in matrix materials. The heterogenious phases are not found out in phosphate glasses with lowered contents of Al_2O_3 (2-3 %) containing ~20 wt.% of plutonium.

With endurance time increase the significant redistribution of plutonium is observed on volume of received glass blocks, resulting in to enrichments by its bottom layers. The thikness and density of layers enriched by plutonium and general contents of plutonium dioxide in top layers of molten mass are strong depends on the composition of initial matrices.

The x-ray phase analysis of received compositions has shown the presence of plutonium dioxide phase in glasses, which precipitation result to significant enrichment of bottom layers of molten mass by plutonium due to high density of PuO_2 dispersed particles in comparison with one of matrix. Other phases, containing plutonium, is not found out.

3. POSSIBLE SCHEMES OF WEAPON PLUTONIUM PROCESSING TO PRODUCE THE GLASS COMPOSITIONS SUITABLE FOR FURTHER SAFE DISPOSAL

The analysis of literature deal with the researches of plutonium and HLLW behaviours during vitrification and study of received final glass properties have allowed to produce the approaches to managements with weapon plutonium and to offer the schemes of it processing.

On fig.2 the offered scheme of plutonium processing with receiption of phosphate glass compositions is presented.

After dismantle of nuclear war-head the fission material is necessary to be pounded and subjected to oxidation with receiption of high chemical-active plutonium dioxide with required physico-chemical properties for simplification further it processing.

Dissolution of metal plutonium in mineral acids is known to be conjugate with significant difficulties and, as a rule, the plutonium is not managed completely to transfer in solution. Furthermore, HCl and HF using for more total dissolution of metal causes the additional difficulties for following processing of solutions containing Cl^- and F^- ions. Using high chemical-active plutonium dioxide for it further processing (dissolution to produce solutions of Pu nitrate, direct vitrification of dioxide by meltering with fluxed additives) permits considerably to reduce the quantity of secondary Pu-content wastes (the precipitates forming at direct dissolution of plutonium metal).

On following stage it is possible to consider two ways of further management with received dioxide of plutonium.

The first variant is realization of “wet” process, where it is necessary to produce Pu-content nitrate solution, required for execution of process by using high level liquid waste (HLLW) vitrification technology. This variant of vitrification permit to receive, in dependence on requirements and method of realization of process, glasses containing plutonium together with fission product (mixing Pu-content and HLLW solutions from processing of nuclear fuel before vitrification) or glasses, containing only the plutonium. On dependence of requirements to final compound the different compositions of fluxed additives, enabling to receive the materials with required properties, are used on stage of high-temperature synthesis. Second variant consists in realization of “dry” process, enabling to avoid the stage of dissolution. On the one hand this way makes more simple the execution of technological process, and on the other hand it permits to receive on following stage or Pu-content semi-product with high content of fission material (for subsequent synthesis of glass composite on it base), or directly the glass-like materials.

Received glass melts are discharged in containers of required form and (after decontamination of containers) departed on temporary storage or disposal.

The offered scheme of weapon plutonium processing permits to execute the process of vitrification with receiption, in dependence on delivered tasks, as the pure Pu-content glasses with contents of solidified component to 20 wt.%, as the materials containing together plutonium and fission product from HLLW formed in nuclear fuel processing. Various phosphate glass compositions and methods of it high-temperature synthesis makes the offerred scheme of processing suitable for receiption of materials with wide interval of Pu content.

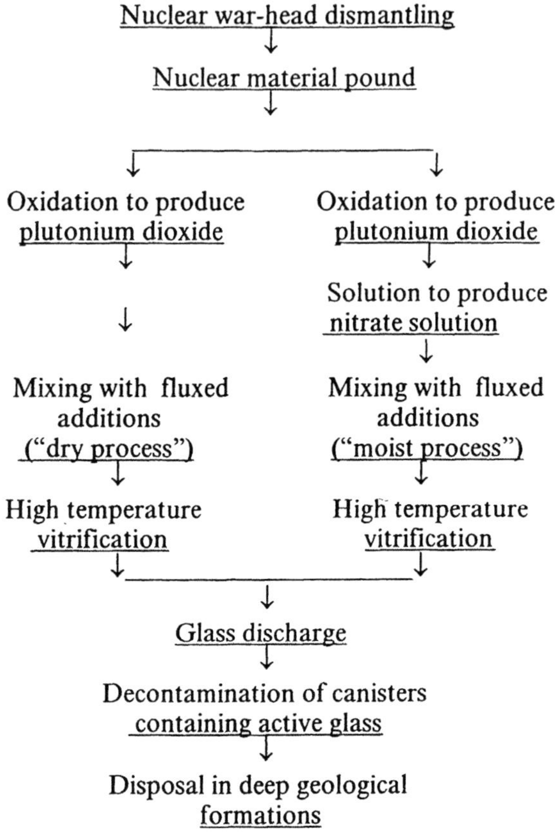

Figure 2 . Proposal sheme of weapon Plutonium process

The additional research and design studies in following directions are required for realization of production on offerred scheme:

- optimization of the methods of plutonium metal oxidation and choosing of optimum regimes of its execution;
- choosing of glass matrix compositions satisfying to requirements to final products of vitrification (level of plutonium inclusion, chemical and radiation stability and mechanical strength of received materials and others);
- research of opportunity of Pu localization in mineral-like compositions;
- comparative analysis of compositions received by using of various technological ways and methods and final choosing of methods for synthesis of Pu-content products;
- research of physico-chemical properties of received compositions with high plutonium contents (radiation, chemical stability of materials and other);
- development, manufacturing and test of equipment for execution of technological processes of plutonium vitrification on offered scheme.

MANAGEMENT OF OPERATIONAL WASTE FROM MOX FUEL PLANTS

FRIEDRICH W. LEDEBRINK
Siemens AG
Power Generation Group KWU
Hanau Fuel Fabrication Facility
P. O. Box 110060, D-63434 Hanau

KEYWORDS

Plutonium / weapons plutonium / disposition / MOX fuel / waste / operational waste / conditioning / waste form / cementation / disposal / disposal volume

ABSTRACT

Disposition of weapons plutonium can be performed following the MOX fuel route. Operational waste from the MOX fuel fabrication has to be conditioned with respect to the safe interim storage and final disposal. The cementation of the waste provides a waste product which fulfil all requirements. A proven technique for the cementation process is available. However, due to criticality criteria for final repositories this process is not suitable for the direct disposition of excess plutonium from nuclear weapons.

1 INTRODUCTION

One frequently discussed possibility for the disposition of weapons-grade plutonium is to reuse it in mixed-oxide (MOX) fuel assemblies to generate electric power in fast or thermal reactors and subsequently store the spent MOX fuel assemblies at suitable repositories. This process of plutonium reprocessing, which has been developed and practiced with great success in Germany and several other countries, is nevertheless not completely free of radioactive waste. This radwaste includes measurable amounts of plutonium from recycling process, and requires special treatment in order to achieve safe and reliable interim and final storage.

The Power Generation Group of Siemens AG, with its mixed oxide fuel assembly fabrication facility in Hanau, Germany, has gained extensive experience in the treatment and conditioning of plutonium waste in the course of the licensing procedures currently underway for the future repository at the Konrad Mine near Salzgitter, Germany. This know-how is the subject of this paper. To date, approximately 8.5 tons of plutonium have

E.R. Merz and C.E. Walter (eds.), Disposal of Weapon Plutonium, 129-142.

been recycled into mixed-oxide fuel, and all of the waste arising from this reprocessing has been conditioned in some 2,900 drums using the Siemens cementation process.

2 ORIGIN AND CHARACTERIZATION OF WASTES CONTAINING PLUTONIUM

By the time fabrication of a MOX fuel assembly is complete, the plutonium, in the form of plutonium nitrate solution, has passed through the following processing stages (see Figure 1):

- Conversion to (U, Pu)0_2 mixed oxide via the AUPuC process
- Blending with UO_2
- Pressing
- Sintering and grinding
- Filling and welding of fuel rods
- Assembly of fuel rods into fuel assemblies.

In practically in all these process stages a small, but unavoidable amount of plutonium-bearing waste in the form of contaminated liquid or solid materials is generated which cannot be used in any further process and which therefore must be disposed of as radioactive waste. Figure 2 presents an overview of the various types of waste arising, although only those wastes from gloveboxes where MOX is processed contain measurable quantities of plutonium relevant to fissile material accountancy. For waste arising outside the gloveboxes, e.g. in the working compartments within the controlled area, plutonium contamination is unlikely, but cannot be completely ruled out. These materials can either be released following measurement or, as a precautionary measure, they can be disposed of as radioactive waste, e.g. by incineration and subsequent final storage of the ashes.

Waste management, however, concentrates on those wastes containing measurable quantities of plutonium:

- Liquid wastes resulting from conversion of U/Pu nitrate to U/Pu oxide (filtrates/distillates) and from wet chemical analysis methods (laboratory wastes).
- Solid wastes from all gloveboxes in the MOX manufacturing plant
- Solid wastes arising during cleaning, repair and maintenance work
- Waste from plant components which have reached the end of their service life.

Process stages following pellet fabrication - fuel rod loading, welding and assembly - generate, if at all, only low-level waste. Measurable plutonium losses do not occur in these latter stages.

Experience with MOX fuel assembly fabrication to date shows that, in an automated plant, plutonium amounts of about 0,2 wt% can be found in the waste. For a MOX plant with an annual throughput of 120 Mg of MOX material containing 5 Mg of

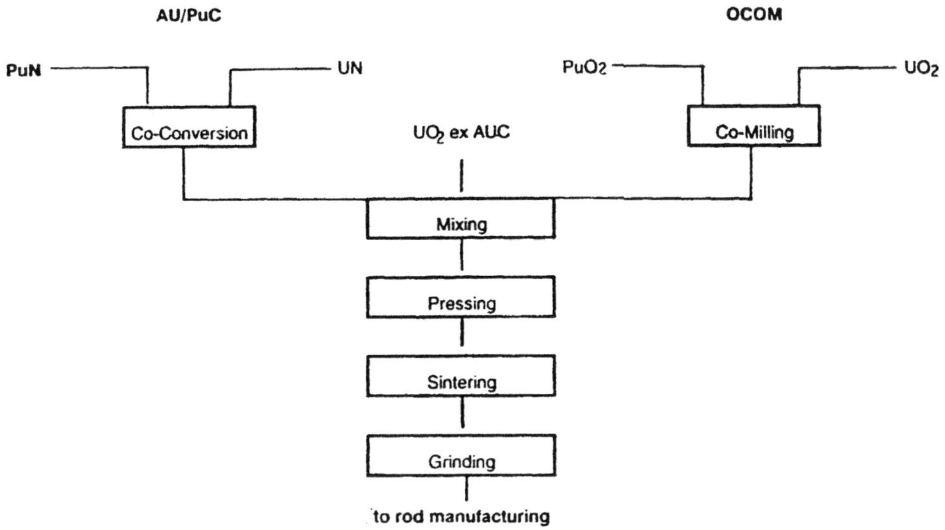

Figure 1: Process steps for manufacturing of MOX fuel

room waste (suspected contamination or very low contaminated)			waste arising in α-tight enclosure		
	combustible	non combustible		soft	hard
solid	surgery-gloves wipes, tissues foils (PVC,PE)	filters	solid	bags (PVC) gloves wipes, tissues bottles (PE) miscellaneous	filters cans glass small plant items tools
liquid		sewage laundry sewage	liquid		aqueous waste from processes and analytics oil

Figure 2: Waste generation during MOX fuel fabrication according to origin and nature

Pu, this would mean an amount of 10 kg of plutonium per year can be measured in the waste or, when processing a total amount of 120 Mg of plutonium fuel, totaling 240 kg Pu in the waste (Figure 3).

Pu - Throughput	5 Mg			20 Mg		
Glovebox waste	Vol (m^3)	Mass (Mg)	Pu (kg)	Vol (m^3)	Mass (Mg)	Pu (kg)
Solid	80	20	9.25	1,920	480	222
- organic	60	12		1,440	288	
- inorganic	20	8		480	192	
Liquid [1)]	75	75	0.75	1,800	1,800	18
Σ			10.00			240

1) Distillates of AUPuC-process (200g heavy metal/l in initial solution) + laboratory water.

Figure. 3: Raw waste generation and material composition

As long as natural uranium or depleted uranium are used for MOX, the amount of uranium mixed with the plutonium is of no consequence to later treatment and final storage of the waste, and is therefore not considered here.

For the waste conditioning the material composition of the waste is of interest. Statistical long-term investigations on glovebox wastes have shown that around 60 wt% of these wastes comprise organic substances and the remaining 40 wt% inorganic substances, mainly metals /1/. This waste is pre-sorted into organic and inorganic wastes in the gloveboxes and collected in transparent cans. These are then removed from the gloveboxes in sealed foil bags

Figure 4 shows such a waste package with multiple barrier system. The transparent container is provided with a welded, double foil packaging. To ensure safe handling up to processing, the waste package is inserted into reusable metal containers.

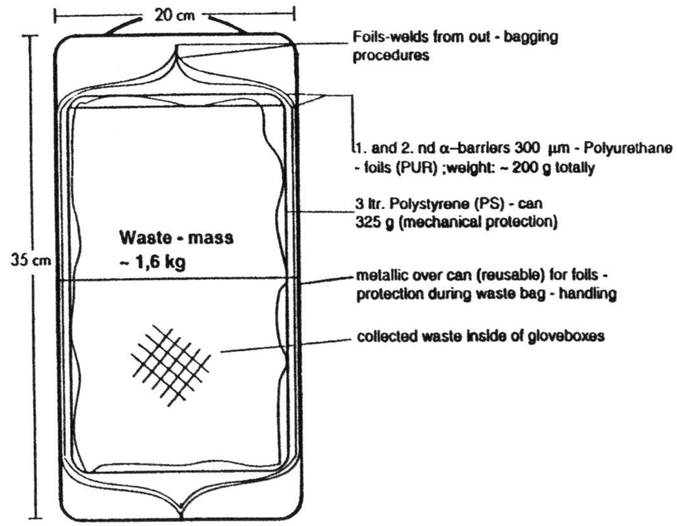

Figure 4: "Standard" waste package with solid operational waste from glove boxes;

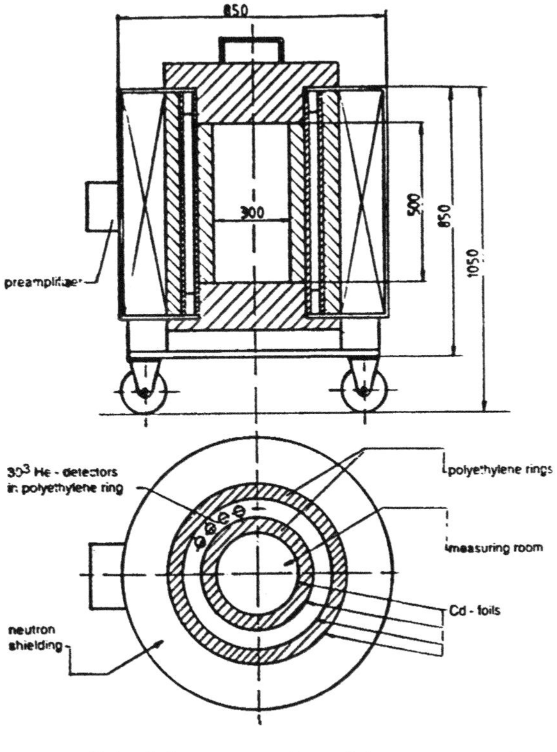

Figure 5: Principle construction of the neutron coincidence measuring device DRC-100

3 DETERMINATION OF PLUTONIUM CONTENT

Determination of the Pu-240 äqui (Σ Pu-238 + Pu-240 + Pu-242) content of the individual waste packages is performed by nondestructive passive neutron coincidence measurement (NDA). The total quantities of plutonium and uranium are calculated with the aid of the known Pu vector for each manufacturing order and the Pu/U ratio.

The widely utilized measuring device DRC-100 (Figure 5) was developed at Los Alamos. The cylindrical measuring chamber, 30 cm in diameter, is fitted with thirty He^3 counting tubes, and the base and cover are equipped with reflectors. The signals are evaluated according to the shift-register method.

The special advantage of measuring the individual packages lies in the applicability of small measuring geometries having high detection capacity and comparatively high measuring accuracy, the detection of Pu concentrations in small quantities of waste.

4 CRITERIA FOR MOX WASTE DISPOSAL AT PLANNED REPOSITORY IN GERMANY

Prior to selecting and specifying a waste conditioning process, the specific requirements should be known with respect to the waste package for final storage. These requirements are always specific to the repository in question and are derived from the given geological formation as well as its retention capabilities. In cases in which repository-specific requirements are not known due to the lack of a definite repository site, the criteria specified for safe above-ground interim storage can suffice as minimum requirements.

In the following, the requirements to apply for waste conditioning are briefly described using Germany's KONRAD repository as an example /2/. KONRAD is a repository site which is currently undergoing licensing procedures and in future is to accommodate, among other radwaste, MOX waste from fuel facilities.

The KONRAD repository is a former iron ore mine having a thick bed of clay as shielding to isolate deposited wastes from the biosphere. Due to the given geological environment, it was determined that in order to ensure safe long-term storage, the temperature differential between the waste store and the surrounding geology shall not exceed 3°C /3/. Moreover, for reasons of criticality safety and accident control, the following requirements were stipulated for a waste product over the repository service life (see Figure 6).

No.	Waste Product Properties	Requirement		
1	Aggregate condition	Solid or solidified waste		
2	Conditioning form	encapsulated in cement/concrete		
3	Activity distribution	uniformed as possible		
4	Minimum compressive strength	10 N/mm^2		
4a	Water-to-cement ratio	≤ 0.45		
4b	Inorganic waste / drum	≤ 100 kg		
4c	Organic waste / drum	≤ 40 kg		
5	Final disposal container	Standard KONRAD container e.g. 10 drums in Type VI-cont. (5,4m^3) /3/		
6	Maximum Nuclide Content	Bq/drum	g/drum	Limitation due to
	Pu-238	1.9E+11	0.3	H
	Pu-239	2.5E+10	11.0	C
	Pu-240	4.3E+10	5.1	H
	Pu-241	2.8E+12	0.7	H
	Am-241	9.3E+10	0.7	H
	U-235 in U_{nat}	3.1E+09	39,000.0	H
	U-235 in U_{an} ≤ 5%	3.4E+06	42.5	C
	U-238	1.1E+10	88,530.0	H

Legende: H= Heat generation
C= Criticality safety

Figure 6: Requirements for waste package (200 l-drum) for final disposal at KONRAD / Germany as example

If more than one radionuclide is present in the waste, the effects of these multiple nuclides must also be taken into consideration, i.e. the allowable activities in a 200-liter drum must comply with the following summation criteria S:

$$S_{W(K)} = \sum_i \frac{A(i)}{G_{W(K)}(i)} < 1$$

where:
- A(i) is the activity of the radionuclide
- G(i) is the maximum allowable activity for a single nuclide
- W = heat, and
- K = criticality.

The criticality limit value for Pu-239, with 11 grams per 200l-drum, presents itself as limiting variable for waste disposal. The waste mass accommodated in a single drum can also possibly be decisive in determining the number of waste drums required, e.g. in cases in which the Pu content in the raw waste is very low.

In order to understand the extraordinarily low criticality limit value it must be noted that a very conservative incident scenario has been selected, namely that plutonium is flushed from the waste product and collects in a spherically-shaped hollow space of the storage cavern. Given the more realistic assumption that the very unsoluble PuO_2 remains in the alkali milieu of the cement and is present inside the drum only in ball form at the least favorable location, a multiplication factor of $K_{eff} = 0.194$ results for an infinitely extended storage arrangement of drums containing waste solidified in cement with 120 g of Pu per drum (95 % Pu-239 and 5 % Pu-240).

5 WASTE CONDITIONING

Requirements 1 to 3 shown in Figure 6 clearly stipulate the cementation of waste as a conditioning process. Furthermore, activity should be distributed as uniformly as possible throughout the waste product, and the minimum compressive strength of the waste package should be 10 N/mm^2. In order to comply with Requirement 4, the waste cement product must meet the derived Requirements 4a to 4c. Compliance with this criterion was not simple due to the fact that plastic foil and other lightweight plastics cannot simply be concrete encapsulated without further ado. Due to differences in density, the waste swims to the top of the cement within a short amount of time, resulting in an undesirable final mix.

This problem was solved by shredding the organic waste to a particle size of approximately 5 mm prior to mixing it with the cement. This shredded material is comparable to a filler material and can be distributed homogeneously throughout the cementation product. Due to the adhesion forces predominanting in such a mixture, buoyancy forces do not come to bear, i.e. the waste particles remain in place as mixed within the compound.

Cementation allows all MOX waste to be conditioned to the desired form for final disposal in a single process (Figure 7).

The organic waste quantity, shredded for a waste drum designed for cementation, is mixed in a continuously running screw blender with cement and liquid waste quantities likewise fed in doses into the blender, and the mixture filled into 200l-drums. The mostly metallic and inorganic lumps of waste scheduled for a particular drum are gradually added manually during the drum filling procedure. Thus, all three types of waste - liquid as well as organic and inorganic wastes - are combined in the waste product (Figure 8).

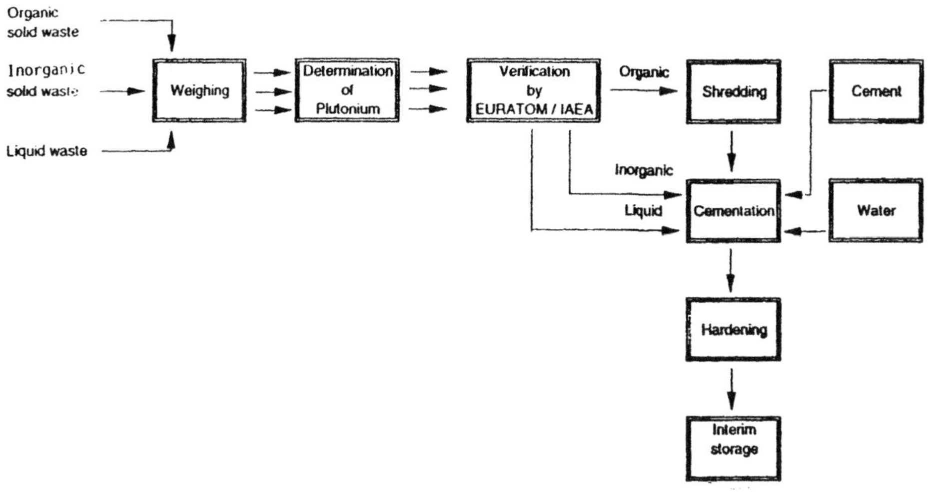

Figure 7: Mass flow of the process for MOX waste cementation

Figure 8: Cementation of operational MOX waste with a continuous screw blender

After completion of the cementation process, the waste product is compacted using a vibratory lance in order to eliminate any cavities which may have formed at waste pieces. Using the double lid technique, the finished waste drum is then disconnected from the glovebox and sealed with the main lid.

6 PROPERTIES OF WASTE PRODUCT

Figure 9 illustrates the appearance of the waste product in a corrosion-protected drum. The organic waste is uniformly distributed throughout the drum. In order to prevent any pressure buildup inside the drum due to a gas generation by alpha-radiolysis, the drum inner and main lids are each provided with a special high-efficiency filter providing 99.997 % retention. Measurements conducted on a large number of drums have shown that in the initial months after cementation approximately 2 l H_2 / (Ci-alpha x year) arise, but that the gas generation rate declines over the long term. For the purpose of final disposal, this characteristic of the waste product does not present any problem since H_2 diffusion through the surrounding geology allows considerably higher gas releases.

As long as the organic fraction of the shredded waste does not exceed 40 kg per drum, the required compressive strength of 10 N/mm^2 will be maintained. Solid pieces of waste have no effect on the compressive strength. However, for volumetric reasons, the content of waste pieces to be concrete encapsulated in a single drum must be limited to 100 kg. Cavities are eliminated through the use of vibratory equipment, waste pieces are completed encapsulated on all sides and porosity is reduced to less than 4 % /4/.

7 FINAL PRODUCT

For logistical reasons, the 200l-drums are not stored individually at the final disposal site, but rather they packed in standard steel containers well adapted to the repository caverns. The free volume between the drums must be filled with concrete in order to withstand the pressure of the surrounding geology.

The Type VI container, which has a gross volumetric capacity totaling 5.4 m^3, can accommodate ten drums. This corresponds to a required final storage volume of 540 l per 200l-drum (see Figure 10).

8 DISPOSAL VOLUME

As shown in Figure 6, requirements governing the criticality safety of the repository limit the Pu-239 content of the waste product. On the basis of these requirements the demand of repository volume can be calculated for the annual waste arisings from processing of weapons Pu with 93,5 wt% Pu-239 as shown in Figure 11.

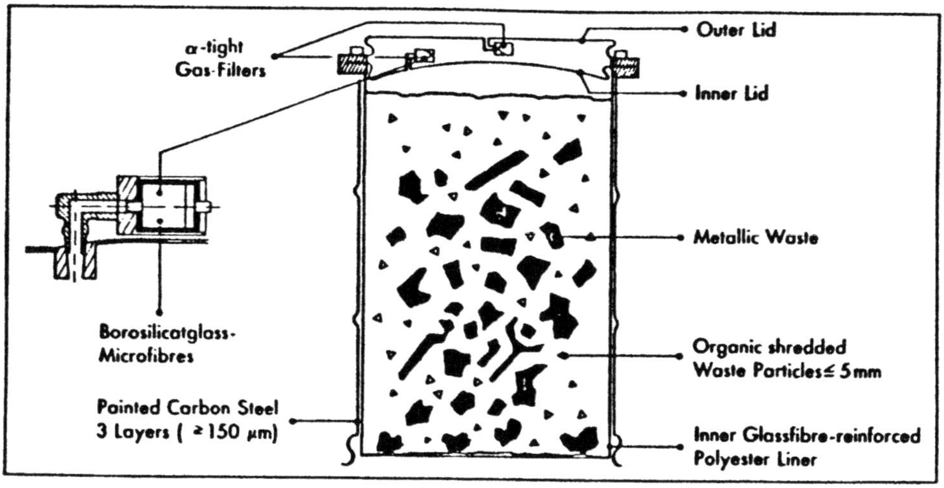

Figure 9: Vented 200 l-drum with cemented MOX waste

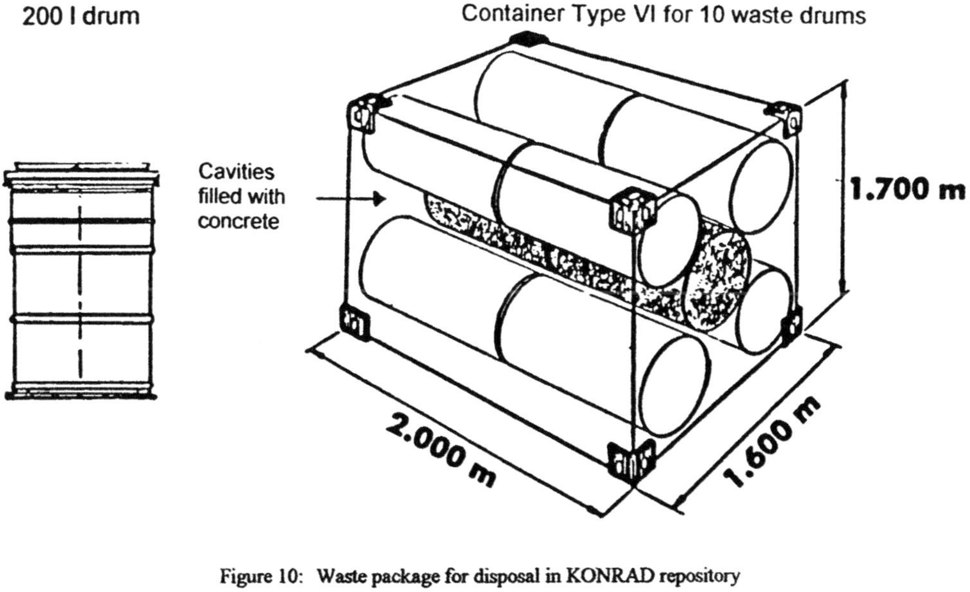

Figure 10: Waste package for disposal in KONRAD repository

Pu-Throughput		5 Mg Pu (120 Mg MOX)				120 Mg Pu			
Waste type Determini	Max. value per drum	Mass (Mg)	200-l drums (Qty)	Type VI cont. (Qty)	Disposal volume (m^3)	Mass (Mg)	200-l drums (Qty)	Type VI cont. (Qty)	Disposal volume (m^3)
Solid organic	40 kg	12	300	30	162	288	7,200	720	3,888
Solid inorganic	100 kg	8	80	8	43	192	1,920	192	1,037
Liquid	90 kg	75	833	84	454	1,800	20,000	2,000	10,800
Pu-239 inventory	11 g	0.01	850	85	459	0.24	20,400	2,040	11,016

Figure 11: Determination of number of waste drums a 11 g Pu-239/drum and of required disposal volume

Determining for the 459 m^3/a disposal volume are the criticality requirements, at 11 grams of Pu-239 per drum, i.e. virtually equal to the required volume for solidified liquid waste (454 m^3/a). Thus, for the processing 120 Mg of Pu, a final disposal volume of approximately 11,000 m^3 would be required for operational waste. Heat production of this waste volume would total a maximum of 600 W, or 55 mW/m^3.

At the KONRAD repository, ten Type VI containers fit into the cross section of the storage area (W x H = 6.4 m x 5.1 m of storage cross-sectional area), taking up 2 m of the cavern length. Final disposal at KONRAD of 2,040 containers resulting from the processing 120 tons of Pu would cover a cavern length of 408 m (Figure 12).

If a repository is available with lower restrictions with respect to criticality safety and heat effects, the disposal volume could be reduced by a factor of 3 to 3,890 m^3 for 120 tons of Pu throughput. In this case the excess liquid waste quantities must be further reduced by evaporation so that, in the end, it would be the quantity of organic, solid waste which determines the number of waste drums.

9 PRACTICAL EXPERIENCE WITH WASTE CONDITIONING

The cementation process for waste conditioning was developed by Siemens MOX fuel fabrication facility in Hanau and has been used since 1984. Over the last ten years, nearly 70,000 waste packages have been processed and cemented in approximately 2,900 waste drums. The peak throughput, achieved in two-shift operation, was 600 drums per year. In this time the cementation facility has proven robust and unsusceptible to malfunction.

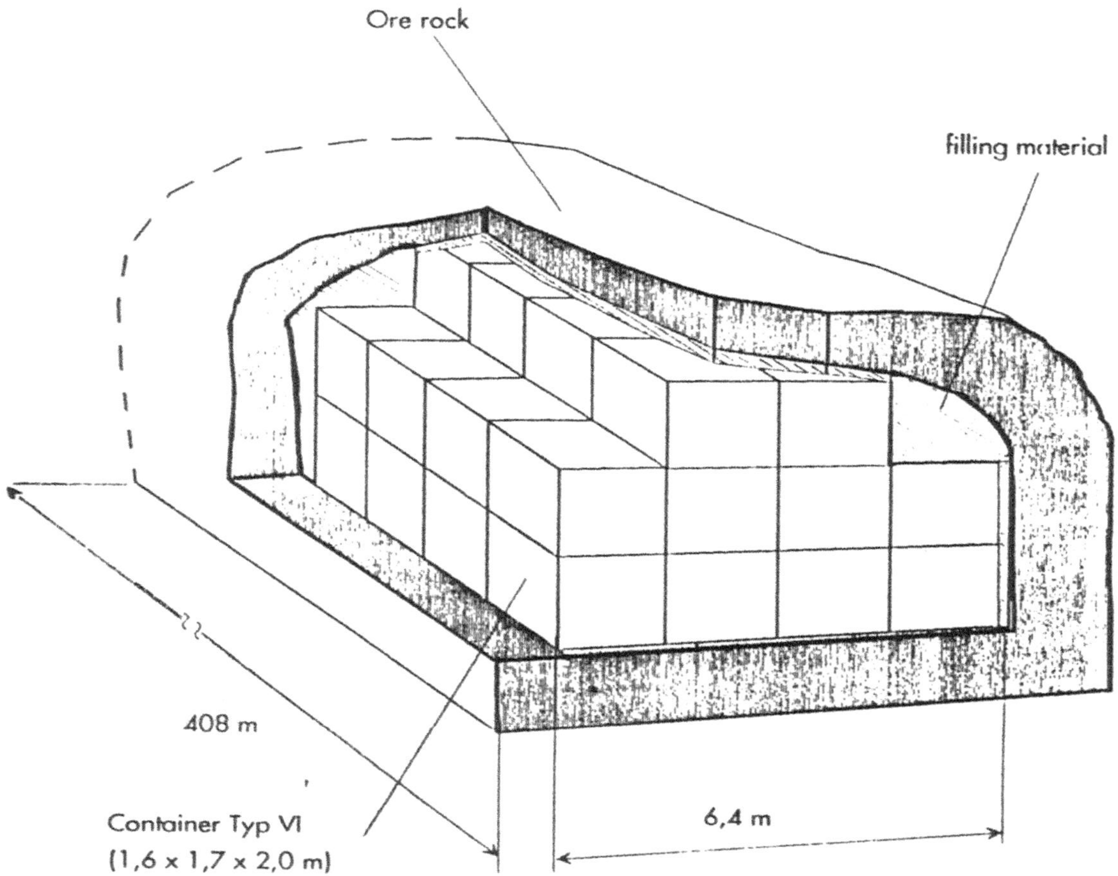

Figure 12: Cavern in the final repository KONRAD for operational waste from processing of 120 Mg Pu

10 CONCLUSION

A proven process is available for conditioning the operational wastes arising from MOX fuel processing which allows the various forms of raw waste to be processed in a single waste product. This process of cement solidification of wastes employs neither high temperatures nor high pressures so that a high degree of safety can be achieved in the operation of such facilities. Components and systems employed are all of proven design used in conventional industrial applications.

Using Germany's KONRAD repository with the stringent acceptance criteria as a basis for calculations, approximately 11,000 m^3 of disposal volume would be required for the disposal of waste generated from 120 t weapons Pu throughput in a MOX fuel plant. In a repository with more realistic acceptance criteria only 4,000 m^3 would be required.

With respect to the gas generation by alpha-radiolysis and criticality limitations the cementation process is not applicable for the direct conditioning of Pu from weapons.

11 REFERENCES

/1/ Ledebrink F. W., Haas E.
"Experience in waste management in a MOX fuel fabrication plant",
Working Document of 14th annual ESARDA Meeting , Salamanca, Spain, Mai 1992, p. 149 - 156

/2/ Anforderungen an endzulagernde radioaktive Abfälle
- Schachtanlage KONRAD - Bundesamt für Strahlenschutz (BfS)
BfS Salzgitter 1994, ET-3/90-Rev. 3
ISSN 0937-4434

/3/ Brennecke P. et al., "Behälter für die Endlagerung radioaktiver Abfälle",
Atomwirtschaft, Nov. 1987, p. 537 - 541

/4/ Schneider V. W., Ledebrink F. W.,
"Cementation of TRU waste by a new process: Properties of the products",
Advances in Ceramics, Vol: 8. 1984, p. 394-400

SRS VITRIFICATION STUDIES IN SUPPORT OF THE U.S. PROGRAM FOR DISPOSITION OF EXCESS PLUTONIUM

George G. Wicks*, J. Mal McKibben, M. John Plodinec and William G. Ramsey,
Westinghouse Savannah River Company
P.O. Box 616, Aiken, South Carolina 29802 USA

1. ABSTRACT

Many thousands of nuclear weapons are being retired in the U.S. and Russia as a result of nuclear disarmament activities. These efforts are expected to produce a surplus of more than 50 MT of weapons grade plutonium (Pu) in each country. As part of this inventory, the U.S. Department of Energy (DOE) has more than 20 MT of Pu scrap, residue, etc., and Russia is also believed to have at least as much of this type of material. The entire surplus Pu inventories in the U.S. and Russia present a clear and immediate danger to national and international security. It is important that a solution be found to secure and manage this material effectively and that such an effort be implemented as quickly as possible.

One option under consideration is vitrification of Pu into a safe, durable, accountable, and proliferation-resistant form. As a result of decades of experience within the DOE community involving vitrification of a variety of hazardous and radioactive wastes, this existing technology can now be expanded to include immobilization of large amounts of Pu. This technology can then be implemented rapidly using the many existing resources currently available. An overall strategy to vitrify many different types of Pu will be discussed. In this strategy, the arsenal of vitrification tools, procedures and techniques already developed throughout the waste management community can be used in a staged Pu vitrification effort. This approach uses the flexible vitrification technology already available and can even be made portable so that it may be brought to the source and ultimately, used to produce a consistent and common borosilicate glass composition for the vitrified Pu. The final composition of this product can be made similar to nationally and internationally accepted HLW glasses.

2. BACKGROUND

In January of 1994, the Committee on International Security and Arms Control (CISAC) of the National Academy of Sciences (NAS) issued a report entitled "Management and Disposition of Excess Pu" [1]. The purpose of this study was to evaluate disposition alternatives for management of excess plutonium resulting from disarmament activities. The treatise covered many important aspects of the Pu disposition question and with respect to long-term management of the excess Pu stated the following:

"The two most promising alternatives for achieving these aims (long-term Pu disposition) are:

- fabrication and use as fuel, without reprocessing, in existing or modified nuclear reactors; or
- vitrification in combination with high-level radioactive waste."

A third option involving burial of Pu in deep boreholes was also mentioned for possible future consideration.

As a result of the NAS study, the vitrification option was elevated to the same level of importance as a reactor option and listed as one of two leading preferences for ultimate disposition of weapons grade Pu.

The Savannah River Site (SRS) has been involved in vitrification of high level waste for several decades and also associated with vitrification of a variety of other types of radioactive and non-radioactive materials [2,3]. As a result of over 20 years of vitrification experience and about 40 years of Pu handling and

E.R. Merz and C.E. Walter (eds.), Disposal of Weapon Plutonium, 143-154.

processing experience, the site has been requested to provide input into this subject by several agencies, including the U.S. Department of Energy. These efforts are directed at addressing a variety of Pu disposition alternatives [4-7], including management of weapons grade Pu and also Pu scrap, residues, etc. that currently exist throughout the DOE complex. An early summary of vitrification options as well as a recent update of the advantages of this technology are presented elsewhere [8,9].

3. POTENTIAL ADVANTAGES OF VITRIFICATION

There are many potential advantages associated with the vitrification option for long-term management of plutonium. These include the following:

3.1 Immediacy of Implementation

Many experts in the field believe it is important to act quickly to immobilize Pu for security, safeguards, safety and environmental reasons. Because of the advanced state of the art of vitrification and as a result of the existing capabilities and experience in this area, there is no other Pu option that can be implemented as rapidly.

3.2 Flexibility

The vitrification option provides a common technology for treatment of almost all forms of Pu. This includes not only weapons grade Pu, but also significant quantities of more complex Pu scrap and residue compositions, currently existing within the DOE community and posing additional problems.

3.3 Technology Availability

As a result of the High Level Waste program and associated waste management efforts, the techniques and procedures for vitrifying radioactive and hazardous components are mature, the equipment and facilities for immobilizing actinides are available, and waste form specifications to assure product quality are well understood. These efforts are directly applicable to Pu vitrification and could be "piggy-backed" upon. A very extensive and capable vitrification infrastructure exists containing experienced and dedicated experts throughout federal and national laboratories, academia, and industry throughout the United States as well as in other countries.

3.4 Waste Glass Performance

HLW glasses have been demonstrated to have excellent chemical durability, mechanical integrity, radiation and thermal stability. Chemical durability is considered to be the most important technical performance property of a waste glass form. It is important to note that actinide bearing glasses exhibit excellent chemical durability. The leaching of actinides is generally 10-100x better (lower leach rates) than modifiers or alkali cations contained in HLW waste glass systems.

3.5 Processing Consideration

The ability to vitrify radioactive materials is not only well developed, but also well demonstrated. In the case of HLW, actual production facilities are in operation world-wide. These include the French process in Marcoule and also La Hague, the German vitrification operation in Mol Belgium, the Sellafield facility in England, the Russian process in Chelyabinsk, and others. Construction of the first HLW vitrification facility in the United States, the Defense Waste Processing Facility (DWPF) at SRS, has recently been completed, is actively making cold glass pours and is scheduled to be in production in about one year. Vitrification of Pu represents an extension of this already available vitrification technology and its many components. It also represents a technology that can be made portable.

3.6 Waste Minimization

Due to the type of equipment and buildings needed to process radioactive materials, and as a result of the vitrification facilities and equipment used to treat other types of radioactive wastes, existing contaminated and non-contaminated facilities could be modified to perform Pu vitrification. These facilities include buildings and equipment designed to receive Pu, store the material, process the Pu, and vitrify it into acceptable products. These facilities currently exist within the U.S. DOE complex and although they would require various degrees of modification, their existence would eliminate the need to build and later D&D additional buildings.

3.7 Ability to Immobilize Pu in Glass

Immobilization of Pu into borosilicate glass has already been demonstrated in HLW programs. This early work involved immobilization of 7 wt. % of plutonium oxide into a glass matrix [10]. Higher Pu loading has also been demonstrated on a laboratory scale [13]. Hence, there is no question whether Pu can be vitrified- the maximum amount immobilized remains to be defined and will most likely not be determined by solubility limits but by criticality considerations.

3.8 Inherent Criticality Control

Due to the composition of borosilicate glasses, boron as well as lithium, which are normally present in HLW forms, can act as poisons to assist in criticality control. This is especially important during handling and processing operations. Other poisons such as gadolinium or erbium could be added to further this effect, which would be especially relevant for long-term repository storage scenarios due to their insolubility. Additional work would be necessary to better assess the effects of potential poisons for criticality control in some of the final disposal options under consideration and for increasing the difficulty in reclaiming Pu from the glass matrix. Criticality control represents the most important consideration in all stages of any disposal option.

3.9 Acceptability/ Waste Form Qualifications

The only waste form which has achieved a degree of national and international acceptance for immobilizing HLW is borosilicate glass. HLW glasses already contain Pu, although in small amounts. It took approximately ten years and thirty million dollars to qualify the SRS HLW waste glass composition. This important and necessary effort could be piggy-backed upon for the Pu vitrification option.

3.10 Proliferation Resistance

There are many ways in which Pu can be immobilized into glass to produce durable, safe, proliferation-resistant forms. These options depend on the degree of proliferation resistance required and are directly proportional to the cost and complexity of the operation. For example, simply immobilizing Pu into glass can be achieved rapidly and most easily and provides the highest degree of flexibility. The Pu-glass product produced would be more proliferation resistant that Pu in its weapons form, but could be reclaimed fairly easily by those reasonably familiar with this field. The degree of proliferation resistance could be increased significantly, however, by either initially mixing the Pu directly or by re-melting Pu-only glass, with fission products or existing HLW. This can be accomplished by "direct vitrification" or by a "staged approach", using a "can-in-canister" technique which will be discussed later. The radiation field associated with the radioactive additives would considerably increase the difficulty in obtaining or handling this material and in subsequent transportation and reprocessing operations to reclaim Pu. Proliferation resistance can be further enhanced by the potential size and weight of the product, chemistry of the form, and most important, by the safeguards that would be necessary for any undertaking of this type.

4. OVERVIEW OF A POTENTIAL Pu VITRIFICATION STRATEGY

Vitrification provides an important option to immobilize and dispose of not only weapons grade Pu, but also many other forms of Pu as well as other radionuclides of concern currently existing in weapons producing nations. Major steps involved in a Pu vitrification strategy include Pu handling and glass preparation, conversion to melter feed, vitrification into intermediate or final Pu-glass products, interim storage of Pu-glass forms and final product dispositon.

A simplified flowsheet summarizing this overall vitrification strategy is shown in Figure 1 and following is a brief description of each of the important phases.

4.1 Preparation

The vitrification strategy applies to Pu in metal form currently contained in weapons as well as Pu in other forms such as oxides, buttons, scrap, solutions, ash, salts, residues, etc. The other forms of Pu are contained at DOE sites including Rocky Flats, Los Alamos National Laboratory, Hanford, Argonne National Laboratory, Idaho National Engineering Laboratory, and the Savannah River Site. Preparation of these materials will involve receipt, disassembly/ separations in some cases, assaying, interim storage and pre-treatment prior to immobilization.

4.2 Conversion

There are two major options that can be used for conversion of Pu from a metallic form to a suitable melter feed. These include (a) *oxidation-* burning it to produce plutonium oxide powder and (b) *dissolution-* dissolving it to yield a plutonium acid solution. Additional preparation involving Pu scrap and residues could be performed, depending on composition and subsequent melting characteristics. Vitrification can handle either liquid or oxide feeds.

4.3 Vitrification

A vast array of electric melting techniques have been developed for vitrification of radioactive wastes over the years and include indirect heating, joule heating, plasma and microwave vitrification. Many of these technologies could be utilized for vitrification of Pu. The proposed strategy considers two main options for vitrification:

- Direct Vitrification; A direct means of vitrifying Pu into glass is shown as Option A in Figure 1. In this case, the treated Pu is melted directly with fission products such as Cs-137 or HLW to produce a highly radioactive Pu glass product. This could be accomplished in several ways and provides the most immediate, highest degree of proliferation resistance. Because of the highly radioactive wastes to be mixed with the Pu, specially contained and heavily shielded facilities would be necessary.

- Staged Vitrification Approach; A two-staged vitrification option is given as Option B in Figure 1. First, a Pu-only glass is produced as an interim product. This step has the advantage of being able to be implemented very rapidly and most easily due to the relatively low amounts of radioactivity present, resulting from the initial absence of fission products or HLW that would be added later. The production of this interim product increases the flexibility of the process in that it can be performed by the use of relatively simple equipment in gloveboxes, and the process can even be made portable, if desired. It also allows use of a wide range of different site-specific vitrification technologies that exist and are already located at sites containing some of the waste. While the resulting Pu-only glass does not produce the most proliferation resistant glass product initially, it does provide the most flexibility. This flexibility allows interim products to be transported to other locations that have more highly radioactive wastes, such as HLW or selected fission products, and later be re-melted or re-combined with this waste as in the "can-in-canister" approach to produce more proliferation resistant forms. Specially contained facilities and equipment similar to the DIrect Vitrification option would also be necessary for this final part of the operation. However, by using the staged approach, this mission involving Pu immobilization could be most easily and efficiently meshed into other existing facilities, programs and subsequent schedules to produce an optimum use of existing resources.

An important part of the Pu vitrification strategy discussed above is that for either option, Direct or Staged Vitrification, a common final waste glass form can be produced with a composition similar to HLW glass [11]. The HLW glass composition has already undergone a very time consuming and expensive process to be certified and made acceptable. A resulting Pu bearing glass can "piggy-back" on this work. The Pu glass compositions would be expected to be as good, if not better, than the already acceptable HLW glass compositions. In general, the observed "leachability" of actinides in glass matrices is significantly lower (better) than most other components found in waste glass systems.

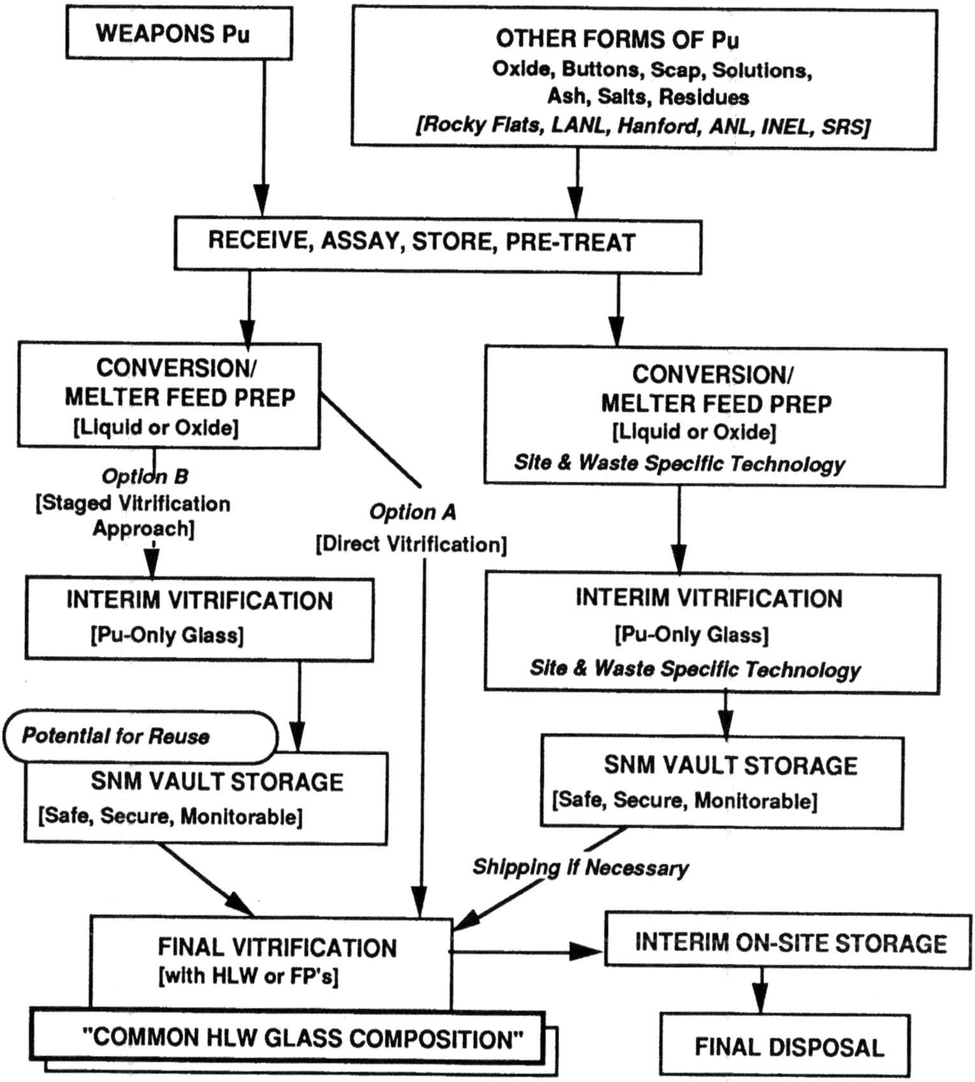

Figure 1. Overall Pu Vitrification Strategy

4.4 Interim Product Storage

A need to store Pu bearing glasses temporarily is important to the vitrification option. The design of the interim storage facility would depend on factors such as the composition and radioactive content of the products and the intended duration of storage before ultimate disposal. Among the most important considerations for this facility are worker radiation exposure, public and environmental protection from radiologic hazards, and material safeguards and accountability.

4.5 Final Product Disposition

The reference concept for ultimate disposition of HLW glass and spent fuel is to dispose of this material by deep burial in carefully selected geologic repositories. This is also being considered for Pu bearing glasses. A very significant challenge for any repository scenario is to demonstrate safe and effective performance of products out to very long time periods (1000 to 10,000 years, and longer). This challenge would be expected to be even more formidable for any immobilization alternative containing large amounts of Pu, due to criticality considerations. The waste form and waste package will be designed to prevent criticality from occurring but this must be demonstrated to a very high degree in a very complex environment. While this would not be expected to be as significant an undertaking for a "Retrievable Surface Storage Facility or RSSF", it would be expected to take much more effort to demonstrate in a geologic repository.

The final form and composition of Pu glass would be tailored for technical and political considerations, involving possible reuse or non-reuse by the weapons producer and for optimizing proliferation resistance, especially towards potential non-friendly nations.

5. VITRIFICATION FACILITIES & CAPABILITIES

5.1 Vitrification of Pu Using New Facilities

One of the options being considered for immobilization of Pu is to design new plants, buildings and equipment tailored for Pu processing. This is sometimes referred to as the "Greenfield Option". A potential main advantage of this alternative is that facilities can be designed for optimum performance to accomplish the mission. Main disadvantages include possible increased costs, time for implementation and the dispositon of facilities which must ultimately be decontaminated and decommissioned after completion of the campaign. A variety of immobilization options involving not only glass but other alternatives, as well as various immobilization techniques and technologies using new as well as existing facilities, are now being evaulated for disposition of surplus fissile materials in the United Statess [12]. A Programmatic Environmental Impact Statement or PEIS is in draft on these options.

5.2 Vitrification of Pu Using Existing Facilities

As discussed earlier, there exists an infrastructure knowledgeable on vitrification of radioactive and hazardous wastes that could be applied to vitrification of Pu. For example, there have already been several technology exchanges between U.S. and Russian scientists on possible joint efforts involving vitrification of a variety of radioactive wastes, including actinide-bearing materials. Further exchanges and interactions are also planned. Immobilization of Pu scrap and residues are of immediate interest within the U.S. DOE complex and these efforts could also be extended to include immobilization of Pu derived from weapons dismantlement activities.

Along with developed and demonstrated vitrification technology, experience and expertise, there also exists buildings, equipment and supporting hardware as well as software for immobilization efforts. In support of the NAS study, the Savannah River Site was asked in 1993 to perform a study examining vitrification options for Pu using facilities and capabilities that already existed. While capabilities exist throughout the DOE complex for this task, those at the Savannah River Site (SRS) were emphasized. Following are some of the results of this study including examples of types of facilities needed [8].

Five major Pu vitrification options, along with variations of most of these options, were identified using SRS facilities for this study [8]. The processing options involved three potential Pu-bearing products that could be made; Pu-only glass (interim product), Pu/Cs-137 glass (final product) or Pu/HLW glass (final product). Examples of types of relevant existing facilities which could be modified and would be important to support vitrification of plutonium include the following:

- Plutonium Storage Facility (PSF); The PSF along with New Special Recovery are facilities that are part of the 221-F Canyon building located at SRS. This complex is shown in Figure 2. The mission of the PSF was to receive Pu materials from offsite that would later be processed. The building contains a fully safeguarded vault for automated transport and stacking, gloveboxes for opening drums and removing and inspecting contents, and instruments for non-destructive assay and computer accountability. The facility also contains a delivery systems for later processing of materials in the adjacent facility, the New Special Recovery (NSR).

- New Special Recovery (NSR); The mission of the NSR was to process plutonium from throughout the DOE complex. This includes dissolving Pu scrap, oxide or metal from various sources to produce purified Pu metal buttons or oxide powder. The facility contains state of the art glovebox trains (Figure 3) for feed preparation, waste handling, dissolution processes, samples and analysis, along with a supporting remote control room. This facility also contains extra room which could be outfitted with glass melters to pursue one of the vitrification options under consideration.

- F Canyon (Including the Multi-Purpose Processing Facility or MPPF); The 221-F Canyon was the world's first PUREX production plant used to dissolve natural and depleted uranium targets and to recover U and Pu. It is a large, heavily shielded facility operated by remote means. The building has the potential of being refitted with a melter and used for vitrification in support of another vitrification option of immobilizing Pu directly with HLW or selected fission products. The MPPF is located within the facility and contains eight modules that are now being used for vitrification of Cm and Am using a bushing melter, and can potentially be used for vitrification and subsequent clean up of Pu scrap on site.

- Defense Waste Processing Facility (DWPF); The DWPF is the first major waste vitrification facility constructed in the United States for immobilization of high level radioactive waste (HLW). It is scheduled to be in production in about one year. The mission of the DWPF, shown in Figure 4, is to immobilize the 34 million gallons of HLW currently being stored at SRS into borosilicate glass. All of the operations necessary to process, vitrify, containerize, seal and decontaminate HLW glass units are present, including supporting capabilities to control the process and verify product quality. This facility and its equipment could be modified in support of various options to vitrify larger quantities of Pu than presently contained within HLW.

Important for DWPF options is not to adversely affect the current, important mission of the facility. There are two primary ways that Pu vitrification could be conducted using the DWPF. First, during the scheduled melter change out of the DWPF, the facility and equipment could be modified and Pu bearing feed introduced either directly as a Pu feed or as a Pu bearing glass frit, to produce Pu-HLW glass forms in 304L stainless steel containers. These units would be similar to HLW glasses in canisters, 2-ft. in diameter and almost 10-ft. high. An important potential advantage of this option is that waste loading to the glass could be increased by 1-2%, which would still result in a highly durable form which could immobilize 50 MT of Pu as part of the HLW program, and without increasing the number of waste canisters produced. Important disadvantages of this option are that the facility is not currently designed nor approved to handle large amounts of fissile material and to do so, additional criticality controls would be needed. This would be expected to be both expensive and very time consuming, and presents the potential of adversely affecting the existing HLW program.

An alternative or second main option is designed to minimize the potential disadvantages discussed above and allow immobilization of excess Pu to be started sooner. This option involves the staged approach discussed earlier, in which a Pu-only interim glass form is first made and later, mixed with fission products or HLW to produce a Pu/ FP or Pu/ HLW final glass product. A means to accomplish this task using the DWPF facility and its extensive capabilties is by the *"can-in-canister"* method. This involves placing cans of interim Pu-only glass produced earlier in more convenient and available facilities into a rack in "cold" HLW canisters and then pouring HLW around them in the DWPF as part of the current HLW vitrification campaign. While this option may not produce as integral a product as the direct vitrification of Pu-HLW glass, it would be expected to produce an excellent form with a high radiation field, very heavy,

Figure 2. Building 221-F, NSR, and PSF Facilities (89-2076-18)

Figure 3. Gloveboxes in NSR (92-2051-6, altered)

Figure 4. Overview of DWPF Complex (89-2079-1)

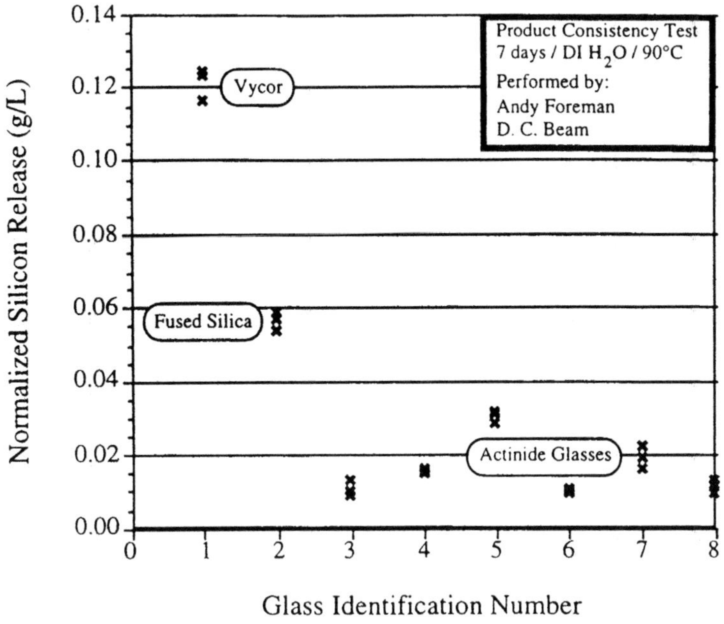

Figure 5. Relative Durability of Löffler Glasses and Common Leach Resistant Glasses.

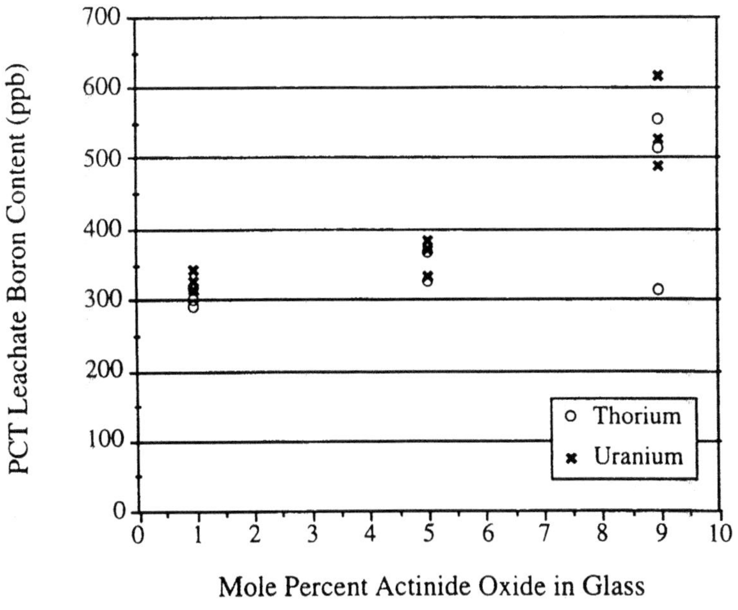

Figure 6. Effect of Actinide Oxide Concentration on Löffler Glass Durability.

cumbersome in size, and subsequently, high proliferation protection. This approach represents the easiest and most rapid to engineer and least expensive to implement. It also represents an option of maximum flexibility that can utilize existing facilities in an efficient manner by allowing Pu immobilization to fit into existing schedules, programs and buildings, without adversely effecting other efforts.

5.3 Current Vitrification Studies on Actinides (Th, U, Am, Cm & Pu)

The Savannah River Technology Center is actively investigating actinide glass formulations as part of current waste management and environmental remediation efforts. Studies are in progress to (a) demonstrate the feasibility of immobilizing actinides into glass, (b) determine the maximum concentration of actinides that can be loaded into a glass and (c) evaluate both the processability and chemical durability of a variety of potential actinide glass formulations. Glass samples have been prepared with Th, U, Am, Cm and Pu [13]. A commercial borosilicate glass composition was seen to accept up to 9 mole percent (approximately 20 weight percent) ThO_2 or UO_2, and PuO_2 has been successfully added to a glass composition at a level of 6.6 mole percent (15 weight percent). Am and Cm glasses have also been fabricated. Characterization and optimization of these systems are in progress.

Löffler glass, the borosilicate composition discussed above, is one of the glass systems under study. This glass has been shown to exhibit very good chemical durability. This is exemplified in Figure 5 which shows a plot of the chemical durability of several Löffler glasses made with rare earth surrogates for the actinides compared to more common glasses of known excellent durability, i.e., fused silica and Vycor®. The addition of the various actinide surrogates appear to have little effect on the excellent chemical durability of the Löffler glass. The durability of glasses with increasing concentrations of actual ThO_2 and UO_2 is summarized in Figure 6. These data indicate that the Löffler glass is both highly compatible with actinides and extremely resistant to aqueous attack.

6. SUMMARY

Vitrification is a technically viable option to immobilize and manage Pu resulting from disarmament activities as well as a wide range of existing Pu scrap and residue compositions. A vitrification infrastructure exists from waste management programs. This includes expertise and experience, personnel, buildings, equipment and supporting capabilities, which could be used to implement vitrification of Pu in a time expedient and cost effective manner.

7. REFERENCES

1 COMMITTEE ON INTERNATIONAL SECURITY AND ARMS CONTROL, National Academy of Sciences, "Management and Disposition of Excess Weapons Plutonium", National Academy Press, Washington DC (1994).

2 M.J. PLODINEC, G.G. WICKS, and N.E. BIBLER, "An Assessment of Savannah River Borosilicate Glass in the Repository Environment", DP-1629, Savannah River Laboratory, Aiken SC (1982).

3 G.G. WICKS, "Nuclear Waste Glasses", in Treatise on Materials Science and Technology, Glass IV, M. Tomozawa and R.H. Doremus, eds., Vol. 26, pp. 57-188, Academic Press, Inc. (1985).

4 G.G. WICKS, "Advantages and Limitations of Disposing of Pu in Glass", presented to the Committee on International Security and Arms Control, National Academy of Sciences, Washington DC (1993).

5. J.M. MCKIBBEN and G.G. WICKS, "Vitrification of Excess Plutonium", presented at the International Policy Forum: Management & Disposition of Nuclear Weapons Materials, Leesburg, VA (1994).

6 G.G. WICKS, J.M. MCKIBBEN and M.J. PLODINEC, "Vitrification of Plutonium", presented to the Interagency Working Group on Plutonium Disposition, Aiken SC (1994).

7 G.G. WICKS, M.J. PLODINEC, and J.M. MCKIBBEN, "Leveraging of Resources for Vitrification of Plutonium", presented to the Immobilization Task Team on Plutonium Disposition, Aiken SC (1994).

8 J.M. MCKIBBEN, R.W. BENJAMIN, D.F. BICKFORD, L.P. FERNANDEZ, W.N. JACKSON, W.R. MCDONELL,E.N. MOORE, P.B. PARKS, M.J. PLODINEC, W.M. RAJCZAK, S.K. SKILES, and G.G. WICKS, "Vitrification of Excess Plutonium (U)", WSRC-RP-93-755, Predecisional Draft, Westinghouse Savannah River Company (1993).

9 G.G. WICKS, J.M. MCKIBBEN and M.J. PLODINEC, "Vitrification for Disposition of Excess Plutonium", presented at Waste Management '95, Tucson, AZ, Feb. 26- Mar. 2, 1995 and to be published in the proceedings (1995).

10 M.J. PLODINEC "Development of Glass Compositions for Immobilization of SRP Waste", DP-1517, Savannah River Site (1979).

11 W.G. RAMSEY and G.G. WICKS, "WIPP/SRL In-Situ Tests: Compositional Correlations of MIIT Waste Glasses", in Ceramic Transactions, Nuclear Waste Management III, G.B. Mellinger, ed., Vol. 9, pp. 257-270 (1990).

12 L.W. GRAY, "Pu Immobilization: Progress, Challenges, and Path Forward", presented at the 2nd International Policy Forum: Management & Disposition of Nuclear Weapons Materials", Lansdowne, VA, March 21-24 (1995).

13 W.G. RAMSEY, N.E. BIBLER and T.F. MEAKER, "Compositions and Durabilities of Glasses for Immobilization of Plutonium and Uranium", presented at Waste Management '95, Tucson, AZ, Feb. 26- Mar. 2, 1995 and to be published in the proceedings (1995).

ON THE FEASIBILITY OF VITRIFYING HIGH-LEVEL LIQUID WASTE CONTAINING HIGH AMOUNTS OF PLUTONIUM

M. DEMONIE, H. CUYVERS
Belgoprocess
Gravenstraat 73
B-2480 Dessel (Belgium)

1. Introduction

The conditioning of high-level liquid waste (HLLW) is a major topic in nuclear waste management. The basic aim of converting HLLW into a solid form is to insolubilize the radionuclides and reduce to a bare minimum the degree of possible pollution of the biosphere.

Glass is one of the solid forms which correspond to these requirements. The main advantage of glass is its capacity to incorporate the large variety of chemical elements present in HLLW into a stable matrix. Moreover, the glass product remains chemically stable over very long periods of time.

In order to produce a glass product of excellent quality, the precise composition of the glass frit to be used must be defined in an extensive glass development programme. The composition of the glass frit can be adapted to the composition of the waste solution, thus ensuring complete incorporation and insolubilization of the radionuclides contained in the waste solution.

Vitrification of high-level liquid waste from the reprocessing of irradiated fuel was carried out on an industrial scale at Belgoprocess in Dessel, Belgium. Between 1985 and 1991 some 900 m^3 of waste concentrates were incorporated into 500 t of glass product (see table 1 for the overall production data of the vitrification programme in the Pamela facility).

Preliminary investigations indicate that the vitrification of a mixture of high-level reprocessing waste and fissile material from the dismantling of nuclear arms is feasible in a facility such as the Pamela plant at Belgoprocess.

2. Process Description

The Pamela vitrification facility at Belgoprocess vitrifies high-level reprocessing waste in a single-step process using a joule-heated ceramic melting unit. Both the liquid waste concentrate and the glass frit are directly introduced into the melter, forming a process zone on top of the glass pool where drying, calcination and melting reactions occur.

E.R. Merz and C.E. Walter (eds.), Disposal of Weapon Plutonium, 155-172.

TABLE 1. Global Pamela Production Data for the Vitrification Period 1985-91

Parameter	Unit	LEWC vitrification 1985-10-01 to 1986-06-12	HEWC vitrification 1986-10-01 to 1991-09-05	Total vitrification period 1985-91
Total liquid waste fed to melter (1)	m^3	62.5	894.6	957.1
Total LEWC and HEWC fed to melter	m^3	47.2	860.1	907.3
Alpha activity in feed	Bq	$1.28x10^{15}$	$2.26x10^{14}$	$1.51x10^{15}$
Beta activity in feed	Bq	$2.78x10^{17}$	$1.64x10^{17}$	$4.42x10^{17}$
Waste oxides	t	(2) 7.97	88.94	96.91
Glass product	t	(2) 81.5	411.7	493.2
Waste oxide content	wt%	9.9	21.60	-
Containers filled				
• 60 l	No.	567	934	1,501
• 150 l	No.		700	700
Total weight filled containers	t	159.28	563.42	722.70
Time efficiency	%	88	93.16	-
Production efficiency	%	69	96.58	-
Alpha activity released to stack	Bq	$4.4x10^3$	$4.08x10^3$	$8.5x10^3$
Beta activity released to stack	Bq	$3.4x10^6$	$6.02x10^5$	$4.0x10^6$
Occupational dose	mSv	40	211	251

(1) inclusive dilution and recycling.

(2) inclusive 0.3 t of waste oxides and 2.6 t of glass product from active start-up and 1.1 t of glass product from melter rinsing at end of LEWC campaign.

Waste is fed at a flow rate of 25 l/h, while glass frit is added at a rate of approx. 15 kg/h. The glass bath temperature is maintained at approximately 1150 °C. The glass product is discharged batchwise, approximately 50 l every 8 hours. Glass can be poured into stainless steel containers either using the central bottom drain or the overflow drain.

Filled containers cool down for approximately five days, before being closed by an air-tight weld. They are decontaminated to a level of maximum 4 Bq/cm^2 beta and 0.4 Bq/cm^2 alpha. All containers are transported to the interim storage facility on-site.

The melter off-gases are cleaned in a dust scrubber, condensed, passed through an NO_x absorber, a scrubber, a glass fiber filter and a series of absolute filters before being released through the stack. As a result of the high efficiency of the off-gas treatment, decontamination factors of ca. 10^{10} for beta and 10^{11} for alpha emitters are obtained.

3. Operational Experience Gained at the Pamela Facility

The main topics of the operational experience in the Pamela facility cover the vitrification of high-level liquid waste from reprocessing, the dismantling of large components and the conditioning of high-level solid waste. Figure 1 gives the overall time schedule of the various working programmes.

3.1 VITRIFICATION

The Pamela vitrification plant was constructed within three years (1981-84). After a year of cold and hot testing with diluted HLLW, hot operation was started on October 1, 1985. Up to September 5, 1991, some 900 m^3 of HLLW from reprocessing were successfully vitrified.

The first phase of the vitrification programme demonstrated the feasibility of the process. Fifty cubic meters of high-level waste concentrates from the reprocessing of 181 t of natural and low-enriched uranium fuels unloaded from power reactors were incorporated into 78 t of glass product containing 2.78×10^{17} Bq beta emitters.

The second phase of the programme was carried out on an industrial basis under the terms of a Belgo-German cooperation agreement. 860 m^3 of high level liquid waste concentrates from the reprocessing of 30.6 t of high-enriched uranium-aluminium alloy fuels unloaded from materials testing reactors were incorporated into 400 t of glass product containing 1.64×10^{17} Bq beta emitters.

The power reactor fuels had a uranium enrichment of max. 4.3%, hence the resulting waste is called low-enriched waste concentrate (LEWC). The materials testing reactor fuels had a uranium enrichment of max. 93%, the resulting waste is called high-enriched waste concentrate (HEWC).

The vitrification programme was completed 9 months ahead of schedule and well within the estimated cost. The progress of the second phase of the vitrification

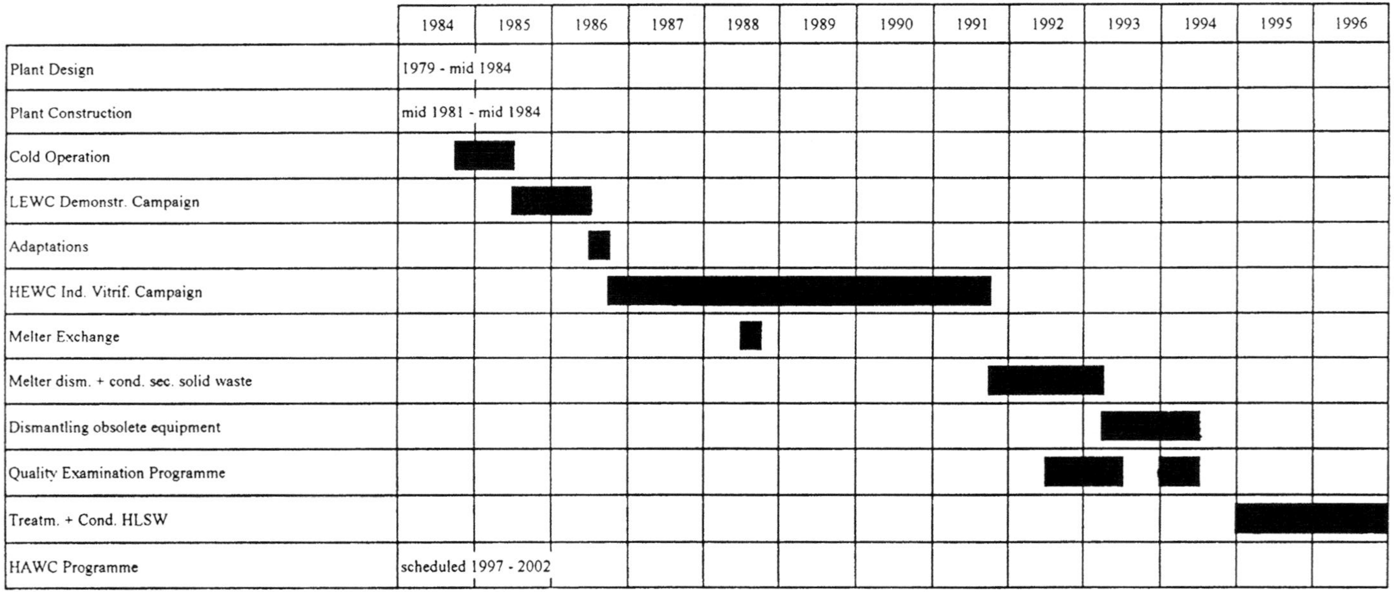

Figure 1. Overall Time Schedule Pamela Working Programmes

programme is shown in figure 2, indicating the gradual depletion over time of the three HEWC storage tanks. It most clearly shows the efficiency of the Pamela facility.

Two ceramic melters were used. After almost three years of operation, the first melter was remotely replaced.

At a feed rate of approx. 16 l/h during the LEWC campaign and 25 l/h during the HEWC campaign, the average throughput of the vitrification facility was respectively 420 and 300 kg of glass product per day, containing resp. 40 and 66 kg of waste oxides.

The average annual activity released through the stack over the total vitrification period of 6 years was limited to 1,400 Bq of alpha emitters and 0.7 Mbq of beta emitters. This emission corresponds to only 0.005 % of the licensed value for alpha activity and 0.01 % for beta activity.

The occupational exposure amounted to 251 mSv for the complete hot operation period. Doses were mainly due to intervention works for repairing remote handling equipment. The average weekly exposure of a staff member working in the controlled area amounted to 0.1 mSv. The maximum accumulated individual annual dose was 20 mSv.

3.2. ANALYTICAL CONTROL ON THE VITRIFICATION PROCESS

Analytical control is of paramount importance when vitrifying high-level waste. The composition of the waste fed to the melter must precisely be known prior to vitrification in order to select the most suitable glass frit and to obtain a qualified glass product.

The composition of the glass frit can be precisely adapted to incorporate all chemical elements present in the liquid feed. The composition of the used glass frits and the average composition of the resulting glass products calculated on the basis of analyses is given in table 2.

Topics of significant influence in the vitrification process and on the quality of the glass product are :

- chemical composition of the HLLW
- noble metals concentration in the HLLW
- the radioactive content of the HLLW
- the presence of insoluble particles in the HLLW
- the composition of the selected glass frit.

Prior to vitrification, every batch of high-level liquid waste is analyzed for: total alpha activity, total beta-gamma activity, density, acid content, calcined residue and nitrate content. The relative standard deviation for the used technique for total alpha determination is approximately 2.0%. The calcined residue concentration in the glass product is based upon the calcined residue concentration of the HLLW feed solution and the feed flow rate. The required amount of glass frit to be added is calculated and adjusted every 8 hours to obtain the reference glass product composition. The glass product was regularly analyzed for total alpha and beta-gamma

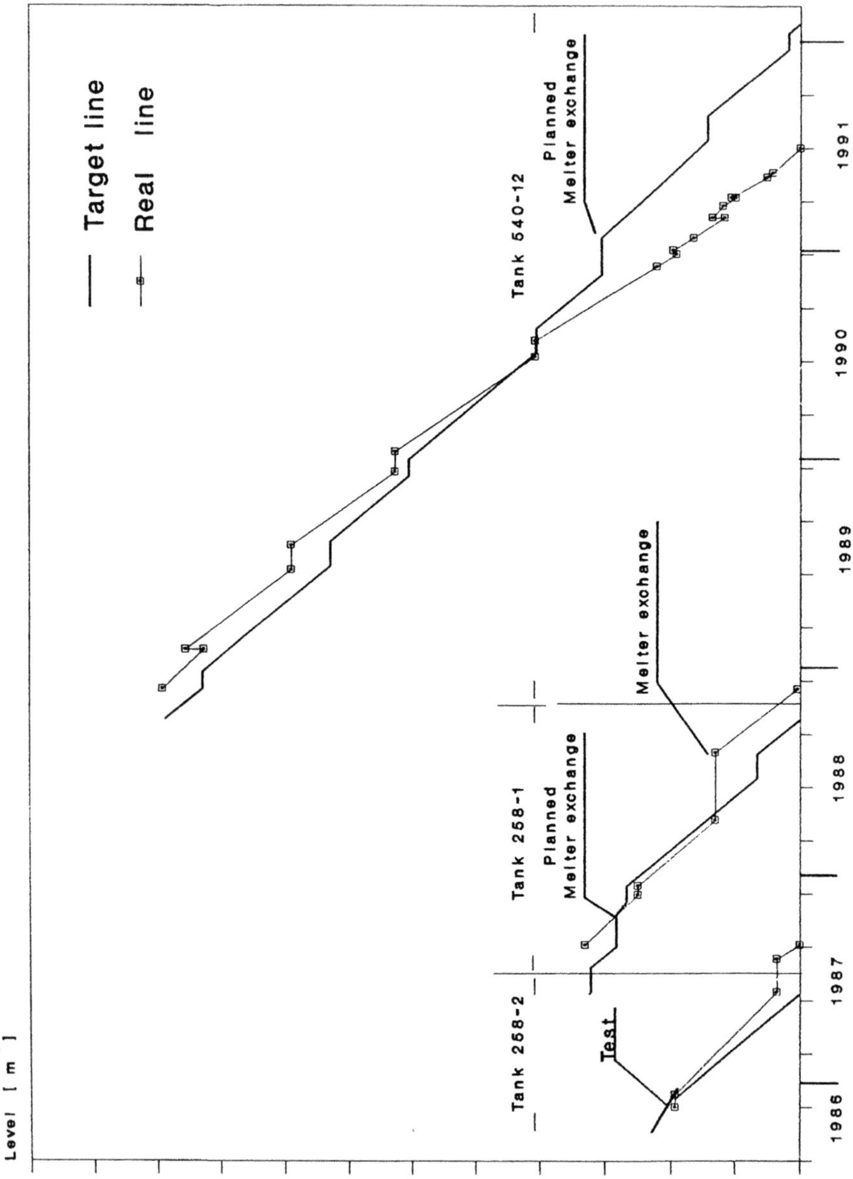

Figure 2. Progress of the HEWC Vitrification Programme of PAMELA

TABLE 2. Standard Glass Frit and Average Glass Product Composition

Chemical Components	Glass frit type: LEWC SM513 FR Wt %	Glass frit type: HEWC SM539FR-Sb Wt %	Glass product: LEWC Wt %	Glass product: HEWC Wt %
SiO_2	58.60	45.50	52.73	35.67
B_2O_3	14.70	33.00	13.24	25.11
Al_2O_3	3.00		4.01	19.19
Li_2O	4.70	4.50	3.82	3.37
Na_2O	6.50	9.70	7.79	8.97
MgO	2.30		2.08	0.12
CaO	5.10	6.50	4.63	5.32
TiO_2	5.10		4.21	0.02
NiO			0.35	0.03
Sb_2O_3		0.80		0.42
Cr_2O_3			0.49	0.08
Fe_2O_3			1.33	0.55
MnO_2			0.52	0.03
ZrO_2			0.5	0.12
U_3O_8			0.06	0.04
Others			4.24	0.96

activity and for calcined residue in order to check correspondence between the real end product and its calculated composition.

Solutions with a higher plutonium content require a more precise analysis such as the isotopic dilution mass spectrometry. Here the relative standard deviation, including dilution and sampling errors, is 0.7%. Moreover the accuracy of the plutonium determination is also depending upon the charactarization of the input solutions whereby a precision of 0.3% in volume measurement and 0.1% in homogeneity is achievable.

3.3. PARTICULAR PROBLEMS ENCOUNTERED IN THE PAMELA PLANT

3.3.1 Noble Metals Behaviour in the Melter

The HLLW vitrified in Pamela contained small amounts of ruthenium, palladium and rhodium (see table 3). During the vitrification of power reactor waste concentrates (LEWC), a highly electrically conductive and viscous layer formed on the flat bottom of the ceramic melter. This layer of noble metals could not be discharged by the routine glass pouring performed through the central bottom drain system. The presence of that layer distorted the electrical field of the joule-heated melter, leading to a loss of power and hence throughput. These noble metals have a very low solubility in the glass melt and have a distinctly higher density.

After 34 months of operation the first melter had to be shut down due to corrosion damage of the lower power electrodes on the one hand, caused by excessively high local current densities, and to corrosion of both drain systems on the other hand.

It is obvious that when vitrifying wastes from commercial reprocessing resulting in a glass product with a noble metals concentration of approximately 1 wt% serious problems can be anticipated.

With a view to vitrifying reprocessing waste with a higher noble metals content in the Pamela plant, the "Institut für Nuklear Entsorgung" (INE) of the nuclear research centre at Karlsruhe developed a new melter design. Instead of a flat bottom, the lower melt tank walls have an inclination of 60 to 75° [1]. If this new melter is operated on a fully continuous basis, with idling times kept to the necessities of maintenance and repair, the sedimentation of noble metals will drastically be reduced.

3.3.2 Blockage of Both Bottom and Overflow Outlets

Blockage of the bottom drain repeatedly occurred in both melters used for the vitrification programme completed at Belgoprocess. The central bottom drain had a relatively small diameter of 18.2 mm. In both melters the bottom drain blocked beyond repair after 11 months of operation. Vitrification operations were continued using the overflow system.

The overflow outlet was also prone to blocking, but could more easily be made free again by means of additional external heating sources. The redesigned bottom drain of the new melter developed by INE will certainly reduce the occurrence of blocking.

TABLE 3. Amount of Noble Metals Fed to the Ceramic Melters K4, K5 and Contained in the Glass Product

Melter	HLLW Type	Original Volume (m^3)	Max. noble metals content (kg) in HLLW			
			Ru	Pd	Rh	total
K4	LEWC	47.2	70.3	10.5	15.9	96.7
	HEWC	218	10.7	11.5	2.2	24.4
	Total	265.2	81.0	22.0	18.1	121.1
K5	HEWC	642.1	26.9	51.4	6.4	84.7
Overall total		907.3	107.9	73.4	24.5	205.8

HLLW Type	Original Volume (m^3)	Max. noble metals concentration (wt%) in the glass product (1)		
		RuO_2	PdO	RhO_2
LEWC	47.2	0.06	0.03	0.025
HEWC	860.1	0.01	0.10	0.01

(1) The analytical errors, especially for the glass analyses, are high due to the low concentrations

3.4. PARTICULAR OPERATIONS CARRIED OUT IN THE PAMELA PLANT

3.4.1 Melter Exchange

After a service life of two years and ten months the first melter was fully remotely exchanged. The required equipment for the remote exchange of the melter was included in the concept of the facility as from the engineering phase. The carefully planned operation only took nine weeks.

First the melter was emptied as completely as possible using the corroded bottom drain. Afterwards all piping and cable connections were remotely disconnected. Subsequently the 20 t melter was hoisted from the melter cell into the crane hall and lowered into the spare melter cell awaiting dismantling.

After decontaminating the crane hall, the new melter was locked into the melter cell and remotely connected. The melter was heated up using a specially developed starting frit and the auxiliary heating elements in the ceiling of the melter. This second melter was operated from August 1988 till September 1991 to vitrify the remaining HEWC.

3.4.2 Dismantling of the Second Melter Used

Upon the completion of the vitrification programme, the secondary solid waste was first conditioned in the dismantling cell of the facility. From June 1992 until March 1994 four large components, including the second melter used, were remotely dismantled and the resulting waste conditioned into cement.

The dismantling operations were carried out in the melter cell, using the existing manipulation systems: master-slave manipulators, a heavy duty manipulator and a 2 t crane. Traditional tools like grinding discs, impact wrenches, vacuum cleaners, jack screws, hammer drills and grab tools were adapted for remote control. The dismantling cell was used concurrently for the conditioning of waste in cement.

The four large components dismantled were: the second melter used, the vitromet equipment, the container lifting and weighing device and the wet dust scrubber. Dismantling operations carried out concurrently with the conditioning of the resulting waste products in cement, took 25,000 manhours over a period of two and a half years. The occupational dose to the eight operators amounted to 47 mSv over the entire dismantling period. Most of this exposure was due to repair of telemanipulation equipment.

The bulk of the dismantling waste from the melter (83.5 %) was conditioned into cement as medium-level waste. The remaining 16.5 % was sent to the appropriate waste treatment facilities onsite as low-level waste.

3.4.3 Decontamination Programme

In view of the quality examination and the conditioning of high-level solid waste, several heavily contaminated cells were decontaminated. Chemical reagents were used in the off-gas and evaporator cells. Remotely operated high-pressure steam jets were used in the melter, container handling and dismantling cells. Those cells were already equipped with remotely operated equipment.

All the cells concerned were decontaminated to a level allowing direct interventions in the cells. The average radiation level is 0.1 mSv/h, with spots up to 0.5 mSv/h.

3.4.4 Quality Examination Programme

After the initial vitrification programme, in which all reprocessing high-level liquid waste from the former Eurochemic plant was vitrified, the Pamela plant was put in safe stand-by. In preparation of future vitrification programmes a quality examination was performed on the off-gas purification and evaporator units.

Although very corrosive off-gas vapours were treated over a period of six years, extensive corrosion in the off-gas purification unit remained limited to the dust scrubber and the off-gas pipe between dust scrubber and condensor. Since then, the used dust scrubber has been dismantled, together with other large components. The off-gas pipe to the condensor will be replaced before a new vitrification campaign starts.

At least one of the valves located in the trays of the NO_x absorption column is damaged. As a consequence the pressure drop behaviour of the column is unstable and the required NO_x absorption efficiency cannot be guaranteed. Therefore a new NO_x absorption column will be installed in series with the existing one. Besides, new self priming off-gas scrubber pumps will be installed in order to reduce the frequency of interventions. Maintenance and repair of the installed scrubber pumps account for a substantial share of the overall occupational dose during vitrification operations.

The corrosion of the evaporator was measured using the ultrasonic technique. The actual corrosion was found to be within the margins expected for six years of operation. The evaporator unit can be used for further programmes without any modification.

3.4.5 Treatment of High-Level Solid Waste

From January 1995 till June 1996, some 40 m^3 of high-level solid waste will be conditioned in cement using various cells of the Pamela facility. End pieces of spent fuel and non dissolved fissile material, kept in a storage pond of the reprocessing plant onsite, will be cemented in 130 stainless steel containers of 150 l. These 25 m^3 of solid waste contain 900 kg uranium and 4 kg plutonium with an overall activity of 5.8×10^{15} Bq.

In separate sub-campaigns, 30 drums of 400 l containing pieces of the thermal shield of the BR3 reactor will be cemented, and 50 containers of 150 l will be filled with pieces of the beryllium matrix of the materials testing reactor BR2.

4. Uranium and Plutonium in the HLLW

4.1 PAMELA EXPERIENCE

The experience with the behaviour of fissile material in the waste feed and resulting glass products was limited. The U and Pu concentrations in the high-level waste concentrates fed to the melter were comparatively low. The highest concentrations of fissile material were found in the LEWC glass product: 0.05 g Pu and 0.57 g U per kilogramme of glass produced. The U and Pu concentrations in the HEWC were roughly 50 % lower. Exact figures are given in table 4.

4.2. IMPACT OF HIGHER U-PU CONCENTRATIONS IN THE HLLW

When the U and Pu concentrations in the HLLW would be significantly increased, three specific problems must be investigated:

- Criticality risks
- Glass product quality
- Safeguards.

4.2.1 Criticality Risk Analysis

In order to control criticality resulting from the fissile material present in the HLLW criticality calculations have to be made. They will refer to the isotopic composition of uranium and plutonium, their concentration and total amount within the solutions and materials handled.

Criticality control is performed by either limitation of the size of the batches processed using mass and/or concentration control; the use of a neutron poison, such as boron, gadolinium or cadmium; or a safe geometry of used vessels and piping. Criticality control can also be achieved by a mixture of the above mentioned techniques.

The basic glass used to vitrify reprocessing waste in Pamela has a comparatively high boron content resulting in a boron content of 4.1 wt% in the LEWC glass product and 7.8 wt% in the HEWC glass product (see table 2).

The feed chain of the Pamela plant is equipped with vertical cylindrical vessels with a capacity of 3.4 and 0.15 m^3 respectively. The input vessel of 3.4 m^3 is provided with a pulsation system for homogenization. Liquids fed to the melter are transferred by air lifts between the different vessels.

If the mixing of Pu and HLLW solutions must be carried out in the Pamela plant, additional equipment is required. The use of borated austenitic stainless steels, enriched with the ^{10}B isotope, is preferred for the construction of new equipment. Precipitation of the plutonium must be excluded, the solution must always remain acidic.

Additional criticality control measures required to vitrify HLLW enriched with plutonium in the Pamela plant refer to the use of increased amounts of boron as a neutron poison and the limitation of the plutonium concentration below the critical concentration of 7.3 g/l ^{239}Pu.

addition of boron solution in the HLLW up to a calculated safe concentration.

The use of a soluble poison to obtain a homogeneous poisoning effect is the most suitable for the Pamela plant. In fact, boron is already present in large amounts in the glass product. On the other hand, the limitations of the existing installation and equipment can thus be taken into account.

TABLE 4. U, Pu Concentration in the HLLW and in the Glass Products

	LEWC		HEWC	
	Feed	Glass product	Feed	Glass product
Uranium	0.93 g/l	0.57 g/kg	0.16 g/l	0.33 g/kg
Plutonium	0.08 g/l	0.05 g/kg	0.01 g/l	0.02 g/kg

U, Pu Mass in HLLW Solutions

	LEWC 47.2 m^3	HEWC 860.1 m^3	Total 907.3 m^3
Uranium	44 kg	135 kg	179 kg
Plutonium	4.0 kg	9.7 kg	13.7 kg

The addition of borosilicate glass raschig rings as neutron absorbers in the concerned vessels is not recommended mainly because the HLLW is a solution containing insolubles. The raschig rings would hinder the proper homogenization of the HLLW in the vessels.

- *limitation of the Pu concentration below the lowest critical concentration of 7.3 g/l or even better: limitation to the safe concentration of 5 g/l.*

It would be preferable if the mixing of the plutonium and HLLW solutions up to an enrichment of for instance 6 g/l should be carried out before being sent to the vitrification plant.

A key factor in an analytical control system, especially when remotely operated, is reliable sampling and analysis. Both must be duplicated for the contents of the process vessels. The plutonium concentration in the two samples should remain within the 20 % margin.

If the discrepancy in mass between physical inventory and accountability balance in a given material balance area, including possible analytical errors, exceeds 1,000 g of Pu, the vessels of the material balance area concerned have to be cleaned according to a well defined rinsing procedure in order to resolve the discrepancy.

The only criticality control in the melter is the poisoning effect of the boron and of the other elements present in the glass product. Based upon a plutonium content of 6 g/l in the HLLW, the concentration of plutonium in the glass can be calculated for different concentrations of calcined residue in the HLLW versus different concentrations of waste oxides in the glass product (see table 5).

The maximum concentration of plutonium in the glass product corresponds with the lowest concentration of calcined residue in the HLLW and the highest concentration of waste oxide in the glass product.

In the given case the glass product would contain 16 g Pu per kilogramme. The Pamela ceramic melter, with a glass bath of approximately 340 l (800 kg) would contain some 12.8 kg Pu. A 150 l container with vitrified waste would contain ca. 400 kg glass product with max. 6.4 kg Pu.

4.2.2 *Glass Product Quality Analysis*

The influence of the higher fissile material concentration on the glass product has to be investigated. The presence of high amounts of fissile material in the glass will significantly increase the alpha irradiation. The alpha radiation from the decay of Pu, U and other alpha emitters converts into helium gas in the glass product.

^{239}Pu finally decays to ^{207}Pb, producing 8 gram-atom He per gram-atom ^{239}Pu. The 6.4 kg Pu present in one container for storage and disposal represents 26.8 gram-atom ^{239}Pu which produces 241.2 gram-atom He or 4,798 l of helium gas over the complete decay chain.

Figure 3 shows the production rate of helium per kg of ^{239}Pu in function of time. The first 10,000 years the helium production rate is the highest. However it is limited to 2.7 ml per kg ^{239} Pu. This corresponds to 17 ml per year and per storage container with vitrified waste. The risk of cracking or change in density of the glass product will depend upon the diffusion rate of helium through the glass product and

TABLE 5. Pu Concentration (g/kg) in the Glass Product with 6 g/l Pu in the HLLW

g/l calcined residue in HLLW	Waste oxide content in the glass product in g/kg			
	80	100	140	160
60	8	10	14	16
80	6	7.5	10.5	12
100	4.8	6	8.4	9.6
120	4	5	7	8

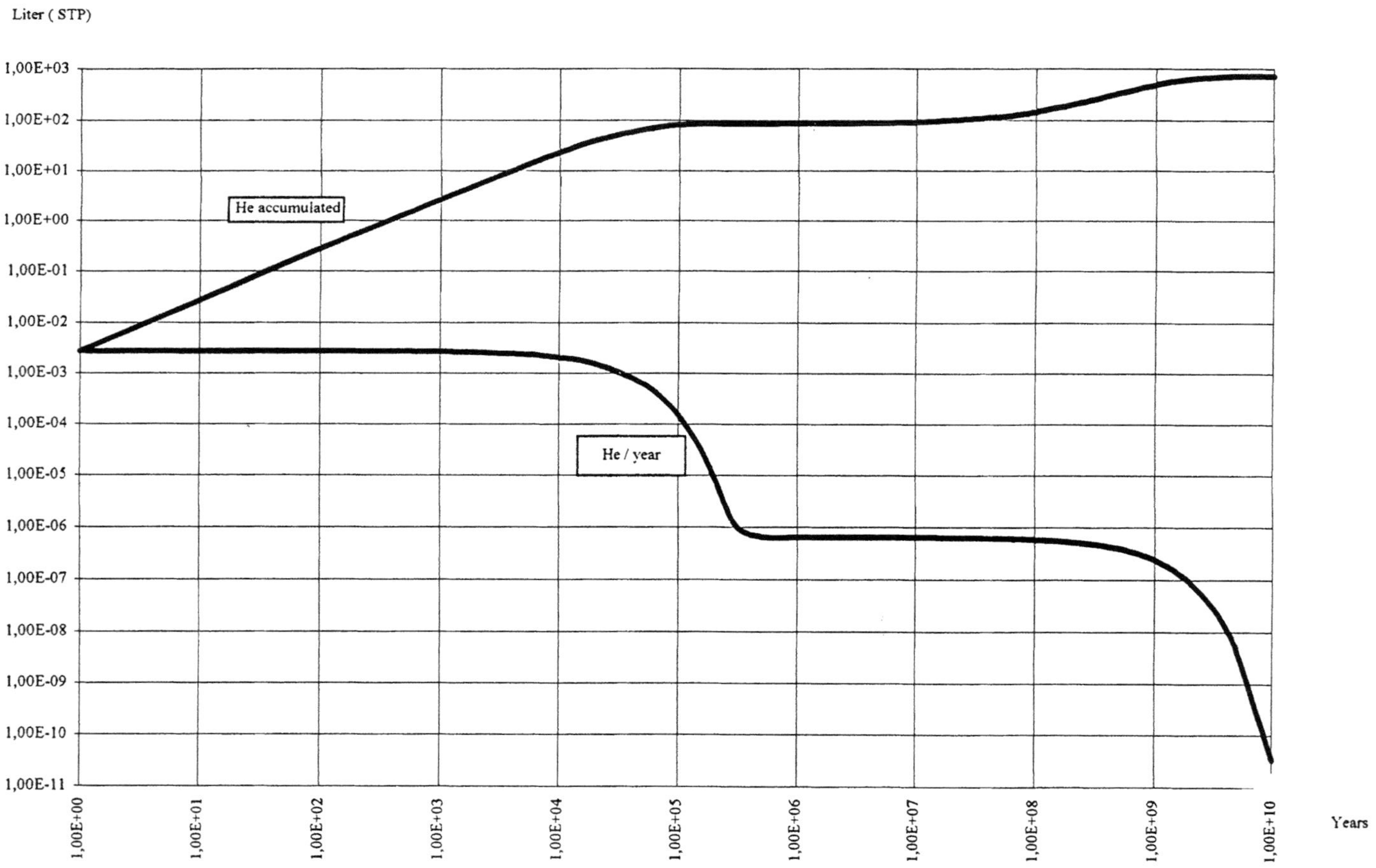

Figure 3. Production of Helium resulting from Decay of 1 kg ^{239}Pu

through the 5 mm stainless steel wall of the container. Pressure build-up in the containers can be avoided by providing the container lid with a carbon filter.

On the other hand, increased radiation damage to the glass as a result of the alpha, beta and gamma radiation has to be investigated too. The input of the higher fissile material content mainly increases the alpha radiation effect which is spread over a long period of time. It significantly contributes to the total accumulated dose over very long periods. During the first thousands of years the alpha radiation effect on the glass product will roughly remain the same as for vitrified reprocessing waste.

Chemically uranium and plutonium behave similarly when present in borosilicate glass. Their distribution in the glass product can be verified by means of scanning electron microscope (SEM) and autoradiography. Experiments demonstrated that concentrations up to 2 wt% U_3O_8 and PuO_2 completely dissolve [2].

4.2.3 Safeguards

Increased analytical control, verification and security will be required when additional fissile material, especially plutonium, will be added to the HLLW. The throughput of fissile material in excess of the normal amounts present in reprocessing waste, will also require a revision of the operating licence.

5. Conclusions

The various programmes carried out so far have demonstrated that the Pamela plant is a multipurpose facility. Any new vitrification programme will require the installation of a new melter, dust scrubber, and container lifting carriage. Besides, the off-gas purification unit will have to be adapted to the new waste stream to be processed. Detailed engineering studies already showed that the facility can be upgraded to vitrify waste concentrates from higher burn-up fuels.

Moreover, process efficiency and throughput will be enhanced. Information and experience from the hot operation of the Pamela plant and from indusrial-scale vitrification tests at the "Institut für Nuklear Entsorgung" at Karlsruhe resulted in a new design and operational mode for the next generation of ceramic melters.

The vitrification of high-level waste solutions enriched with 6 g Pu per litre in a facility of the Pamela type, with a feed rate of 25 l/h would result in a daily throughput of 3.6 kg plutonium, corresponding with an annual throughput of 1 t plutonium (280 days of operation).

Before embarking on such a programme however, the problems related to criticality risks, glass product quality, especially over very long periods of time, and safeguards must be thoroughly investigated.

A criticality control study has to be made for all solicited process units: i.e. feed preparation, melting unit and interim storage facility for conditioned waste.

If the mixing and homogenization of the HLLW with Pu solution has to be carried out in the Pamela plant, additional, critically safe equipment will be required. The use of borated stainless steel as a construction material is recommended.

For each specific vitrification programme a specific glass development programme must be carried out to select the proper glass frit for the waste to be vitrified. In the case of a plutonium-enriched waste stream, additional investigations must determine the impact of increased alpha radiation on the glass product quality in the long term.

Procedures for increased analytical control, verification and security must be determined in concert with the Belgian safety authorities and in compliance with the international rules of safeguarding

6. References

1. Roth, G. (1995) *INE's HLLW Vitrification Technology.* Atomwirtschaft, No. 3, p. 174-7.

2. Walker, C.T., and Riege, U. (1979) *Ceramics in Nuclear Waste Management*, Cincinnati, CONF-790420, p. 198

VITRIFICATION OF WEAPONS PLUTONIUM: A HYPOTHETIC OPTION OR A REAL WAY TO SOLVE THE PROBLEM?

A.S.ALOY, E.B.ANDERSON
RPA " V.G.Khlopin Radium Institute"
St.Petersburg,Russia

1. Introduction

When choosing the technology for the disposition of accumulated weapons plutonium, the basic and essentially the only criterion is the safety in a wider sense of the notion - the safety without borders and over geological time-scales.

Taking into account this fact, may one consider the process of plutonium vitrification together with high-level wastes (HLW) from radiochemical plants and subsequent burial as meeting this criterion? Any impartial analyst would very likely reply in the negative. What are reasons for saying this?

Firstly, the technology of HLW vitrification up to its realization on industrial scale was developed using solutions with a very low content of transuranium elements (TUE). Moreover, the efforts of radiochemists were directed to a more complete separation of plutonium as a valuable strategic product. Any specialist daring to consider it as a waste would look a "white crow".

This results in a high degree of uncertainty in physical, chemical,radiation, thermal and other properties of intermediate and final products of HLW vitrification when adding to them even insignificant amounts of plutonium.

Another paradox resides in the fact that the most suitable form for plutonium storage - PuO_2 is a ceramic material having a very low solubility in water, particularly after annealing at high temperatures. When incorporating plutonium in a glass matrix its leachability increases as compared to oxide, and the probability of its spreading outside the solidified form increases respectively.

Nevertheless, in a publication of U.S. National Academy of Sciences [1] the option of plutonium vitrification with HLW and their subsequent disposal is mentioned as an alternative deserving attention for realization in the near future.

Having discussed some scientific and technological aspects of this problem, we shall attempt to answer the question put in the title of our report.

2. Scientific Aspects

Among scientific aspects affecting the proper choice of an option we shall give prominence to criticality, radiation stability and homogeneity of composition. For a long-term prediction even small errors and inaccuracies in the evaluation of these parameters can significantly change the values corresponding to a longer period of time.

E.R. Merz and C.E. Walter (eds.), Disposal of Weapon Plutonium, 173-177.

2.1. CRITICALITY

Plutonium-239 is a fissile material with a relatively small critical mass, therefore it is necessary to carry out a careful analysis and the investigation of all factors having an influence on feasibility of local rise of its concentration both in the process of vitrification and during the storage and disposal.

Calculations show that the theoretical maximum of plutonium incorporation into a borosilicate glass matrix is 8-10 wt.%, and it depends on isotopic composition [2]. This relatively high content is primarily defined by the presence of boron acting as a neutron absorber which, in particular, fulfils the role of an additive decreasing the radiation hazard [3].

This circumstance leads essentially to a higher degree of incertainty in future, since it is known that boron is the most readily leached from the glass network. Moreover, the influence of time factor on the criticality change should be studied taking into accound the accumulation of uranium-235 and americium - 241 and their possible redistribution in the matrix volume with time.

This circumstance causes also a particular anxiety, since with the rise of plutonium concentration in glass only to 1 wt.% its content in a canister of European standard will reach nearly 4 kg, whereas according to the specification it should not exceed 110 g for plutonium and 90 g for curium [4].

Owing to the migration of daughter uranium-235 and plutonium outside a glass block, zones with a higher content of these elements may appear on burial sites in media saturated with water, which promote the moderation and capture of neutrons. The dimensions of such zones and their geometry defy all calculations and predictions taking into account time-scales and possibilities of changes in geological and climatic conditions [5].

2.2.RADIATION STABILITY

Until quite recently it was thought that the ionizing radiation from radionuclides in vitrified HLW could not have a critical effect on structural, mechanical and chemical properties of a matrix. These conclusions were based on the results of accelerated experiments with the of electronic or other kinds of beams and with incorporation in glass of short-lived TUE in concentrations exceeding treir content in real HLW. But in 1985 it was shown that under action of a rather low-power photon flux of cobalt-60 even at a dose comparable with 10-year storage period the oxygen microbubles were formed in borosilicate glasses randomly distributed over the glass volume [6]. According to the data given in [7], the free oxygen concentration in PNL 76-68 glasses was $3,78x10^{20}$ cm^{-3} at D = 35 MGy. The effect of irreversible destruction of boron-oxygen hole center during glasss irradiation with rather low radiation dose rates was also registered in [8].

It is obvious that the oxygen generation (in addition to helium accumulation) is accompanied by the growth of internal pressure, the change in chemical composition and the augmentation of surface. In actual practice this can cause a dangerous degradation of the glass block isolating property.

Although at present there are no reasons for a radical revision of materials technology priorities in the HLW management, the necessity for a long and comprehensive study of the discovered effects with due to account of the TUE content increase is beyond question.

This necessity is also confirmed by results of the work carried out to study the influence of damages caused by alpha-recoil on the leaching of optical glass containing 4 wt.% of ThO_2 and aged for 17 years [9]. As result of fine experiments the authors using the difference in Th-228/Th-232 yield could find out that alpha-recoil tracks in glass represent local sites with a completely damaged structure. These damages increase significantly the leachability of alpha-decay products from the glass when it is exposed to the action of aqueous solutions.

A distinctive feature of glasses with a higher plutonium-239 content will be a rapid accumulation of daughter americium-241, which leads to a sharp growth of radiation effect on the matrix caused by high-energy alpha-particles (4-6 MeV) and accompanied by the increase in neutron emission by (α,n) reaction.

2.3. HOMOGENEITY

It is known that durind HLW vitrification the phenomena of immiscibility, segregation and formation of insoluble precipitates may be observed. The last phenomenon applies particularly to platinoids as well as to some rare earth and transplutonium elements having a limited solubility in such media [10].

These phenomena have a profound effect not only on conditions of technological processes, but also on properties of the glass block produced.

Components of HLW can be involved in creating the matrix structural network, they can play the part of modifiers or form microinhomogeneous zones enriched with one or another chemical element.

According to the data of [11] obtained using optical and Mossbauer spectroscopy, plutonium (IV) and americium (III) at small concentrations enter into the glass network, whereas uranium (VI) is most likely to be a modifier, which leads to its higher mobility.

Unfortunately, fundamental investigations in the field of TUE spectroscopy in borosilicate glasses are practically lacking,and as yet this does not allows their chemical state to be determined depending on waste composition, TUE concentration, redox potential and so on.

Such investigations would eliminate the discrepances in values of TUE real solubility limits in glasses determined by different investigations.

3. Technology Aspects

Only vitrification technology of HLW has been developed to the industrial stage stage after long time and expensive RAD. The detail review of IAEA [12] observed all current vitrification processes.

In general, two continuons processes have been developed to meet the requirements of today's reprocessing plants (Russia, France,UK) or to dispose of a very large volume of HLW accumulated over the years (USA).

These are:

- the rotary calciner/metallic melter (France,UK)
- the liquid or slurry fed ceramic melter (Belgium,Japan,Russia,USA)

In the described units the low plutonium concentration no present the criticality hazard. At the high plutonium feed concentration this typical vitrification operation involving a pool of melt for homogenization is doubtful to be acceptable.

Firstly, the mass of plutonium in the melt would exceed criticality limits and, secondly, the molten pool is more likely to give separation of PuO_2-rich phases.

The metallic melter has an advantage over the ceramic one because of possibility to mix the glass melt before the pouring into canister.But the capacity of such unit is a significally lower than ceramic melter has.

It will be need to constract a numerous vitrification plants like as UP3 at La Hague to vitrify all 100 t of weapons plutonium during 15-20 years.

Ceramic melters have a higher capacity (DWPF in Savannah River, for example, has a nominal capacity approx 100 kg/h). The residence time of glass melt in the melter is about 60 hours without any agitation.

Probably, nobody could answer now at what level of plutonium in the melt and under what conditionss the formation of precipitates with plutonium will occur in the pool.

4. Conclusion

The world-wide experience in the vitrification of HLW containing mainly fission products relies on more than 30-year research and development work according to initial requirements. Therefore, increasing the content of plutonium in HLW by several fold and perhaps by an order of magnitude necissitates the performance in accordance with recomendation of IAEA [13] of new investigations during many years and requires high expenses.

The lack of reliable safety guarantees for such a process makes it inexpedient to convert into wastes the product having a large energy potential.

The IAEA has expressed [14] its opinion that the safety of excess plutonium storages is mainly depend on safequarding control, where the various surveillance equipment can provide a very high level of assurance against proliferation and diversion.

As it is known, the conception of nuclear power development in Russia is based on the postulate of closed fuel cycle. This implies the use of accumulated weapons plutonium both in fast reactors, and in PWR of new generation [15], and one cannot but agree with this conception considering the foregoing.

References

1. *National Academy of Sciences Management and Dispositions of Excess Weappons Plutonium.* (1994) National Academy Press, Washington, DC.
2. *The treatment and packaging of waste plutonium and waste actinides for disposal.* (1988) Report DOE/RW/88.089.
3. Rimskii-Korsakov A.A. Aloy A.S, Dushin D.N.(1993) Evalution of neutron emission from vitrified HLW compositions, *Proc. of the 4-th Annual Intern. Conf.*, Las-Vegas, Nevada,USA, April 26-30, **2**, 1388-1391.
4. Yvinec J.M., Delahaga P., Baillif L.(1993) Qualication methods and quality control of vitrified HLW in France, *Proc.of the Intern. Conf. of Nuclear Waste Management and Environmental Remediation*, Praque, Chech. Rep., September 5-11, **1**, 387-396.

5. *Geochemistry of long lived transuranic actinides and fission products.* (1992) IAEA - TechDoc-637, Vienna.
6. DeNatale J., Howitt D.(1985) The gamma-irradiation of nuclear waste glasses, *Radiat. Effects.* **91**, 89-96.
7. DeNatale J., Howitt D.(1986) Radiation damage in silicate glass. *Radiat. Effects.* **98**, 229-236.
8. Nechaev A.F., Aloy A.S, Suvorov O.A.,Faddeev I.S. (1993) EPR investigation of the gamma-irradiated borosilicate glass with cesium and strontium oxides, *Atomnaya Energia,* **75**, 295-299 (in Russia).
9. Eyal Y., Ewing R.C. Impact of alpha-recoil damage on dissolution of thoriated glass, In [4], 191-196.
10. Lutze W. (1988) Silicate glasses, in W. Lutze and R.C. Ewing (eds.) *Radiactive waste forms for the future,* North-Holland Physics Publishing, Amsterdam. pp. 29-48.
11. Karraker D.G.(1982) Actinide Valences in borosilicate glass, *J.Amer. Ceramic Society,* 65, **1**, 53-55.
12. *Design and Operation of High-Level Waste Vitrification and Strorage Facilities.* (1992) TRS 339, IAEA, Vienna.
13. *Quality assurance requirements and methods for high level waste package acceptability.*(1992) IAEA TECDOC-680, Vienna.
14. International Nuclear Fuel Cycle Evaluation *"Reprocessing, Plutonium Handling, Recycle"* (1980) report INFCE Working Group 4, IAEA, Vienna.
15. Egorov N.N., Murogov V.M., Kagramanyan V.S. et.al. (1994) Management with Plutonium in Russia, *Report presented at the NATO Advanced Research Workshop*,Obninsk,16-19 October.

VITRIFICATION OF PLUTONIUM FOR DISPOSAL

J.A.C. MARPLES

AEA Technology,
Harwell Laboratory
Oxon, OX11 0RA, UK

Abstract.

The possibility of immobilising plutonium from dismantled nuclear weapons by incorporation in glass is considered. It is concluded that:

(a) Immobilisation of Pu by vitrification would be a feasible process.

(b) Co-processing the Pu with High Level Waste (HLW) would make recovery of the Pu extremely difficult but would mean that vitrification had to be carried out at a reprocessing plant.

(c) The best way of incorporating the plutonium in the glass is probably by dissolving the metal in nitric acid to which HF and, possibly, hydrazine has been added and feeding the acid solution to the vitrification plant. Electrolytic dissolution is a promising alternative and may have the advantage that it may also dissolve any coating on the plutonium.

(d) Borosilicate glass, which is used in all existing and planned vitrification plants in the West, has a limited solubility for PuO_2 which would probably restrict the PuO_2 content to 5 wt.%. In the former Soviet Union, phosphate glasses are used for HLW vitrification and in these the solubility of PuO_2 is greater.

(e) Because of the high neutron absorption cross-section of boron, borosilicate glass would be preferred to phosphate glass for criticality control.

(f) As an additional safeguard, it might be preferable to incorporate gadolinium oxide or a mixture of rare-earth oxides in the glass along with the plutonium oxide.

(h) After disposal, the leach-rate of plutonium from the glass will be very low, as long as the pH of the ground-water contacting the glass is above about 9 i.e.the water is alkaline. In that event the release rate will be dependent on the solubility of Pu which is extremely low at high pH.

E.R. Merz and C.E. Walter (eds.), Disposal of Weapon Plutonium, 179-195.

(i) The heat emission from the Pu-containing glass is sufficiently low that there is no reason why it should not be disposed of immediately.

(j) If the glass were cast in the same size containers as are currently used for vitrified HLW, which hold between 0.4 and 1.4 te of glass, recovery of the plutonium would require very large scale, contained facilities. If the Pu were vitrified along with HLW, the facilities would need to be shielded. After the glass had been dissolved in acid, the silica content of the solution would interfere with either a solvent extraction or an ion-exchange process and an additional flocculation step would be required to remove it.

1. Introduction.

The reductions in the nuclear weapons programmes the USA and in the former Soviet Union has meant that there is a current surplus of weapons grade plutonium. One strategy to deal with this is to store it pending its use in nuclear reactors. This is a very attractive option because 1 kg of plutonium contains as much energy as 2000 tonnes of coal and it seems likely that mankind will need this energy sooner or later. However, the amount of Pu involved, a few hundred tonnes, is small compared to the world's coal reserves and also small compared to the amounts contained in civil reactor fuel.

The drawback to this option is that the plutonium will have to be stored for several decades. At present some plutonium is being used as mixed uranium-plutonium oxide, "MOX" fuel in thermal reactors, but the capacity to manufacture this is small at present, only equivalent to a few tens of tonnes of Pu per year. Furthermore, it is more efficient to use the plutonium, again probably as mixed U,Pu oxide, in fast reactors but only a few of these have been commissioned. When plutonium is used as MOX fuel in thermal reactors, the build up of higher actinides is greater because the ratio of neutron capture cross sections to fission cross sections is greater for thermal neutrons than for fast. This has drawbacks in High-Level Waste (HLW) management, leading to greater radiation damage to the solidified waste.

Storing weapons-grade plutonium for several decades means that, in principle, it is still available for diversion by governments, other organisations or individuals. One way round this problem is to incorporate the plutonium in some material from which it is difficult to recover it. This material could then be stored with less safeguards or disposed of, if it were decided that mankind could reasonably forego the energy content of the plutonium.

One option to render the plutonium unavailable is to incorporate it into a glass, either on its own or with HLW. If the latter is done, recovering the plutonium within the next few centuries would need heavily shielded facilities. When the Sr-90 and Cs-137 have decayed away (both have half-lives of about 30 years), less shielding would be required. However, even if the plutonium were incorporated into glass on its own, a large-scale glove-boxed facility would be required to dissolve the glass and recover the plutonium for re-use.

From the processing point of view, there would be several advantages in adding the plutonium to the glass independently of HLW:

(a) The plutonium would not have to be transported to the reprocessing plant where the HLW is stored and the vitrification plant could be built at the weapon dismantling plant.

(b) The vitrification plant would not have to be fully shielded.

(c) It would be possible to optimise the composition of the glass for plutonium incorporation, without it needing also to accommodate the HLW.

Two distinct types of HLW vitrification equipment are used world-wide. In France and Britain, the AVM (Atelier de Vitrification à Marcoule) system is used, developed by the French [1 - 3]. In this, the HLW, which is in solution/suspension in 6M nitric acid, is mixed with some lithium nitrate and fed into a rotary calciner in which the liquid is evaporated and most of the nitrates calcined to oxides. These then fall from the end of the calciner into an induction heated metallic melter, operating at ca 1150°C, to which small pieces ("frit") of pre-melted glass are also added. When the melter is full, a freeze-valve in the bottom is opened and the glass is run out into a steel storage vessel. The size of the latter is not crucial to the system but in the production plants at Cap La Hague in France and at Sellafield the storage vessels are 43 cm in diameter and 1.3 m high, containing ca 400 kg glass.

In Belgium [4] and in Russia [5], the JCM (Joule-heated Ceramic Melter) system is used, and this is also planned to be used in the USA [6,7]. Here the glass is contained in a furnace lined with a ceramic refractory. The glass is heated by passing a current through it from electrodes let into the sides of the furnace or suspended from the roof. Special arrangements are necessary at start up. The melters are usually "liquid-fed" i.e. the liquid waste is slowly fed onto the top of the molten glass where it is calcined in situ and becomes incorporated into the glass. As with the AVM, glass frit is added separately. Glass is cast into storage vessels, either through a freeze-valve in the bottom of the furnace or by tilting the furnace so that the glass flows over a weir or by sucking it out through a vacuum-operated siphon.

2. Initial Processing of the Plutonium.

Most of the plutonium will initially be in the form of δ-phase metal, stabilised by a small percentage of, probably, gallium or, possibly, aluminium. The Pu metal will probably also be coated with a thin layer of inert metal. For co-vitrification with HLW, it would be desirable to get the Pu into solution in nitric acid. Plutonium metal does not however dissolve in nitric acid, unless HF is added. Even then the dissolution rate is slow, ca 12 $mg.cm^{-2}.min^{-1}$ in 3M HNO_3 to which 0.13 M HF had been added [8]. This is equivalent to ca 0.5 $mm.h^{-1}$. The addition of hydrazine increases this dissolution rate by about a factor of six [9]. Another technique which is worth investigating is electrolytic dissolution [10]. This could have the advantage that it might also dissolve any coating that had been applied to the plutonium. The dissolved coating would probably also be incorporated in the glass.

Plutonium metal does dissolve readily in hydrochloric acid, but using this would add complications to the vitrification plant due to the presence of chloride ions.

Another possible route would be to calcine the plutonium metal by heating it to ca 850°C in air to convert it to the oxide before adding this directly to the glass or first dissolving the oxide in nitric acid plus HF. Adding the oxide directly to the glass might lead to problems with the plutonium oxide settling out because of the large difference in density between them - PuO_2: 11.0; borosilicate glass: ca 2.8 $g.cm^{-3}$: phosphate glass: ca 3.3 $g.cm^{-3}$. If the PuO_2 is added as a nitrate solution, this will lead to greater homogeneity. In AVM-type plants, some of the lithium used in the glass formulation is added as nitrate and co-calcined with the waste. If this practice were followed, the Pu oxide would be intimately mixed with the lithium oxide before melting into the glass. Also, decomposition of the nitrates is incomplete in the calciner, and a stirring action is produced when the decomposition is completed beneath the surface of the glass melt. A similar phenomenon will occur if a JCM is used.

3. Glass Compositions and Plutonium Solubility.

In the West, the glass compositions that are being used [6], or are planned to be used [7] for HLW vitrification are borosilicates In the former Soviet Union, phosphate glasses are used [5]. Early work in the West (for example [11]) lead to the conclusion that some phosphate glass compositions were prone to devitrification (crystallisation) above about 400°C which increased their leach-rates. This seemed serious at the time, when early vitrification of the HLW, giving high glass centre temperatures, seemed a likely scenario, but might not be considered such a drawback nowadays. However, phosphate glasses were also considered to be more corrosive to, particularly, metallic plant components. On the other hand, phosphate glasses are more tolerant of sulphate in the HLW than are borosilicate glasses. In the present context, borosilicate glasses will have a distinct advantage in criticality control due to the high neutron capture cross-section of boron.

3.1 BOROSILICATE GLASSES.

Four of the borosilicate glass compositions that are being or are planned to be used in HLW vitrification plants are given in Table 1. The choice of composition is partly a compromise between low leachability, high HLW solubility in the glass and a reasonably low formation temperature. Typically the glasses contain about 25 wt.% calcined HLW oxides, comprising ca 10 wt.% total fission product oxides with 15 wt.% of oxides of inactive elements. These latter are either added originally to the fuel or are added during processing, for example to condition the Pu valency, or are corrosion products of the fuel element can.

Table 1

Representative Borosilicate Glass Compositions (Weight %)

Oxide	MW [2]	M5 (189) [14]	M22 (209) [14]	R7T7 [1]	Pamela (SM513-LW11) (a)	DWPF [7] Min	DWPF [7] Max
Glass formers							
SiO_2	46.3	41.5	51.0	45.5	52.1	44.6	54.4
B_2O_3	16.2	21.9	11.0	14.0	13.1	9.9	10.2
Na_2O	8.3	7.7	8.3	9.9	5.8	8.2	12.1
Li_2O	4.2	3.7	4.0	2.0	4.2	3.1	4.6
K_2O						2.1	4.6
ZnO				2.5			
CaO				4.0	4.5	0.8	1.2
ZrO_2				1.0			
Al_2O_3				4.9	2.7	2.9	7.1
MgO					2.0		
TiO_2					4.5	0.6	1.0
Waste oxides							
FPO_x	10.6	9.6	9.8	11.3	2.5	0.2	1.1
$ActO_x$	0.2	0.2	0.2	0.8	0.1	0.5	4.0
Al_2O_3	5.2	5.0	5.0		0.9		
MgO	5.6	6.2	6.2			1.3	1.5
Na_2O					3.3		
Fe_2O_3	2.7	2.7	2.7	2.9	1.7	7.4	12.7
MnO						1.6	3.1
Misc	0.7	1.7	1.8	1.2	2.6		
Total waste	25.0	25.4	25.8	16.2	11.1	11.0	22.4
TOTAL	100	100	100	100	100	N/A	N/A

(a) Schiewer, E. (1986). The borosilicate glass for 'Pamela', Radioactive Waste Management and the Nuclear Fuel Cycle 7 121- 138.

Little work has been done on the solubility of PuO_2 in the glasses since the concentration of Pu in the HLW is normally very low (< 0.1 wt.%). Marples et al [12 - 14] doped various glass compositions with $^{238}PuO_2$ in order to test their radiation stability. In the original work, 5 wt.% PuO_2 was added but it was concluded that some of the oxide had crystallised out of the glass. In subsequent experiments [13,14], the additions were restricted to 2.5 wt.% PuO_2 but even then, autoradiography showed that some crystallisation of the PuO_2 had occurred in some of the glass compositions. The PuO_2 was present as small cubes, suggesting that it had originally dissolved and then recrystallised. The International Standard on methods of assessing the radiation stability of vitrified HLW recommends restricting $^{238}PuO_2$ additions to 3 wt.% [15].

Cerium is also primarily tetravalent, and the ion, Ce^{4+} behaves in a similar manner and is similar in size to Pu^{4+}: the lattice parameters of the oxides are: CeO_2: 5.411Å; PuO_2: 5.396Å. Dalton et al [16] conducted a series of experiments on the crystallisation of M5 and M22 glasses which contained a full spectrum of inactive isotopes of the fission products, including 1 wt.% CeO_2 in the glass. They delineated sets of T-T-T (Time, Temperature, amount of Transformation) curves, which showed that ceria dissolved at over ca 1080°C but started to crystallise in less than 100 seconds at 900°C. Thus it will always be precipitated in production-sized glass blocks (ca 500 kg) which of necessity cool slowly. PuO_2 with its similar valency, lattice parameter and somewhat greater stability is expected to behave in the same way. For immobilisation purposes, it may not matter if the PuO_2 crystallises as long as it does not settle out. This latter is improbable as the cubes were small (ca 5 µm) and the glass is fairly viscous below the crystallisation temperature. A deleterious effect might occur if the expansion-coefficient mismatch caused micro-cracking round the crystals and this increased the effective surface area and hence the amount of glass leached.

3.2 PHOSPHATE GLASSES.

The composition of the phosphate glass that is used in Russia is given in Table 2. Little is known about the solubility of PuO_2 in this glass, again because this is not normally a requirement. Marples et al [14] as part of their study of radiation stability, doped a phosphate glass with 5.0 wt..% $^{238}PuO_2$ and found that there was no evidence of crystallisation. This suggests that the solubility of plutonium oxide is higher in phosphate glasses than in the compositions of borosilicate glass normally used for HLW vitrification.

3.3 GLASS CENTRE TEMPERATURES.

For a glass block 43 cm in diameter, containing 5 wt.% weapons-grade Pu, the centre temperature would be 35°C above ambient. This compares to 250°C above ambient for a glass containing 25 wt.% PWR HLW, cooled for 10 years. The latter would drop to 80°C above ambient after 100 years but the temperature of the Pu containing glass would not be much affected for many thousands of years.

Table 2

Representative Phosphate Glass Compositions (Weight %)

Oxide	Pamela 78/7	Russian	Lead-iron Phosphate
Reference	[14]	[5]	(a)
Glass constituents			
P_2O_5	48.5	51	38
Fe_2O_3	15.2		10
Al_2O_3	4.9	18	
Na_2O		23	
PbO			45
Waste			
FPO_x	28.1		
Fe_2O_3	0.5		
Cr_2O_3	0.2		
NiO	0.2		
U_3O_8	2.4		
Total waste	31.4	8	7
Overall total	100	100	100

(a) Sales, B.C. and Boatner, L.A. (1988) Lead-iron phosphate glass in "Radioactive Waste Forms for the Future", W Lutze and RC Ewing, Editors, North Holland, 193 - 231.

3.4 AMOUNTS OF GLASS TO BE PRODUCED AND THE TIME TAKEN FOR VITRIFICATION.

An arbitrary amount of 100 te of Pu, incorporated in glass at the 5 wt.% level, would require 2 000 te of glass. The two-process-line Windscale Vitrification Plant (WVP) at Sellafield in the UK produces three containers per day, a total of 1.4 te of glass. This plant would therefore take about 4 years to vitrify the plutonium, assuming continuous operation and no down-time. With the same assumptions, the Defence Waste Processing Facility (DWPF) at Savannah River, which is designed for a throughput of ca 100 $kg.hr^{-1}$, would take just over 2 years for the task.

4. Radiation Damage.

Radiation damage, expressed as atomic displacements per unit volume, in vitrified HLW is dominated by the α-decays of the incorporated actinides [17,18]. The α-particle typically displaces ca 150 atoms whilst the recoiling actinide nucleus displaces ca 1500. A typical β-decay will only displace < 1 atom so that, although β-decays are much more numerous shortly after vitrification, they do much less damage to the glass.

The damage to the glass has been simulated on an accelerated time-scale in various ways [17,18]. The most realistic is to incorporate a quantity of a short half-life α-emitter into the glass so that the glass experiences in a few years the number of α-decays that the real waste will receive in centuries or millennia. The best isotopes for this purpose are Pu-238 (half-life: 87.8 y) and Cm-244 (half-life: 18.1 y). The latter is probably a better choice because it appears to be more soluble in the glass but it also emits spontaneous fission neutrons, which makes handling more hazardous, and it is also more expensive. More work has therefore been done with Pu-238.

The effect of radiation on glass has been reviewed by Burns et al [17], by Weber and Roberts [18] and by Marples [19]. In general, glasses are little affected by internal α-decays. Their leach-rates are sometimes slightly affected. Most increase, after a dose of about 2×10^{18} α-decays.g^{-1}, by a factor which is usually less than two, but can be as high as three. However, the leach-rate of MW glass appeared to decrease (improve) with radiation dose [20], reaching half its initial value after a dose of 2×10^{18} α-decays.g^{-1}.

The change in density of the doped glasses followed closely an equation of the form:

$$\Delta\rho = \Delta\rho_{\infty} \;(1 - e^{-\alpha D})$$

where: $\Delta\rho_{\infty}$ is a fitted constant, equal to the saturation density change at large radiation doses. This had a maximum value of 1.2% for all the glasses studied and was generally ca 0.5%.
D is the radiation dose (α-decays.ml^{-1})
α is a fitted constant, equal to the volume (ml) damaged by each α-decay.

Most HLW glasses expand under irradiation but some contract i.e. their densities increase. Marples, in a programme sponsored by the European Commission [21], tested six glass compositions suggested by members of the European Community. Four of these expanded and two contacted, including the one phosphate glass, but the maximum predicted density change at saturation was just over 1%. Typically, the density changes reach saturation at a dose of about 5×10^{18} α-decays.ml^{-1} i.e. at about 2×10^{18} α-decays.g^{-1}.

On the grounds that glass is only a metastable phase, it has been suggested that one effect of radiation might be to devitrify the glass. No sign of this has ever been seen: although glasses containing HLW-simulant are black because of the iron content, slight devitrification gives them a "milky" appearance and this is readily recognisable. In addition, devitrification seems unlikely anyway since the effect of irradiation on most crystalline substances is to render them metamict i.e. to destroy their long-range order.

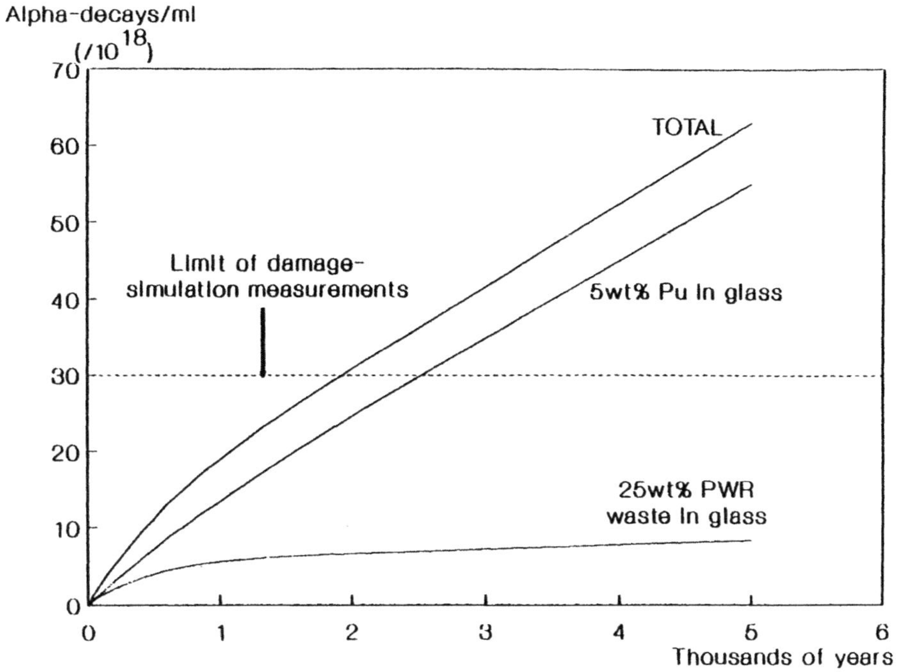

Figure 1. Alpha-decays per ml of glass.

Some energy is stored in the glass due to the displacement energy of the defects but the amounts are small and the energy is released over a wide temperature range so that a temperature excursion is not possible.

After it has come to rest, each α-particle becomes a helium atom but the resulting gas remains mainly in solution in the glass at ambient temperatures and does not form bubbles or pressurise the container.

The largest dose at which measurements have been made is about 3×10^{19} α-decays.ml^{-1} (1.2×10^{19} α-decays.g^{-1}). Calculating the time for vitrified HLW to accumulate this dose requires assumptions to be made as to the percentage of waste in the glass, the amount of plutonium and of higher actinides left in the waste and the length of time before reprocessing and the delay before vitrification. However, for PWR waste, reprocessed after 5 years cooling and vitrified immediately, it is equivalent to ca 500 000 years.

A glass containing 5 wt.% weapons grade PuO_2 would reach 3×10^{19} α-decays.ml^{-1} in about 2000 years (see Figure 1). However, some samples of Pu-238 doped glass have received higher doses than this : a small piece of glass containing 5 wt.% $^{238}PuO_2$ has been stored for 20 years, accumulating a dose of 4×10^{19} α-decays.ml^{-1}, and shows no obvious ill effects.

No radiation stability work has been done on phosphate glasses apart from the CEC-sponsored programme by Marples [21] mentioned above. This composition was taken to a dose of ca 10^{19} α-decays.ml^{-1} (3.2 x 10^{18} α-decays.g^{-1}). By this dose, the glass had contracted by ca 0.45 % i.e. the density had increased and the volume change appeared to have almost reached saturation. The Soxhlet leach-rate at 100°C had remained almost constant at 5 x 10^{-5} g.cm^{-2}.d^{-1}.

In addition however, it may be significant that, as Boatner and Sales [22] have pointed out, monazites, mixed lanthanide orthophosphates, $LnPO_4$, which often contain a proportion of U and Th, have sustained extremely high radiation doses, over geological time-scales, while remaining crystalline and apparently undamaged. Eyal and Olander [23] studied a crystalline monazite specimen containing 0.2 % U-238 and 10.4 % Th-232 that was about 1400 million years old. It had received an irradiation dose resulting in ca 12 displacements per atom (dpa), approximately equivalent to 1.2 x 10^{20} α-decays.g^{-1}. Leach rates for damaged material remained low (ca 2 x 10^{-7} g.cm^{-2}.d^{-1} at 90°C [22]).

5. Leaching of Plutonium from Glass.

Although vitrified HLW contains little plutonium, it is recognised that the plutonium is potentially one of the most hazardous components of the glass, particularly once the Sr-90 and Cs-137 have decayed [24]. Hall et al [25] studied the leaching of Pu from the "MW" glass composition that BNFL are using to vitrify the HLW at Sellafield. The leach tests were in distilled water at 25 and at 75°C with a specimen surface area to leachant volume ratio of 10 m^{-1}. The leachates were acidified on completion of the tests and removal of the specimen so that all Pu released from the specimen was included in the analysis. The normalised Pu leach-rates were similar at 25 and at 70°C at ca 0.3 x 10^{-6} g.cm^{-2}.d^{-1}. A similar result was obtained for the French R7T7 glass [24]. This was about half the mass loss leach-rates measured in the same experiments. The Pu leach-rates measured in a Soxhlet experiment at 100°C were only slightly greater (0.5 x 10^{-6} g.cm^{-2}.d^{-1}) although the mass-loss leach-rates were much greater (2 x 10^{-4} g.cm^{-2}.d^{-1}), showing that much of the Pu was retained in the "gel-layer" that forms on the surface of the glass. In experiments in static leachants at 50 and at 90°C, the Pu leach-rates were much smaller, ca 0.04 x 10^{-6} g.cm^{-2}.d^{-1}.

6. Disposal of the Glass.

The general strategy for the disposal of the vitrified plutonium would be similar to that for vitrified HLW - the "multi-barrier" concept. The glass (first barrier) would be cast into a steel cylinder (second barrier) which would be encased in a metal over-pack (third barrier) and the whole emplaced in a tunnel or borehole several hundred metres underground. The excavation would be "backfilled" with a material (fourth barrier) whose function would be to retard water flows and, possibly, to condition the chemistry of the water contacting the vitrified waste. The surrounding geology is the outer barrier. Among disposal geologies that have been considered are salt, clay, granite and tuff.

For disposal of vitrified HLW, the function of the overpack would be to keep the water away from the glass until most of the fission product activity has decayed away. The latter is dominated by Sr-90 (half-life: 28 years) and Cs-137 (30 years) and a period of 300 years is generally considered adequate for this isolation. Two types of overpack are considered, either a thick (ca 150 mm) steel or cast iron ("corrosion allowance") or a thinner (ca 10 mm) layer of titanium ("corrosion resistant") [26]. An additional function of the corrosion allowance overpack is to provide reducing conditions in the neighbourhood of the glass (the "near-field") which will slightly reduce the solubility of the Pu and also reduce its diffusion coefficient [27]. However, the lifetime of a "corrosion allowance" overpack, in particular, will not be long enough for all of the plutonium to decay. It may therefore be considered desirable to use a thick metallic copper overpack as proposed by the Swedish KBS project for the disposal of spent fuel [28]. Their calculations suggested that, when buried 500 m deep in granite, 10 mm thick copper would last 10^5 years under "unfavourable" conditions and 10^6 years in favourable conditions [28].

It is generally assumed in the UK that vitrified HLW will be kept in a surface or near-surface retrievable store until it can be disposed of such that the temperature at the backfill/rock interface does not exceed 100°C. With an ambient temperature of 35°C at the normally assumed repository depth, this means that the thermal output from each cylinder does not exceed ca 1 kW [29]. If the Pu is vitrified independently of the HLW, then, for 5 wt.% Pu in the glass, the thermal output from the COGEMA / BNFL size glass cylinders would be only ~120 watts. Thus, to satisfy the above-mentioned temperature criterion, there is nothing to prevent glass where Pu is the only active material being disposed of immediately after vitrification.

In a salt dome, the boreholes or tunnels would be backfilled with salt. The safety arguments rely on the fact that the salt dome has remained undisturbed for millions of years in the past and is therefore likely to remain so in the future. Even if water does break through into the salt dome, it will take many millennia to dissolve it. The behaviour of glasses containing Pu, after disposal in salt, has been discussed by Grambow [30]. He showed that one thousand times higher concentrations of Pu in the brine leachant resulted if iron was present so that it would be sensible to dispense with the overpack in that case or possibly to use titanium. In the absence of iron and at pH 4, the solution concentrations of Pu in the vicinity of the glass were 2 x 10^5 times the ingestion limits.

In a repository in clay, the tunnel or borehole would probably be backfilled with the clay itself. Some work has been done on the behaviour of HLW glasses containing Pu in the "Boom" clay that might be used for a repository in Belgium [31]. The release of Pu into the clay pore-water was very slow, giving concentrations of Pu in close proximity to the glass of 10^{-9} molar.

In granite, the tunnels or boreholes in which the overpacked glass cylinders would be emplaced, would probably be backfilled with bentonite although cement might be considered. Grauer [27] has shown that if the bentonite were compacted into blocks before dry emplacement, the swelling produced when water seeps into the repository would further compact the bentonite and the diffusion coefficient, D_a, for Pu would be $\leq 10^{-14}$ $m^2.s^{-1}$ [32]. For a 1m thick layer of backfill, this means the breakthrough time for Pu would be in excess of 300 000 years, well over ten half-lives.

Furthermore, Boult et al [20] have shown that the concentration of Pu in pseudo-equilibrium with the glass in the presence of bentonite was ca 2 x 10^{-13} M after filtering the leachate through a 25 000 Molecular Weight Cut-Off (MWCO) filter which has a pore size of ca 2 nm - larger particles would not penetrate through the pores in the bentonite. For comparison, a concentration of 25 x 10^{-13} M would result in a committed effective dose equivalent of 1 mSv to someone drinking 2 litres of water per day: a typical natural background radiation dose is 2 mSv per year.

If cement (e.g. a 1:3 mix of Ordinary Portland Cement (OPC) and Blast Furnace Slag (BFS)) was used as a backfill in these equilibrium concentration experiments, the results were even lower than with bentonite [20]. This was probably due to the higher pH produced by the OPC decreasing the solubility of the Pu [33,34]. Surprisingly, this higher pH did not increase the overall rate of attack on the glass specimens [20].

In the USA, the present plan is to build a repository for HLW and spent fuel in volcanic "tuff" at Yucca Mountain in Nevada. The water from a well there ("J13") [35] has a pH of ~7, giving Pu solubilities of 2 x 10^{-6} to ~10^{-5} M depending on the speciation. With the pH adjusted to 8.5, which is still rather below the value to which borosilicate glass would buffer the ground-water, the solubility was 2 x 10^{-7} M at 25°C, decreasing to ca 10^{-8} M at 90°C [36]. Leach-rates for proposed DWPF glass compositions at high S/V ratios were < 10^{-7} $g.cm^{-2}.d^{-1}$ for Na, Li and B [37] and would be lower for Pu.

In fact, vitrified waste emplaced at Yucca Mountain, would probably be above the water table. However, the glass will "weather" in the damp air in the repository and may be exposed to intermittent flooding events. The weathered layer on the glass surface has been shown to result in Pu releases in colloidal form when flooding occurs [38].

7. Criticality Considerations.

The problem of criticality will have to be considered at several points in the processing, vitrification and disposal cycle. If a dissolution step is used, probably the best practice is to use "ever-safe" vessels. This is a well-tried and proven technology and should present no problems. i.e. if the process vessels are small enough or narrow enough, criticality is impossible whatever the concentration of the solutions.

During dry processing, e.g. oxidation of the metal, the criticality problem is not so acute. Good plutonium accountancy practices will be necessary for criticality control and also to prevent diversion of the plutonium. The possibility that the plant may be flooded due to water leaking from cooling systems must be considered since the minimum critical mass is much smaller in moderated systems. The same precautions against accidental criticality that were enforced during manufacture of the plutonium will be applicable during the dry stages of processing.

If a borosilicate glass is used for vitrification, there should not be any problem with criticality during the vitrification stage although care will have to be taken if the cubic plutonium oxide crystals mentioned in Section 3 show a tendency to settle out. If a phosphate glass is used, without co-vitrification with HLW, it would probably be advisable to add Gd oxide or a rare-earth oxide mixture for criticality control.

After disposal, two problems might arise in the very long term. If plutonium were to be leached out of the glass by acidic groundwaters, which then flowed into an alkaline rock stratum, the plutonium might be precipitated and build up to a critical concentration. Alternatively, in alkaline conditions, if most of the glass were to be leached away, leaving the plutonium, again a critical concentration might result. Neither scenario seems very likely, but like all negatives, it is difficult to prove rigorously that they cannot happen. Perhaps the best way of guarding against them is to co-process the plutonium with a Gd oxide or a rare-earth oxide mixture as mentioned above. The protection afforded by this co-processing depends on the Pu and the rare-earths leaching and precipitating at the same rate as the plutonium. Although the rare-earths are largely tri-valent, their properties may be sufficiently like those of Pu particularly under reducing conditions to provide a useful safe-guard.

8. Recovery of Plutonium from Glass.

A BNFL / COGEMA-size cylinder would hold ca 400 kg glass, a DWPF cylinder 1700 kg. With 5 wt.% Pu in the glass, these would contain ca 20 and 85 kg of Pu respectively. To recover this, the cylinder would have to transferred into a "boxed" facility or a shielded cell if the Pu had been co-vitrified with HLW. The simplest way of removing the glass from the steel canister would be to melt it out, requiring a large furnace capable of ca 1100°C. It would be preferable to run the glass out into a tank of water in order to break it up into fragments. The glass would then have to be crushed to fine powder and dissolved in HNO_3 to which a small amount of HF had been added. To recover a significant amount of Pu, a large volume of acid would be required. A possible method to reduce the volume would be to circulate the acid over the glass and then through a solvent extraction plant or, possibly, through an ion-exchange column. However, the presence of large amounts of siliceous material, associated with the Pu, would tend to blind ion exchange columns and interfere with phase disengagement in solvent extraction processes. Flocculation and centrifugation would thus probably be required before Pu extraction in order to remove most of the silica.

Although the extraction process would not be too difficult with small amounts of glass, carrying it out on a large scale would require a very large facility. Co-vitrification of the Pu with HLW would make recovery even more difficult.

9. Research Requirements.

Prior to research commencing, it will be necessary to decide whether the Pu is to be co-processed with HLW as this affects many aspects of the vitrification. If the decision is deferred, parallel research on processing with and without HLW would be necessary.

The following topics would need to be researched:

1. What is the best way to get the Pu into the glass - oxidation and direct incorporation of the oxide or dissolution (in nitric acid ?) and calcination? What is the best way to dissolve the, probably coated, metal components? (It is possible this is already known to workers in weapons establishments.) Is electro-dissolution a useful method?
2. Is a phosphate or a borosilicate glass to be used?
3. What glass composition is to be used? This would vary if the Pu is to be co-processed with HLW or if rare-earth element neutron absorbers are to be added.
4. What is the leach resistance of the glass under likely disposal conditions? If neutron absorbers are added, do they leach out at the same rate as the Pu?
5. Is the glass stable to radiation? Is the leach-rate affected by radiation?
6. What is the best technique for making the glass - JCM or AVM?
7. How difficult would it be to recover the Pu?

10. Acknowledgements.

The author would like to acknowledge helpful comments and suggestions from many colleagues in AEA Technology including: John Berry, Richard Bush, David Harris, Les Hutton, Guy Owen, Peter Roberts, Peter Sutcliffe and Simon Wisbey.

11. Conclusions.

(a) Immobilisation of Pu by vitrification would be a feasible process.

(b) Co-processing the Pu with High Level Waste (HLW) would make recovery of the Pu extremely difficult but would mean that vitrification had to be carried out at a reprocessing plant.

(c) The best way of incorporating the plutonium in the glass is probably by dissolving the metal in nitric acid to which HF and, possibly, hydrazine has been added and feeding the acid solution to the vitrification plant. Electrolytic dissolution is a promising alternative and may have the advantage that it may also dissolve any coating on the plutonium.

(d) Borosilicate glass, which is used in all existing and planned vitrification plants in the West, has a limited solubility for PuO_2 which would probably restrict the PuO_2 content to 5 wt.%. In the former Soviet Union, phosphate glasses are used for HLW vitrification and in these the solubility of PuO_2 is greater.

(e) Because of the high neutron absorption cross-section of boron, borosilicate glass would be preferred to phosphate glass for criticality control.

(f) As an additional safeguard, particularly if it is decided to use phosphate glass, it might be preferable to incorporate gadolinium oxide or a mixture of rare-earth oxides in the glass along with the plutonium oxide.

(h) After disposal, the leach-rate of plutonium from the glass will be very low, as long as the pH of the ground-water contacting the glass is above about 9 i.e.the water is alkaline. In that event the release rate will be dependent on the solubility of Pu which is extremely low at high pH.

(i) The heat emission from the Pu-containing glass is sufficiently low that there is no reason why it should not be disposed of immediately.

(j) If the glass were cast in the same size containers as are currently used for vitrified HLW, which hold between 0.4 and 1.4 te of glass, recovery of the plutonium for re-use, clandestine or otherwise, would require very large scale, contained facilities. If the Pu were vitrified along with HLW, the facilities would need to be shielded. After the glass had been dissolved in acid, the silica content of the solution would interfere with either a solvent extraction or an ion-exchange process and an additional flocculation step would be required to remove it.

12. References.

[1] Sombret, C.G. (1985) The vitrification of high level radioactive wastes in France. Nuclear Energy 24 85 - 98.

[2] Larkin, M.J. (1986) Development of highly active waste conditioning at Sellafield. Nuclear Energy 25 343 - 354

[3] Woodhall, A and Maillet J. (1987) Solidifying Sellafield's high level waste. Nuclear Engineering International, 32, 44 - 47

[4] Wiese, H and Demonie, M. (1992) Operation of the Pamela high-level waste vitrification facility. Nuclear Engineering and Design 137 147 - 151.

[5] Polyakov, A.S. et al. (1994) Experience in operating the EP-500/1R ceramic melter for vitrification of liquid high-level wastes. Atomic Energy 76 181 - 185.

[6] Lutze, W (1988) Silicate Glasses in "Radioactive Waste Forms for the Future", W Lutze and RC Ewing, Editors, North Holland 1 - 152.

[7] Cunnane, J.C. et al. (1994) "High Level Waste Borosilicate Glass: A Compendium of Corrosion Characteristics" USDOE Report. DOE EM 0177, Vol 2, p308.

[8] Miner, F.J. et al. (1969) Dissolution of plutonium in dilute nitric acid. I & E C Product Research and Development 8 402 - 405

[9] Karraker, D.G. (1986) Mechanism of plutonium metal dissolution in HNO_3 - HF - N_2H_4 solution. J Less Common Metals 122 337 - 341.

[10] Fox, R.D. (1971) Electrolytic dissolver for plutonium metal. Atlantic Richfield Hanford Co Report No ARH 2203

[11] Grover, J.R. (1973) Glasses for the fixation of high-level radioactive wastes in "Management of Radioactive Wastes from Fuel Reprocessing, Paris 1972" CONF-721107, OECD-NEA, Paris, 593 - 612.

[12] Hall, A.R. et al (1976) Development and radiation stability of glasses for highly radioactive wastes in "Management of Radioactive Wastes from the Nuclear Fuel Cycle", IAEA Symposium, Vienna, Volume II, 3 - 14.

[13] Malow, G et al. (1978) Testing and evaluation of the properties of various potential materials for immobilising high activity waste. Commission of the European Communities Report, EUR 6213

[14] Marples, J.A.C et al. (1981) Testing and evaluation of the properties of various potential materials for immobilising high activity waste. European Applied Research Reports 3 395 - 484.

[15] International Standards Organisation. (1982) "Standard method for testing the long term alpha radiation stability of solidified high-level radioactive waste forms" ISO 6962.

[16] Dalton, J.T.et al. (1981) The influence of metal oxides on the leach rate and crystallisation behaviour of waste glasses in "International Seminar on Chemistry and Process Engineering for High-Level Liquid Waste Solidification" Editors, R Odoj & E Merz. Jul Conf-42, Vol 1, p 515 - 532.

[17] Burns, W.G. et al. (1982) Effects of radiation on the leach rates of vitrified radioactive waste. J Nuclear Materials 107 245 - 270.

[18] Weber, W.J. and Roberts, F.P. (1983). A review of radiation effects in solid nuclear waste forms. Nuclear Technology 60 178 - 198

[19] Marples, J.A.C. (1988) Dose rate effects in radiation damage to vitrified waste. Nuclear Instruments and Methods in Physics Research B32 480 - 486

[20] Boult, K.A. et al. (1991) Radioactive release from solidified high level waste. Commission of the European Communities Report EUR 13604.

[21] Marples, J.A.C et al (1984) Alpha-radiation stability in Commission of the European Communities Report EUR 9268, Ch Engelmann, Editor. 178 - 185.

[22] Boatner, L.A and Sales, B.C. (1988) Monazite, in "Radioactive Waste Forms for the Future", W Lutze and RC Ewing, Editors, North Holland 495 - 564.

[23] Eyal, Y and Olander, D.R. (1990) Leaching of uranium and thorium from

Monazite: I Initial leaching. Geochimica and Cosmochimica Acta 54 1867 - 1877.

[24] Vernaz, E.Y. and Godon, N. Leaching of actinides from nuclear waste glass: French experience. in Scientific Basis for Nuclear Waste Management XV. CG Sombret, Editor, Materials Research Society (1992), 37 - 48

[25] Hall, A.R.et al. (1987) Evaluation of solidified highly active waste: Leaching of MW glass containing Pu and Am. AERE Harwell, UK, Report AERE R-12580.

[26] Smailos, E et al. (1991) "Corrosion of HLW Containers and Repository Structures" in "Radioactive Waste Management and Disposal", L Cecille, Ed. Elsevier , 331 - 349

[27] Grauer, R (1994). Bentonite as a backfill material in a high-level waste repository. MRS Bulletin, XIX, 43 - 46.

[28] SKBF/KBS (Swedish Nuclear Fuel Supply Company) "Final storage of spent nuclear fuel - KBS-3. Vol iii, Barriers. (1983) p10.1

[29] Hodgkinson, D.P., Lever, D.A. and Rae, J, (1983) Thermal aspects of radioactive waste burial in hard rock. Prog Nucl Energy 11 183 - 218

[30] Grambow, B. (1994) Borosilicate glass: Future research requirements or "What we don't know" MRS Bulletin, XIX, 20 - 23

[31] Marples,J.A.C, Godon, N, Lanza, F and van Iseghem, P. (1991) "Radionuclide Release from High-level Waste Forms in Clay or Granite" in "Radioactive Waste Management and Disposal", L Cecille, Ed. Elsevier , 287 - 301.

[32] Torstenfelt, B. (1986) Migration of the actinides thorium, protactinium, uranium, neptunium, plutonium and americium in clay. Radiochem Acta 39 105 - 112

[33] Ewart, F.T. et al (1986) The solubility of actinides in the near field in Scientific Basis for Nuclear Waste Management, IX, LO Werme, Editor, Materials Research Soociety, 701 - 708.

[34] Ewart, F.T. and Tasker, P.W. Chemical effects in the near-field (1987) in Waste Management '87, Vol II, RG Post and M Wacks, Editors,

[35] Nitsche, N. and Edelstein, N.M., Solubilities and speciation of selected transuranium ions. A comparison of a non-complexing solution with a groundwater from the Nevada tuff site. Radiochem Acta 39 23 - 33

[36] Nitsche, N. (1991) Solubility studies of transuranium elements for nuclear waste disposal: Principles and overview. Radiochem Acta 52/53 3 - 8.

[37] Bates, J.K. et al. (1992) Issues affecting the prediction of glass reactivity in an unsaturated environment. J Nuclear Materials 190 198 - 227

[38] Cunnane, J.C. et al, (1994) "High Level Waste Borosilicate Glass: A Compendium of Corrosion Characteristics" USDOE Report. DOE EM 0177 (1994), Vol 2, p 170

FRENCH APPRAISAL OF THE VITRIFICATION OPTION

Weapons plutonium – or how to get rid of it!

Antoine Jouan
Commissariat à l'Energie Atomique (CEA)
Rhône Valley Research Center, BP 171
30207 Bagnols-sur-Cèze Cedex, France

INTRODUCTORY NOTE

"One trembles at the thought of what it might be used for if it were not in the hands of enlightened people... ".

This line, spoken emphatically by Jean Gabin in the film version of *Les Grandes Familles*, written by a contemporary member of the French Academy*, referred to the dangerous or even dramatic consequences of money when improperly used. The same phrase could be applied today to plutonium, with different but certainly graver and more far reaching consequences than those evoked in the film. May the members of this group contribute – even to a modest extent – to nuclear nonproliferation and thus to world equilibrium.

1. INTRODUCTION

When he isolated plutonium-239 for the first time in 1941, Glenn Seaborg probably never imagined he would change the face of the earth, and that the fission properties of this element would have a decisive effect on the world order for the next fifty years – not to speak of the coming fifty years.

The United States is the only country to have used plutonium for wartime purposes. The shock caused by the use of such a weapon was apparently sufficient to dissuade the other countries that subsequently acquired the same capability from ever doing likewise. Nevertheless, the world stockpile of weapons-grade plutonium now represents some 250 tons (Figure 1), and we are facing an increasing risk of proliferation.

To avoid a worst-case scenario in which our planet becomes an explosive garbage dump, it is up to us to make our contribution, even with limited means, to the nonproliferation edifice – and the various ways of making plutonium unrecoverable are one of the means that technicians can make available to statesmen.

---

* *Maurice Druon, former government minister and member of the* Académie française.

E.R. Merz and C.E. Walter (eds.), Disposal of Weapon Plutonium, 197-207.

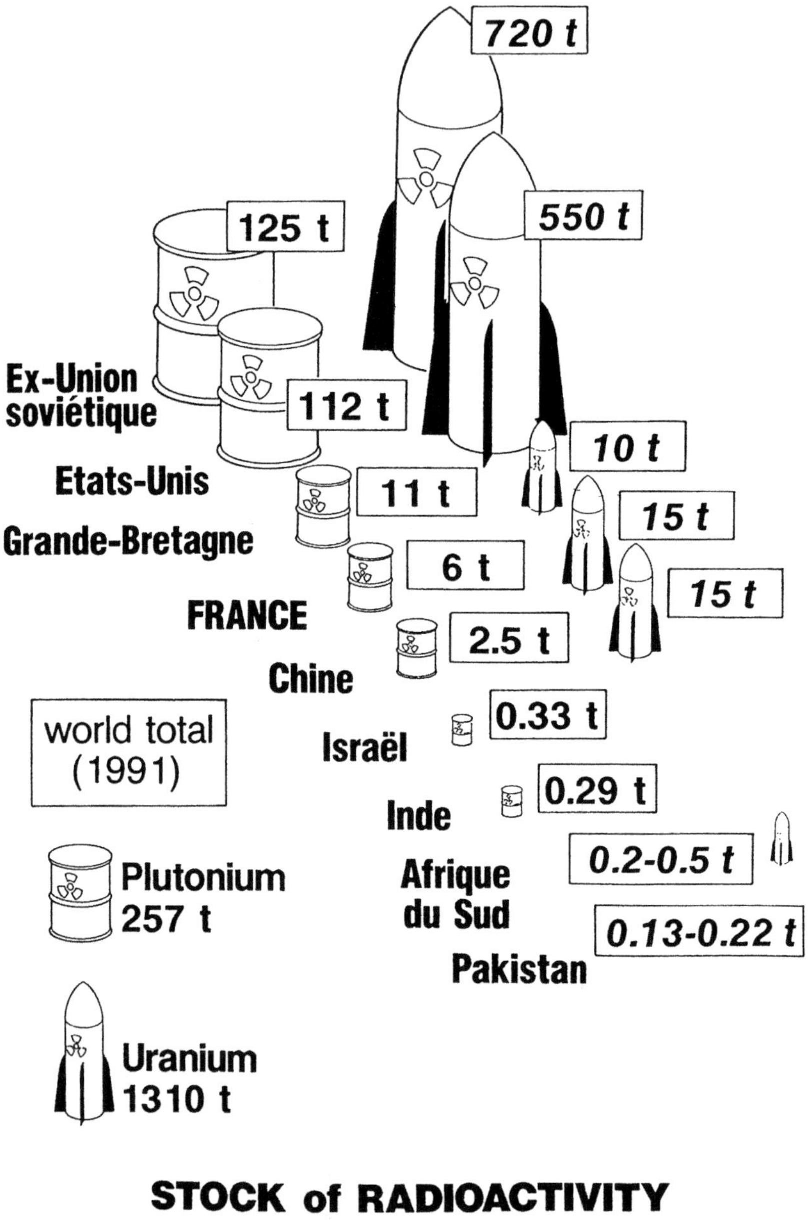
720 t
550 t
125 t
112 t
Ex-Union soviétique
Etats-Unis
11 t
10 t
Grande-Bretagne
6 t
15 t
FRANCE
2.5 t
15 t
Chine
world total (1991)
Israël
0.33 t
Inde
0.29 t
Plutonium 257 t
Afrique du Sud
0.2-0.5 t
Pakistan
0.13-0.22 t
Uranium 1310 t
STOCK of RADIOACTIVITY

Figure 1

2. WHAT TO DO WITH EXCESS WEAPONS PLUTONIUM?

Vitrification of excess weapons plutonium, incorporating it in a relatively insoluble glass network from which recovery would be difficult, is a technical possibility, even if many years of development would be necessary before it could become an industrial solution.

Is this the best solution? A number of alternatives have already been discussed. They are illustrated in Figure 2, taken from a USDOE report[1], showing disposal options (direct or after containment) as well as energy recovery options. The second alternative, namely utilization of weapons plutonium to produce energy, is the option promoted today by France.

It seems only reasonable to benefit from the energy released by plutonium fission: one gram of plutonium, after all, is equivalent to a ton of oil – with no impact on global warming. Proven technology already exists in which Pu is used to fabricate mixed oxide (MOx) fuel for use in power reactors (Figure 3). The MOx fuel fabrication units throughout the world are indicated in Table I[2].

MOx fuel was first used in 1963 in Belgium, with the BR3 demonstration PWR at Mol, and in Germany at Obrigheim in 1970. In France, MOx fuel experience has developed progressively since the first irradiation in the Chooz plant in 1974, followed by the MOx program in light water reactors operated by the French electrical utility EDF at Saint Laurent des Eaux in 1987. Seven French nuclear power stations are currently operating with MOx fuel, and EDF intends to extend its use in 20 to 28 units by the end of the century. The present policy of utilizing MOx fuel is thus solidly based on over three decades of experience[2]. As with commercial plutonium, the processes now available would be capable of fabricating MOx fuel from weapons plutonium. Of course, it would also be necessary to decide in which reactors it would then be burned.

Table I. MOx fabrication plants

Country	Name and Location	Operator	Capacity (tHM/year)	Status
Belgium	P0, Dessel	Belgonucléaire	35	In operation
	P1, Dessel	Belgonucléaire	40	Planned
Germany	Hanau	Siemens	35	Shut down
	Hanau	Siemens	120	Under construction
France	CFCa, Cadarache	COGEMA	20–30	In operation
	MELOX, Marcoule	MELOX	160	Startup
UK	MDF, Sellafield	BNFL	8	In operation
	SMP, Sellafield	BNFL	100	Under construction
Japan			100	Planned

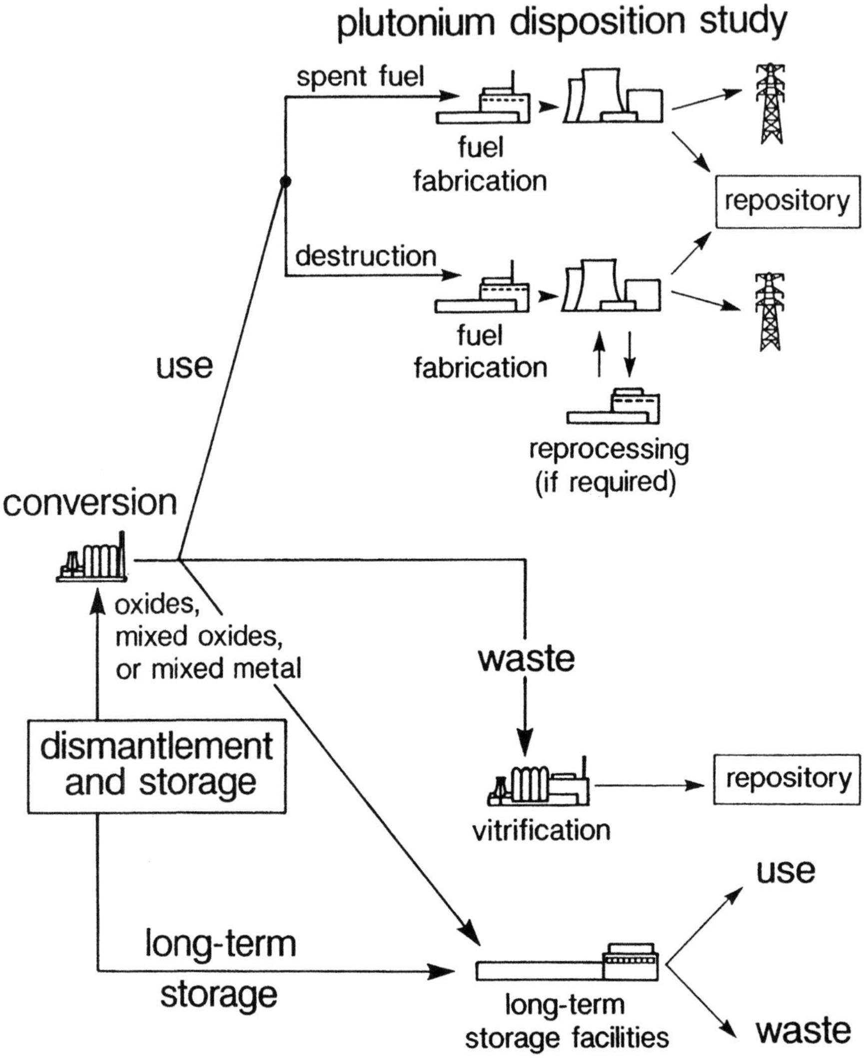

DISPOSITION
of WEAPONS-GRADE PLUTONIUM

Figure 2

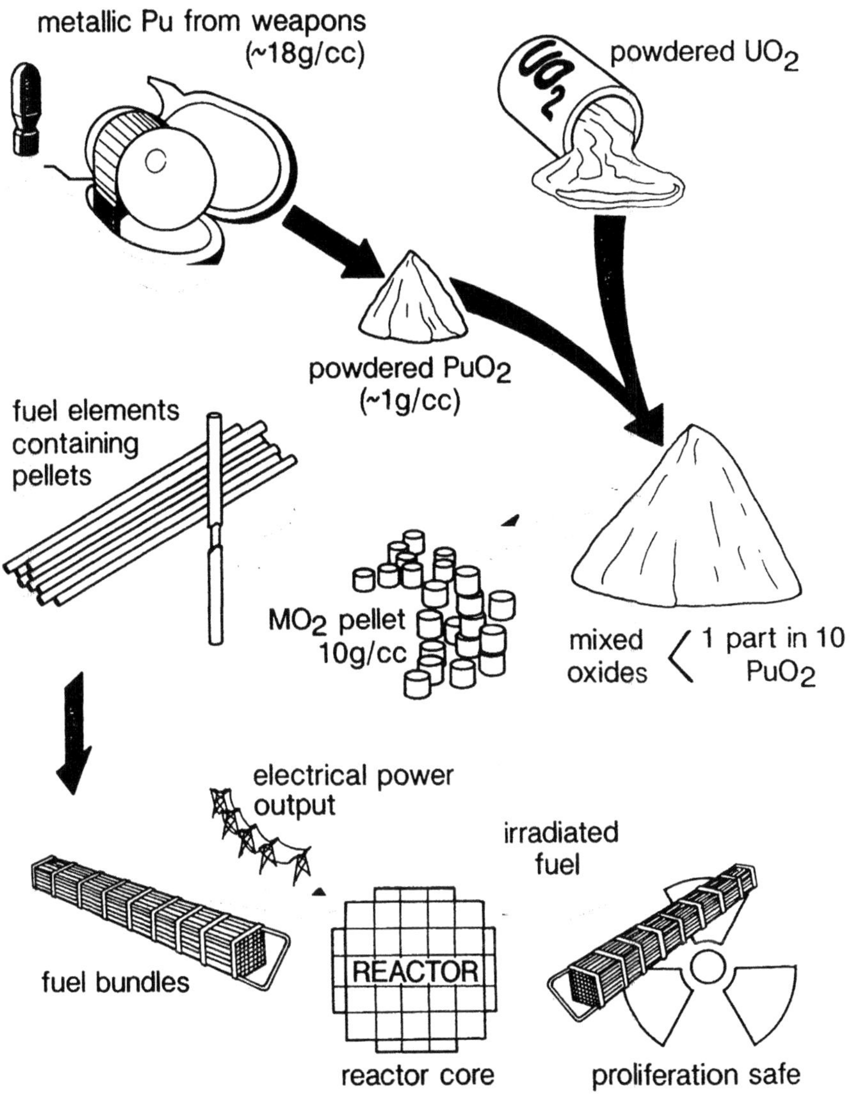

TYPICAL Pu DISPOSITION OXIDE FUEL PROCESS

Figure 3

3. PLUTONIUM VITRIFICATION

3.1 Glass Formulation

Plutonium vitrification has not been a subject of major research in France to date; only a few fission product containment glass blocks were fabricated in the 1970s with a few percent of PuO_2 to dope them with α-emitters in order to investigate α irradiation damage in the glass. The incorporation of PuO_2 in sufficient quantities proved difficult because of segregation problems, and PuO_2 was replaced by α-emitters with shorter half-lives such as AmO_2 or CmO_2.

Vitrification is routinely used in the COGEMA reprocessing plants at Marcoule and La Hague to produce containment glass for concentrated fission product solutions, but the glass contains only negligible amounts of Pu. Glass containing PuO_2 has been fabricated by KfK in Germany, but Pu vitrification has only been seriously investigated in the USA at Savannah River.

Various options are available for PuO_2 vitrification: one would be to determine the optimum formulation for incorporating PuO_2 alone into a glass matrix; another would be to incorporate it into fission product containment glass. In either case, it should be noted that the maximum PuO_2 concentration in the glass would not only depend on its solubility in the glass, but could be limited by the risks of criticality and/or heat release in interim storage or after final disposal. The behavior of α-emitting glass (helium production, leaching resistance) and its long-term behavior in repository conditions must also be investigated.

It is difficult at the present time to produce fission product glasses at industrial scale containing more than a few percent of PuO_2.

3.2 Vitrification Process

3.2.1 French Experience

France has undeniably acquired considerable experience in the vitrification of fission product solutions.

- The continuous vitrification technique[3] (Figure 4) employed in the COGEMA facilities at Marcoule (AVM) and La Hague (UP2 and UP3) is also implemented in the United Kingdom, and is designed to convert high-level liquid waste from Purex reprocessing plants into borosilicate glass.

  The HLW solution in nitrate form is fed at a metered rate into an electrically heated rotating calciner where the salts are evaporated and the nitrates are partially calcined to oxides; the calcine is then supplied together with glass additives into an induction-heated metal crucible. The process thus converts the nitrate solution into a dry oxide form suitable for glass production. The current R7/T7 melter design has a total volume of 200 liters and weighs less than 500 kg; glass is poured about 70 liters at a time every 6 to 8 hours.

  The results logged by the French vitrification units are summarized in Table II. Over 5000 canisters have been produced representing nearly 2000 tons of glass containing more than 1500 million curies of fission products.
- Nearly 30 years ago, in 1968, the CEA began operating a vitrification unit called PIVER using a batch or pot vitrification process in which a radioactive fission

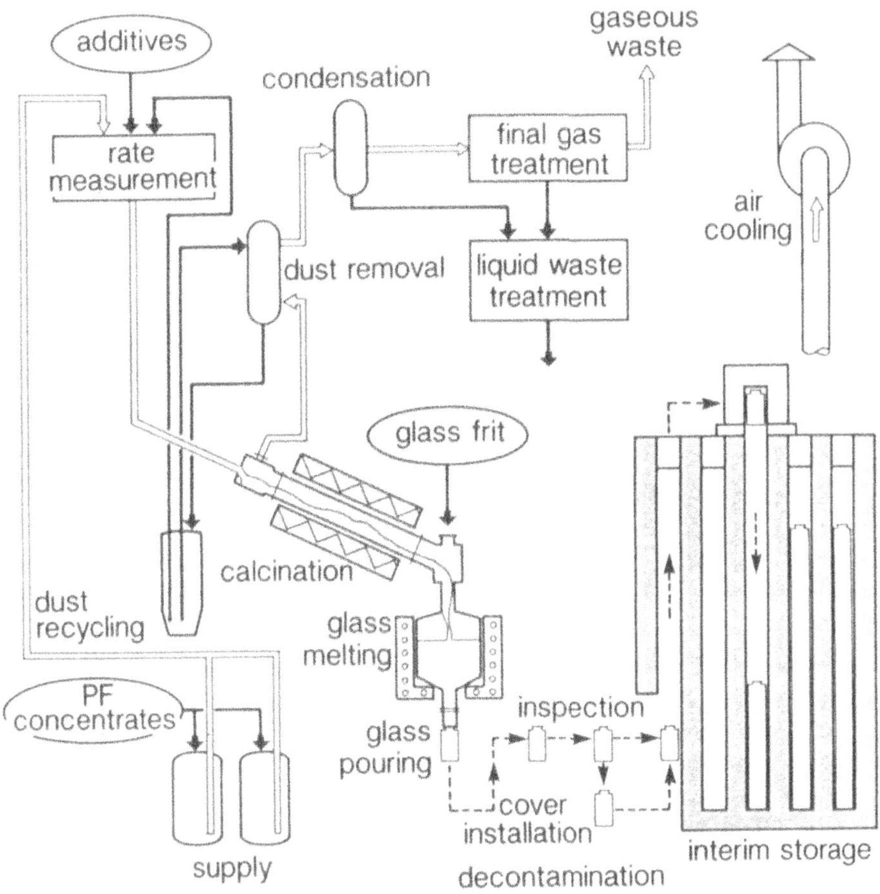

FRENCH TWO-STAGE CONTINUOUS VITRIFICATION PROCESS

Figure 4

Table II. Operating statistics for French vitrification facilities (as of December 31, 1994)

Operating Statistics	Reprocessing Plant		
	UP1 (Marcoule)	**UP2 (La Hague)**	**UP3 (La Hague)**
Vitrification facility	AVM	R7	T7
Startup date	1978	1989	1992
Evaporation capacity ($l \cdot h^{-1}$)	40	60	60
Glass fabrication rate ($kg \cdot h^{-1}$)	10–15	20–30	20–30
Vitrified solution volume (m^3)	1800	1750	587
Glass production (metric tons)	806	806	344
Number of canisters produced	2275	2048	869
Volume reduction factor	5.7	5.6	4.5
Vitrified $\beta\gamma$ activity (MCi)	384.3	745.7	380.4

product solution was converted into glass in a single induction-heated metal melter, used for the evaporation, calcining and vitrification steps (Figure 5). About 200 glass blocks weighing about 100 kg each were fabricated. Twenty years after it was commissioned, PIVER was named a Nuclear Historic Landmark by the American Nuclear Society on November 15, 1989: "a high-level waste vitrification prototype and its storage facility both began operating in 1969. PIVER successfully demonstrated one of the essential safety conditions of the nuclear power option: the safe disposal and storage of long-lived radioactive waste".

- The *Commissariat à l'Energie Atomique* has been developing a new glass melting technique in recent years, using an induction-heated cold crucible melter (Figure 6). This highly efficient heating technique could no doubt be used to vitrify plutonium; compared with existing processes, it can produce more refractory glass or vitrocrystalline formulas with even better containment properties than the current fission product glasses if required. This process has two major advantages: temperatures of 1500°C can easily be reached, and the melter is corrosion-free.

3.2.2 Plutonium Vitrification Process Options

Any of the processes described above could be adapted and developed to vitrify weapons plutonium in solid or liquid form, either alone or together with γ-emitters to ensure greater inaccessibility.

Both the continuous and batch processes could be used for plutonium in solid or liquid form. If the plutonium is obtained as PuO_2 – or even as Pu metal – in granulated or powder form, it could be metered directly with prefabricated glass frit additives into an induction-heated metal pot as in the PIVER or R7T7 processes; a melting pot with subcritical geometry could even be designed.

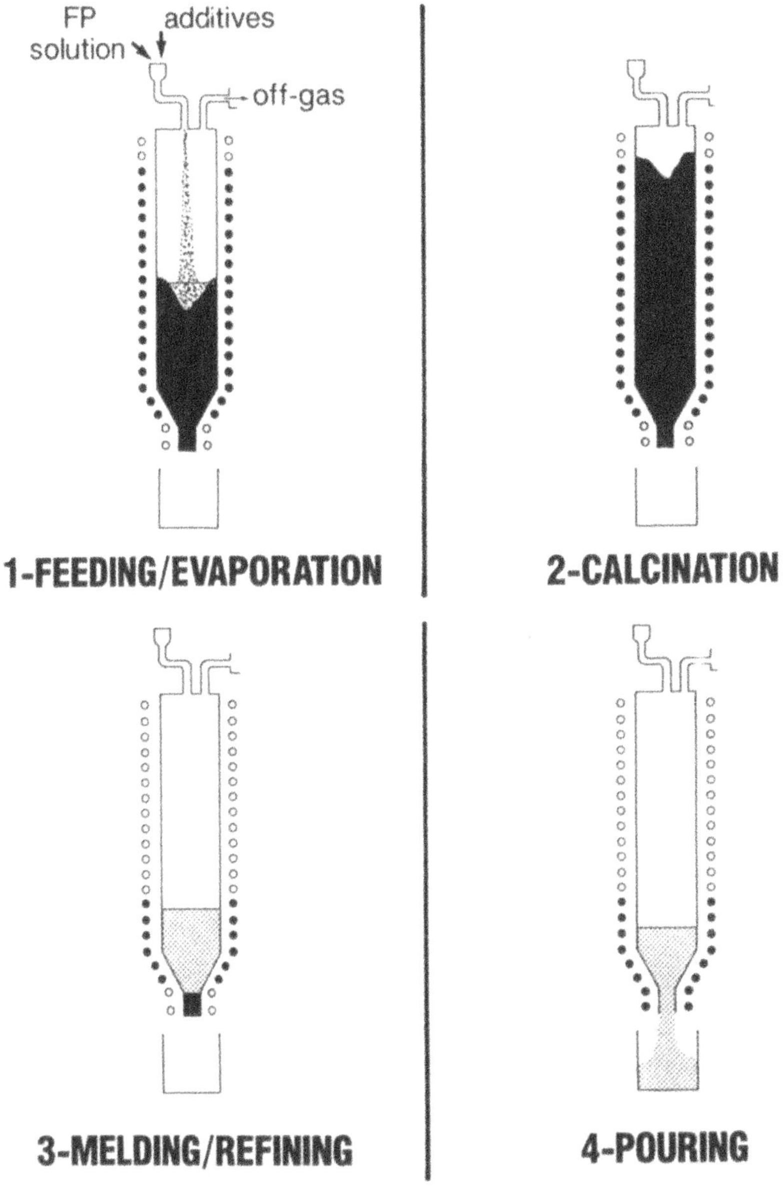

POT VITRIFICATION PROCESS

Figure 5

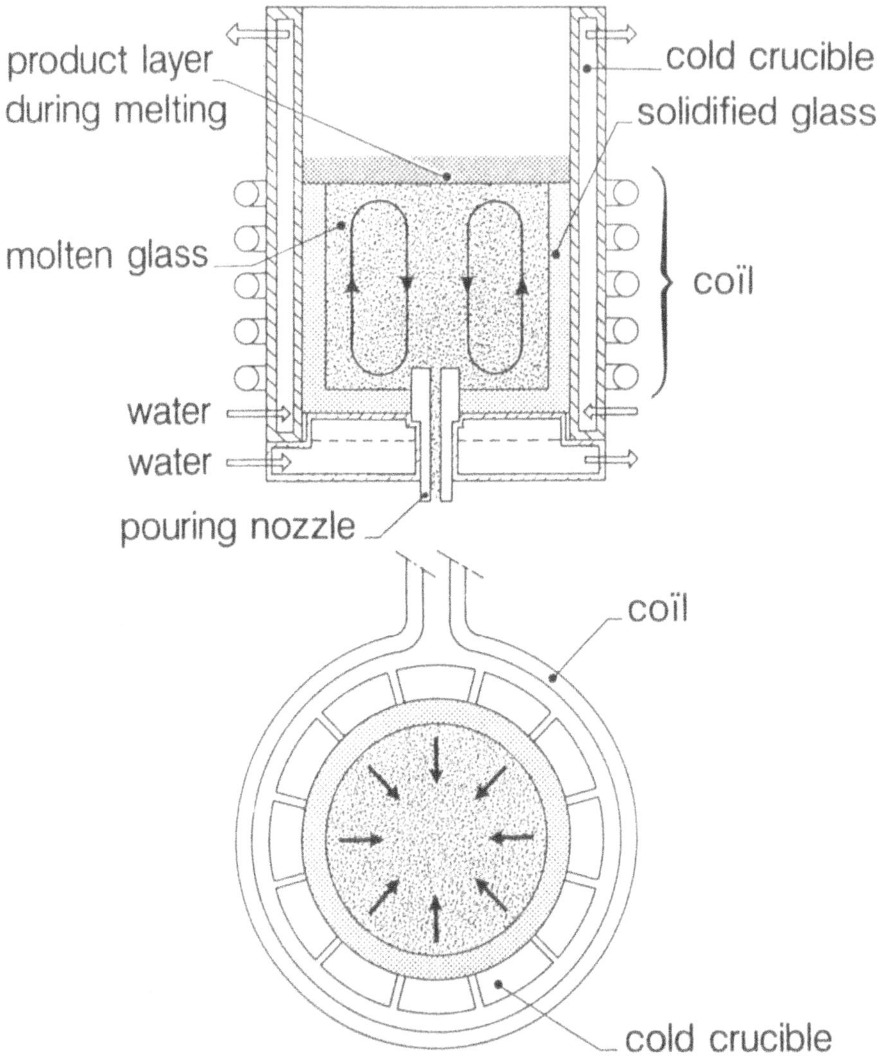

COLD CRUCIBLE MELTING PRINCIPLE

Figure 6

- Vitrification without fission products would obviously be the simplest and least costly solution. Shielded structures would be unnecessary, and a glove box line would suffice, but, the result would be on a par with the cost: the glass could then be handled directly unless, for example, a ^{137}Cs capsule were placed in the canister immediately before pouring the glass (the idea has already been proposed, and some research has been done on this option). Vitrification without fission products, however, would probably not be sufficient to meet nonproliferation requirements.
- It would also be possible, of course, to vitrify the weapons plutonium together with fission product additives in shielded cells. The facility would resemble those at Marcoule and La Hague, or at Tokai Mura in Japan, or at Savannah River in the USA – not only by their design but also by their cost and complexity. In a unit such as those operated today by COGEMA at La Hague, plutonium could be fed as a liquid together with the fission product solution, or as a solid directly into the glass melt.

  This is only a technical approach, however, with no consideration for the political aspect – which is far from negligible. To simplify matters, the plutonium and the fission products should belong to the same country; if not, serious complications are not doubt inevitable.

4. CONCLUSION

From a purely technological standpoint, various technologies are available to vitrify liquid or solid plutonium oxide or plutonium metal, although further design work and testing would be required. A dry process has significant advantages and largely simplifies the criticality hazard; a liquid process would probably allow easier metering of the feed solution. Vitrification with fission products is technically feasible, but will never be a simple process – especially where safety is concerned.

From a more mundane point of view, man has been extracting energy sources from the earth since the beginnings of time, but this is no doubt the first time he has ever considered getting rid of them by burial. Would we not be better inspired to put this fabulous metal to good use?

REFERENCES

1. *Plutonium Disposition Study*, USDOE Technical Review Committee Report, vol. 1 (July 1993).
2. W. Fournier and G. Dalverny, "Le combustible et l'usine de fabrication MELOX", *Revue Générale Nucléaire*, 1995 No. 1.
3. A. Jouan, J.P. Moncouyoux and S. Merlin, "The Development of Nuclear Waste Vitrification in France", *Waste Management*, Tucson (1995).
4. C. Ladirat, R. Boen, A. Jouan and J.P. Moncouyoux, "French Nuclear Waste Vitrification: State of the Art and Future Developments", *Annual Conference on Glass Problems*, Columbus, Ohio (1994).

SAFETY ASPECTS WITH REGARD TO PLUTONIUM VITRIFICATION TECHNIQUES

LEONARD W. GRAY AND TEHMAU KAN
Lawrence Livermore National Laboratory
P.O. Box 808, L-592, Livermore, CA 94551 USA

Substantial inventories of excess plutonium are expected to result from dismantling U.S. and Russian nuclear weapons. Disposition of this material should be a high priority in both countries. Various disposition options are under consideration. One option is to vitrify the plutonium with the addition of ^{137}Cs or high-level waste to act as a deterrent to proliferation. The primary safety problem associated with vitrification of plutonium is to avoid criticality in form fabrication and in the final repository over geologic time. Recovery should be as difficult (costly) as the recovery of plutonium from spent fuel.

1. INTRODUCTION

In the aftermath of the Cold War, significant quantities of enriched uranium and weapons-grade plutonium have become surplus to national defense needs in both the United States and Russia. These excess stockpiles pose a danger to national and international security, not only in the potential proliferation of nuclear weapons but in the potential for environmental, safety, and health consequences if these fissile materials (FMs) are not properly managed.

If agreed reductions are implemented, perhaps 100 metric tonnes of Pu will no longer be needed for military purposes by the Nuclear Weapons States. Continued implementation of arms reduction agreements will result in further dismantling of weapons and increasing stockpiles of surplus weapons-usable materials.

With the transformation of the Soviet Union into the Commonwealth of Independent States and the economic and social challenges faced by those newly formed states, there is a serious risk of nuclear proliferation from the resulting growing stockpiles. Nuclear weapons or fissile materials could fall into the hands of terrorists or non-nuclear nations through theft or diversion of FMs. The U.S. National Academy of Sciences report (NAS) [1] on the management and disposition of excess weapons plutonium characterized this as a "clear and present danger." This nuclear danger is, in many ways, more diffuse, harder to manage, and more dangerous than the nuclear tensions of the Cold War era.

To aid in selecting long-term management options for weapons surplus Pu, the U.S. Department of Energy (DOE) has undertaken a multifaceted study to select options for storage and disposition in keeping with the national policy that excess

E.R. Merz and C.E. Walter (eds.), Disposal of Weapon Plutonium, 209-220.

Pu must be subjected to the highest standards of safety, security, and international accountability. Disposition is defined as a process of use or disposal of materials that results in the remaining material being converted to a form substantially and inherently more proliferation-resistant than the original form. Disposition options must take into account technical, nonproliferation, environmental, and economic considerations. As a part of the overall disposition program, Russia and other nations with relevant interests and experience will be invited to participate in the overall disposition study.

The disposition process (Fig. 1) can be divided into three distinct but overlapping phases—dismantling, intermediate storage, and long-term disposition. Dismantling of U.S. and Former Soviet Union (FSU) weapons and storage of resulting fissile materials (FM) are already under way. Conversion of residue materials and long-term disposition of all FM will take far longer to accomplish.

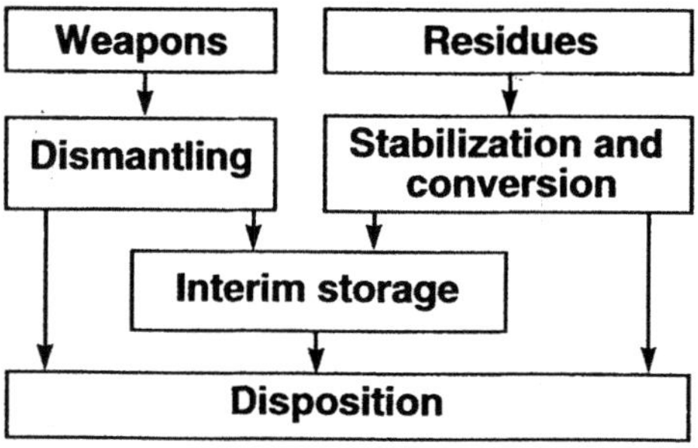

Figure 1. Steps in control and disposition of surplus fissile materials.

One disposition alternative considered for Pu is immobilization. In support of DOE's Materials Disposition (MD) Project Office's Programmatic Environmental Impact Statement (PEIS) for Disposition of Pu, Lawrence Livermore National Laboratory was selected as Lead Laboratory to study and recommend methods for transforming Pu into long-term immobilized forms meeting environmental, safety, and security objectives; to provide appropriate input to other Disposition Tasks Teams so as to assess the technical feasibility of immobilization as a long-term disposition option; and describe infrastructures required to dispose of Pu. Support laboratories include Westinghouse Savannah River Technology Center (South Carolina), Argonne National Laboratory (Illinois), Oak Ridge National Laboratory (Tennessee), Pacific Northwest Laboratory (Idaho), Rocky Flats Environmental Technology Site (Colorado), and Westinghouse Hanford Company (Washington).

1.1. IMMOBILIZATION OPTIONS

Immobilization would embed Pu, alone or mixed with radioactive fission products, in a tailored ceramic, borosilicate glass, or other suitable material to produce a suitable disposal form. To be economically viable, the Pu concentration must be in the range 0.4 to 10 wt%. To arrive at suitable forms, published information on radioactive high level waste (HLW) immobilization technologies was reviewed; 72 uniquely named forms were identified. After pre-screening, candidate forms were further screened using multiattribute utility analysis to determine the most promising technologies for Pu immobilization. Promising immobilization families—glass, ceramics, and metals—were then evaluated to identify inherent chemical, engineering, environmental, safety, and health problems and to seek solutions for them before making technical decisions as to the viability of using any of the families for long-term disposition of Pu. We are also assessing modifications required to existing U.S., British, and French HLW immobilization approaches with respect to environmental, safety, and health implications, as well as costs and schedule. All data, analyses, and reports are being provided to the DOE/MD to support the Record of Decision that is anticipated in the fourth quarter of FY96.

1.2. SPENT FUEL STANDARD

To clarify the issue, the NAS [1] coined the term "spent fuel standard" as follows: "Options for long-term disposition of plutonium should seek to meet a 'spent fuel standard' in which the plutonium is inaccessible for weapons use similar to the much larger and growing quantity of plutonium that exists in spent fuel from commercial power reactors."

The standard consists of four parts: radiological, physical, chemical, and nuclear properties. The first three directly influence requirements, regulations, and practices for applications of domestic and international safeguards and security; therefore, these properties must satisfy the standard. Plutonium isotopic compositions have only secondary effects and thus are not considered as a criterion in meeting the spent fuel standard.

2. SELECTION OF WASTE FORMS FOR IMMOBILIZING PLUTONIUM

A literature search identified 72 waste forms [2] by unique names (but only 45 unique forms) that have been considered for immobilizing radioactive wastes. Individual forms can be grouped into eight families that share common chemical and physical characteristics. These families are calcine, cementitious, ceramic, glass, glass-ceramic, metallic, multibarrier, and polymeric. Distinct waste forms are distinguished by unique physical and chemical properties. A screening process was implemented to select the best waste forms for the Pu disposition application. A more detailed discussion of the screening process is given elsewhere [3], a summary of the process is given below.

2.1. PRESCREENING PROCESS

A two-stage approach (Fig. 2), based on decision analysis techniques, was adopted for screening. This allowed more rigorous selection techniques to be used as options became more closely matched. Stage 1 applied a small set of criteria to a large number of forms; this quickly removed forms that were clearly inappropriate for Pu immobilization. Stage 2 more closely evaluated remaining forms with the goal of selecting a small set (≤3) of the best forms. Stage 2 involved a more formal comparison of waste form characteristics using multiattribute utility analysis techniques from decision analysis principles.

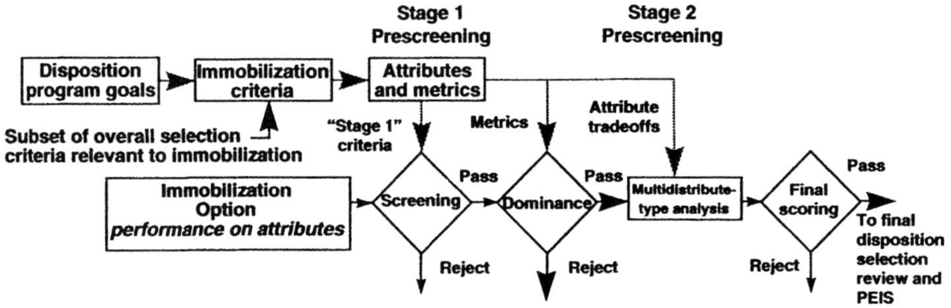

Figure 2. Pu immobilization prescreening process.

The results from individual and combined technical assessments and value and tradeoff assessments are shown in Table 1. "Utility" is the overall figure of merit for a form.

Form	Utility
Borosilicate glass	0.89
Synroc	0.66
Phosphate glass	0.55
Monazite	0.49
Metallic alloy	0.47
High-silica glass	0.44
FUETAP concrete	0.40
Hot-pressed concrete	0.25
Phosphate-bonded ceramic	0.17
Silicon-zirconium phosph.	0.17
Ceramics in concrete	0.13
Iron-enriched basal	0.13
Ceramic pellets in metal	0.13
Supercalcine	0.08
Glass-ceramic monoliths	0.03
Cermet	0

TABLE 1. Ranking of forms according to weightings and utility curves.

2.2. IMMOBILIZATION OPTIONS

Seven immobilization options comprising three technologies are being evaluated in the PEIS process:

- Vitrification
  - Internal radiation barrier
    - New vitrification facilities
    - Defense Waste Processing Facility (DWPF) adjunct melter
  - External radiation barrier
  - Glass Material Oxidation and Dissolution System (GMODS)
- Ceramics
  - Internal radiation barrier
  - External radiation barrier
- Electrometallurgical Treatment

3. VITRIFICATION OF PLUTONIUM IN BOROSILICATE GLASS

Criticality control is much easier in preparing crystalline ceramics than in preparing glass monoliths. Reference [3] discusses crystalline ceramics in more detail. In this paper, only borosilicate glass is discussed as the immobilization medium.

3.1. CONVENTIONAL VITRIFICATION TECHNOLOGY

The technical basis for plutonium vitrification in borosilicate glass derives from nearly 40 years of international development and decades of commercial production of borosilicate glass to stabilize HLW. Borosilicate glass has gained international acceptance in the Western World as the waste form for HLW [4]. Preliminary work has shown that plutonium can be incorporated in borosilicate glass, providing an indication that plutonium-glass can be developed into a relatively safe, durable, accountable, and monitorable form [5–8].

Internal Radiation Barrier: This vitrification option immobilizes Pu in a borosilicate glass. The product of the vitrification facility is a homogeneous glass monolith (Fig. 3a) within a Defense Waste Processing Facility (DWPF) type canister. The glass contains both the Pu and the ^{137}Cs or other radiation source uniformly distributed in the glass matrix.

External Radiation Barrier: This immobilization alternative vitrifies Pu in glass. The plutonium-glass is poured into small stainless-steel cans, seal-welded, and placed into interim storage. The small Pu-glass cans are then inserted into standard DWPF-type waste canisters, and borosilicate glass containing HLW is poured around the cans to form a radiation barrier (Fig. 3b). Since the plutonium-bearing glass does not have to incorporate the volatile cesium, higher-temperature (more-durable) glass compositions can be used to immobilize the plutonium.

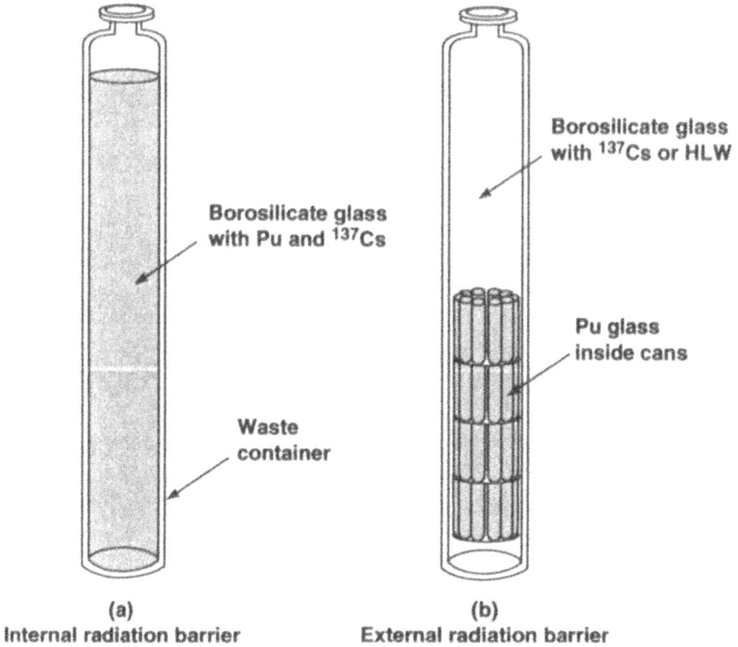

Figure 3. Two options for vitrification of Pu.

World-wide, various equipment and processes are in commercial use to produce borosilicate glass for an assortment of commercial applications, both with and without the incorporation of radioactive elements. These technologies could be adapted to produce plutonium-glass. Although differences exist among applications and the associated equipment—including feed compositions, melter designs, scale of equipment, and heating methods—the basic compositions and properties of the final glass products are similar.

The choice of a heating method depends on the particular application, the scale of the equipment, feed composition, final form requirements, and other considerations. Heating methods in common use include joule, radiant energy, induction, microwave, and plasma.

The most thoroughly demonstrated process for HLW vitrification is the one developed at Marcoule, France [9], which (with minor alterations) is being used in the United Kingdom, France (LaHauge), and Japan. This two-step process first calcines an aqueous acidic waste. The resulting dry, calcined material is fed to an induction glass melter. In the United States, most HLW are alkaline, containing significant quantities of sodium and aluminum. The United States uses a single-step vitrification process in which the waste is in the form of a liquid slurry and is fed directly to the melter. The Defense Waste Processing Facility (DWPF) at the Savannah River site (SRS) is scheduled to begin "hot" (with radioactivity) operation late this year. Other vitrification plants for HLW soon will be operating in Germany, the United States, and Japan. Vitrification plants for low-level waste

(LLW) are even more commonplace.

Figure 4 is a block flow diagram of a generic process to vitrify plutonium in borosilicate glass. In this process, the gamma radiation field is assumed to be provided by the isotope ^{137}Cs in the form of $^{137}CsCl$ from the inventory presently stored at the Hanford Site [10].

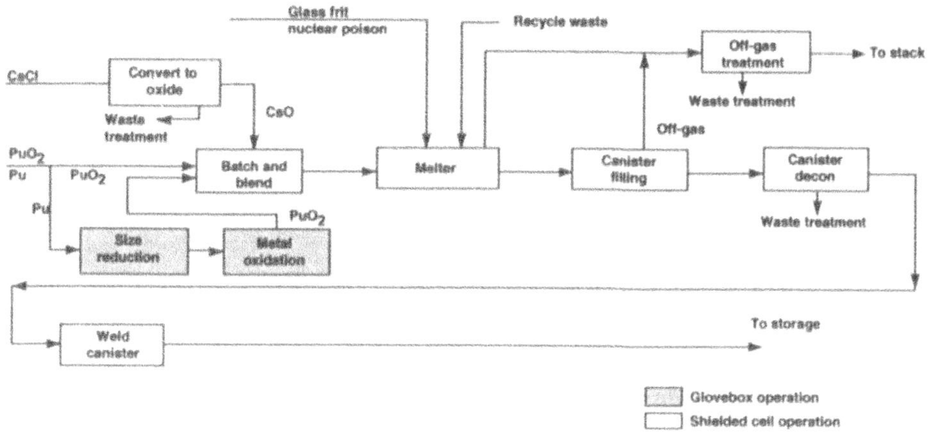

Figure 4. Vitrification block flow diagram.

Although HLW vitrification is relatively well developed, Pu vitrification requires investigation of several issues before a production plant can be built. In the field of glass chemistry, we must determine the optimum glass composition to yield a durable product with an acceptably low Pu leach rate. In addition, process and equipment development must be done for the various plant unit operations to reduce risk in deploying the production plant. Finally, but not least, the issues associated with criticality must be addressed for each step of the immobilization process, from preparing the input feed material to producing the final waste form.

3.2. NUCLEAR CRITICALITY

It is assumed that the facility will process about 25 kg of Pu per operation day. This amount of Pu represents more than 50 times the minimum critical mass for Pu/water systems and is greater than the maximum critical mass for dry, spherical, PuO_2 systems. Therefore, criticality safety is clearly a major concern in designing vitrification processes.

3.2.1. *Process Criticality Safety*

A nuclear criticality accident is the result of inadvertently producing conditions that allow a self-sustaining or divergent neutron chain reaction. These conditions are present when the effective neutron multiplication factor (k_{eff}) for a system of fissionable material is ≥1. The k_{eff} for any system of fissionable material is a complex function that depends on

- The mass and distribution of all fissionable materials within the system.
- The mass, distribution, and nuclear properties of all other materials associated with the system.

Operations with fissionable materials always present a risk of nuclear criticality accidents. Nuclear criticality safety is achieved by controlling one or more of the parameters that affect k_{eff} within subcritical limits. These controls may be achieved through

- Geometry restraints such as limiting the diameter of tanks that contain fissionable materials within the subcritical limit.
- Use of instrumentation, such as devices to limit the concentration of fissionable material in a process.
- Administrative controls, such as procedures to limit the inventory of fissionable material in a process.
- Use of nuclear poisons to reduce k_{eff} .

DOE Order 5480.24 requires application of the double contingency principle:

> "Process designs shall incorporate sufficient factors of safety to require at least two unlikely, independent, and concurrent changes in the process conditions before a criticality accident is possible. Protection shall be provided by either (a) the control of two independent process parameters (which is the preferred approach, if practical) or (b) a system of multiple (at least two) controls on a single parameter. In all cases, no single failure shall result in the potential for a critical accident."

DOE Order 5480.24 also ranks, in order of preference, the different types of control that may be used to ensure criticality safety. Process designs are to rely on passive geometry controls whenever possible. When these are not possible, the order of preference is for other passive engineered controls, active engineered controls, and finally, administrative controls.

ANSI/ANS-8.1 allows the use of neutron absorbers for criticality control. Extreme care is required with solutions of absorbers because of the difficulty of exercising this type of control.

ANSI/ANS-8.1 requires that subcritical limits shall be established with adequate allowances for uncertainties. Methods of calculation used to determine k_{eff} for a system or used to establish subcritical limits shall be validated. The bias in the results shall be determined. The calculation shall incorporate sufficient margin to ensure subcriticality. The margin of subcriticality shall include allowances for the uncertainty in the bias.

No criticality analysis of the vitrification process has yet been done. The criticality issues for each stage of the immobilization process are discussed below. These issues are assessed based on engineering judgment and extrapolation from similar processes. These assessments produce criticality safety assumptions used to size the facility. They also indicate that it will be possible to produce facility, process, materials-handling, and waste-form designs that satisfy all applicable criticality safety requirements. However, all of these criticality safety assumptions and assessments require confirmation that can be provided only by a detailed criticality safety evaluation.

3.2.2. *Vitrification Specific Criticality Issues*
Nuclear criticality avoidance is ensured by three mechanisms: process design, process instrumentation, and nuclear safety systems. Each of these is examined below.

Process Design. The locations of buildup of FMs in the processing system depend upon the process chemistry; thus, the first level of nuclear criticality safety is a detailed understanding of the fissile material process chemistry. A number of processing decisions have not yet been made, and a number of critical parameters have not been determined.

Experiments to date on the solubility of plutonium in glass primarily have focused on the low concentrations of plutonium in the various waste streams instead of the solubility of high-concentration feed stocks. As such, experiments have focused on whether these low concentrations are soluble in the various glass formulations—very little experimentation has therefore been done on plutonium concentrations in glass above about 2 to 3 wt% Pu. Experimental data, which have been done above the 2- to 3-wt% concentration level, are sparse and conflicting even when the experiments were done under nearly the same conditions. As a result of the conflicting data, an experimental program is under way to determine the solubility of various Pu feed stocks under varying process conditions and cycle times. These experiments, however, will not be done in time to support the PEIS.

Because of this, a two-step approach has been taken to all vitrification flowsheets. The first step is the conversion of PuO_2 to Pu-borosilicate glass in small batches. To satisfy the double contingency principle, a batch process based on a bottom-draining, geometrically favorable melter such as the Bushing Melter (wrapped with a neutron-loaded insulation, if necessary) and the ever-safe, water-reflected PuO_2 sphere mass limits are being considered for the PEIS flowsheets. For weapons-grade plutonium (i.e., about 6% ^{240}Pu), the ever-safe PuO_2 sphere mass is 4.53 kg. Preliminary calculations suggest that, until the plutonium is dissolved into the borosilicate glass, the safe mass in a melter may be as high as 27 kg; provided the water content is held to less than 3% of the plutonium content. More definitive calculations will be performed as the design firms up; at the present time, for the PEIS, it is sufficient to know that the batch size will be <27 kg.

Once the Pu has dissolved into the borosilicate glass to give a homogeneous melt, then the k_{eff} drops to a very low value (<<0.1). Therefore a larger melter containing much more Pu can be used for the second stage of vitrification. Glass frit added to the second stage also will have an additional dose of neutron poison dissolved in it beyond just the boron inherent in the glass—gadolinium or other rare earth, hafnium, and other neutron poisons have been suggested.

Once the kinetics of dissolution and the solubility of plutonium in the specific glass formulation have been determined, and the melter design decided upon, then full-scale criticality calculation will be done for the specific process chosen. Until that time, the conservative approach will be taken for the PEIS.

Process Instrumentation. At the level of analysis required for the PEIS, actual instrumentation need not be specified—just that the type of instrumentation required is available. Avoidance of nuclear criticality also will be guarded against by on-line process control systems that measure the concentrations and quantities of FMs in the various parts of the plant process. This capability is required for process-control purposes independent of nuclear criticality considerations, but the process-control system also has the capability to confirm that concentrations of FMs are not building up within the system. Product requirements necessitate a uniform glass-product composition, starting with highly heterogeneous feed stocks. This requires multiple measurement and control systems to ensure production of a homogenous glass product.

3.2.3. *Nuclear Safety Systems*

The nuclear safety systems are those items required to ensure criticality control. Therefore, the system is a combination of facility and equipment design, instrumentation, and process control. To the extent possible, engineered design features will be designed into the system to prevent criticality. Only where absolutely necessary will administrative control be used.

3.2.4. *Repository Criticality Safety*

Criticality control can be maintained in the repository by using a combination of fissile concentration, immobilization matrix, neutron poisons, overpacks, backfills, etc. The three items within the control of the Immobilization Task are fissile concentration, the immobilization matrix, and neutron poisons.

Fissile Concentration. The range of plutonium concentrations being examined for immobilization is about 0.4 to 10 wt%. Below that range, the cost is considered to be prohibitive. Above that range, the immobilized form is considered to be too attractive as a target for a potential proliferant. Within this range, however, there is some flexibility for criticality considerations. For the purposes of the PEIS, the external radiation cases have been arbitrarily set at 10 wt% Pu and the internal radiation cases at 5 wt% Pu. The final average concentrations in both cases is 5 wt%. After the PEIS is complete, more detailed calculations will be performed to verify or change the original assumptions on plutonium concentration.

Immobilization Matrix. For criticality safety to be maintained over time in the repository, the glass formulation must optimized so that it will

- Not undergo phase devitrification;
- Minimize the formation of crystalline phases that have a negative effect on the chemical durability of the final form;
- Minimize inhomogeneity, amorphous phase separation, devitrification, etc.;
- Have a reasonable high glass transition temperature;
- Have low leach rates of both the FMs and the neutron poison; and
- Have similar leach rates for plutonium and the neutron poison.

An R&D program is underway to determine the glass formulation that will be acceptable from a processing perspective as well as durable enough to prevent migration and separation of the plutonium and the neutron poison into the host rock strata.

4. CONCLUSIONS

Three primary immobilization forms have been identified for possible use in the Fissile Material Disposition Program. These are borosilicate glass, crystalline ceramics, and metals. Borosilicate glass has been the subject of this paper. The primary safety problem for any of the immobilization forms is criticality control during form fabrication. Through the PEIS process, initial assessments indicate that the primary method of criticality control should, conservatively, be mass and geometry control. As the design for the process continues, more detailed calculational methods will be employed to determine if it is reasonable to relax the mass control somewhat. Mass and geometry control have dictated a two-step vitrification process. If the mass controls can be relaxed sufficiently, the two-step process will be rejected in favor of a cheaper one-step process.

5. ACKNOWLEDGMENT

This work was performed under the auspices of the U.S. Department of Energy by the Lawrence Livermore National Laboratory under Contract No. W-7405-Eng-48.

6. REFERENCES

1. National Academy of Sciences (1994) Management and Disposition of Excess Weapons Plutonium, National Academy Press, Washington, DC.
2. Screening of Alternate Immobilization Candidates for SFM Disposition, Pre-decisional Draft, Lawrence Livermore National Laboratory, Nov. 1995.
3. Gray, L.W., et al. (1995) Immobilization as a route to surplus fissile materials disposition, accepted for presentation at ASME's Fifth International Conference on Radioactive Waste Management and Environmental Remediation, Sept. 3–7, 1995, Berlin, Germany, and to be published in the Proceedings.
4. Lutze, W., and Ewing, R.C. (eds.) (1988) Radioactive Waste Forms for the Future (North-Holland, Amsterdam).
5. Wicks, G.G., McKibben, J.M., and Plodinec, M.J. (1995) Vitrification of excess plutonium, presented at Waste Management '95, Feb. 26–Mar. 1, 1995, Tucson, AZ, and accepted for publication in the Proceedings.
6. McKibben, J.M., and Wicks, G.G. (1994) Vitrification of excess plutonium, presented at the International Policy Forum: Management and Disposition of Nuclear Weapons Materials, Leesburg, Va.
7. McKibben, J.M., Benjamin, R.W., Bickfor, D.F., Fernandez, L.P., Jackson, W.N., McDonnell, W.R., Moore, E.N., Parks, P.B., Plodinec, M..J., Rajczak, W.M., Skiles, S.K., and Wicks, G.G. (1993) Vitrification of Excess Plutonium, WSRC-RP-93-755, Predecisional Draft, Westinghouse Savannah River Company, Aiken, So. Car.
8. Wicks, G.G. (1985) Nuclear waste glasses, in *Treatises on Materials Science and Technology, Glass IV*, M. Tomozawa and R.H. Doremus, Eds., Vol. 26 (Academic Press, Inc., New York), pp. 57–188.
9. Sombret, C.G. (1977) The vitrification of high-level radioactive wastes in France, *Nucl. Energy* **24**(2), 85.
10. Jackson, R.R. (1977) Hanford waste encapsulation: strontium and cesium, *Nucl. Technol.* **32**, 10.

DISPOSITION OF RADIONUCLIDES IN DEEP GEOLOGICAL FORMATIONS EITHER IN CONTAINERS OR IN VITRIFIED FORM

Tatyana A. Gupalo
All-Russian Research and Design
Institute of Production Engineering
VNIPIPromtechnologii, Kashirskoye Shosse, 33
115409 Moscow, Russia

It is universally recognized that at present and from a technical point of view the only realistic method to dispose of radioactive materials not destined to further utilization consists in their solidification and isolation in the bowels of the Earth. Depending on the concept for the management of fissionable materials, applying concrete waste forms, for either their temporary storage and subsequent utilization or their final disposal in geological formations must satisfy definite requirements. Lately, plutonium from excess weapons dismantlement has attained special attention. Two points of view have been formulated concerning plutonium. One of them considers plutonium as waste from nuclear energetics; on the other hand, plutonium is considered a national property. Both points of view require scientific substantiation of the possibility for its safe isolation both for long-term storage, subsequent utilization and for disposal.

Underground storage in repositories for solidified radwaste present multibarrier systems of isolation in which engineered barriers play only a limited role as time-concerned factor (up to hundreds of years). Natural barriers (enclosing rocks) represent the main isolating system keeping its properties for longer periods of time.

The problem for the choice of most stable blocks of the earth`s crust for disposal or long-term storage of fissionable materials requires the solution of scientific and engineering tasks concerning geology, geomechanics, hydrogeology, and radiochemistry.

At present, existing approaches to the choice of geological sites for the disposition of ecologically dangereous substances are based on comparative consideration of blocks in the earth crust with low tectonic activity, possessing structural stability, low permeability and low water content. As an assessment criteria, relative characteristics of the geologic massifs are of importance (rock type, tectonic disturbance state, thickness of confining layer). However, it is evident that even the same type of rocks may possess significant inhomogeneity of properties, disturbance and permeability, i.e. the adoption of the solution on the priority of the place for isolation of radwaste should be made on the basis of numerical criteria and analysis of ecological safety.

It will allow:

- to investigate the probabilities for emergency situations and their sequences;

E.R. Merz and C.E. Walter (eds.), Disposal of Weapon Plutonium. 221-236.

- to assess the risk of underground isolation of fissionable materials with concrete compositions as the solidification matrix, and quantity on the investigated geological site;
- to determine most vulnerable points in the multibarrier system of underground isolation of fissionable materials;
- to adopt scientifically substantiated solutions of sizes, quantity, and characteristics of barrier materials.

For a substantiation of ecological suitability of chosen geological formation it is necessary to carry out the following stages:

1. Assessment of properties, structure of rocks, physical processes, and range of change of the massif characteristics under the action of natural and technogenic factors and influences.
2. Conduct mathematical modelling of massif state for the required period of time.
3. Investigate scenarios of possible emergency situations which can emerge under the considered geological conditions.
4. Determine the probabilities of emergency situations and their sequences on the basis of quantitative parameters of destruction and water permeability of the rock massif.
5. Conduct probability safety analyses for normal and emergency situations with the assessment of risk for radionuclide migration and determination of most vulnerable points in the multibarrier system for fissionable materials encapsulation.

The main physical processes defining the state of the nearest and distant zones of a storage area or a repository are as following: geomechanical, thermal mass transfer, and hydrogeological processes. It is expedient to note the main difficulty for prognosticating of parameters of those processes during the period when waste presents a potential hazard. That is the reason why all quantitative criteria have been oriented for maximum values of parameters concerning destruction of rock massifs. For example, concrete configurations with a law and intercommunications of parameters and processes will be further used. Their characteristics have been received during long-term operation of a set of underground facilities located in gneisses to the depth of 300 m from the surface.

1. GEOMECHANICAL STATE OF MASSIFS

1.1 Prognosis of potential possibility for disturbance of natural isolating properties of rock massif

It is well known the the main mechanism of radionuclide migration from an underground repository is due to a transport through the fractures in rocks after leaching from the matrix materials. Analysis of mining and geological conditions, properties and disturbances of hard rocks reveal that the main reasons for underground water penetration through underground repositories and its storage formations can be attributed to the following effects:

- opening of natural fractures and zones of tectonic disturbances and breaks induced by formation of zones with relative tension and movement of rock massif;
- formation of rock destruction zones and new main fractures on those areas of rock massifs where the prognosis for stresses approach or exceed the strength limit of the types of rocks investigated;
- formation of structural disturbances of the contour part of the massif under the action of technogenic loads and long periods of time.

Those reasons were considered from the point of view of the assessment for critical value when the risk of underground waters penetration to isolated radioactive materials emerged. Objects for prognosis while considering geomechanical criteria, i.e. the maximum values for required period of time taking into account concrete mining and geological conditions are as follows:

- zone of relative tension of rock massif formed at the expense of deformations and rock shifts (prognosis carried out by deformation values which exceed critical values for investigated rocks) $\varepsilon > \varepsilon_{cr}$ = 4 mm/m;
- zones of concentration of compression stresses approaching the limits of rock strength in the interval $\sigma < \sigma_{cr}$ (depending on the grade of rock disturbance);
- zones of increased structural disturbance of rock massif which execute the role of initial channels for underground waters filtration (prognosis can be carried out by exceeded values of some threshold of structural disturbance which corresponds to maximum geodynamic shift. Figures 1 and 2 show mathematical models and graphical results as well as calculations for stress-strain state of contour zones of massif and processes movements of thickness of the layer over underground facilities supposed for isolation of materials containing long-lived radionuclides.

Figure 3 shows graphs of dependence of time for rock destruction on different values of destruction factors characterising critical values of stresses when the process of rock disturbance begins.

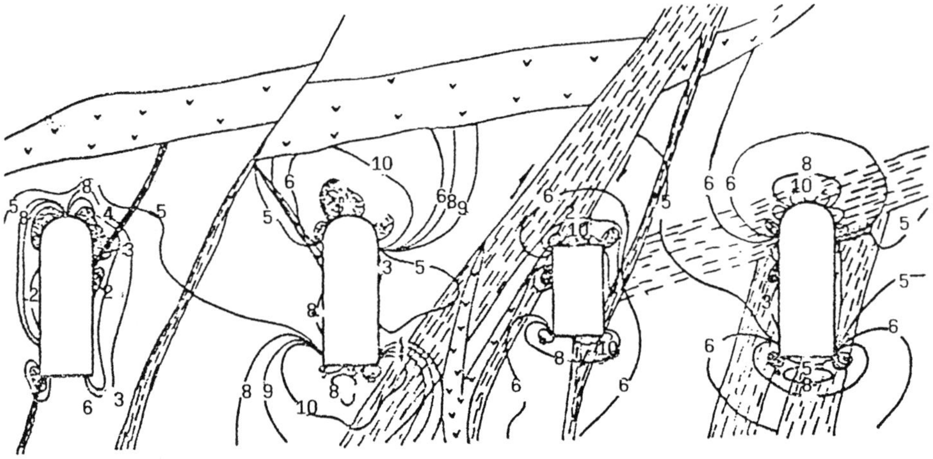

Fig. 1: Prognosis for potential areas of rock massif destruction

isolines for distribution of stress fields

zones of critical stresses - potential areas of destructed rocks

tectonic disturbances

$\varepsilon > 0$ = tension

$\varepsilon < 0$ = compression

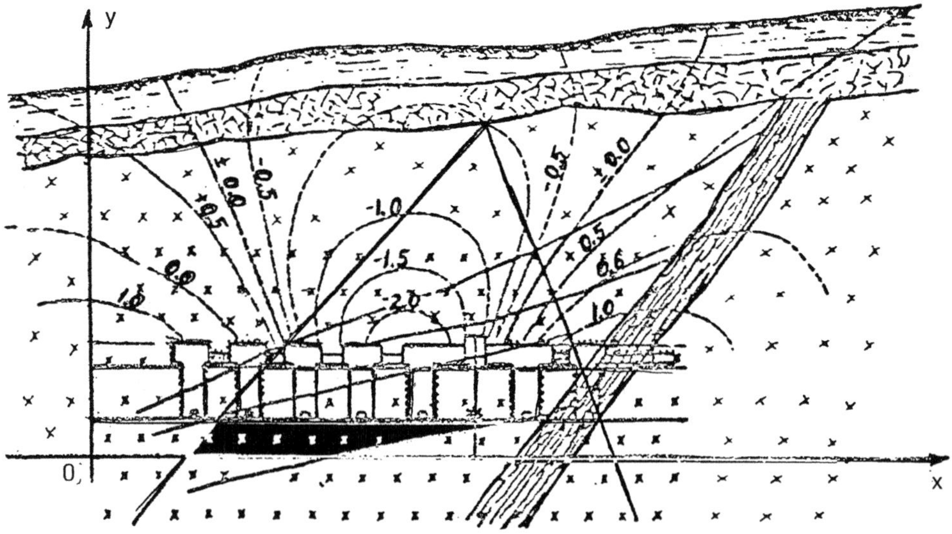

Horizontal deformations in massif

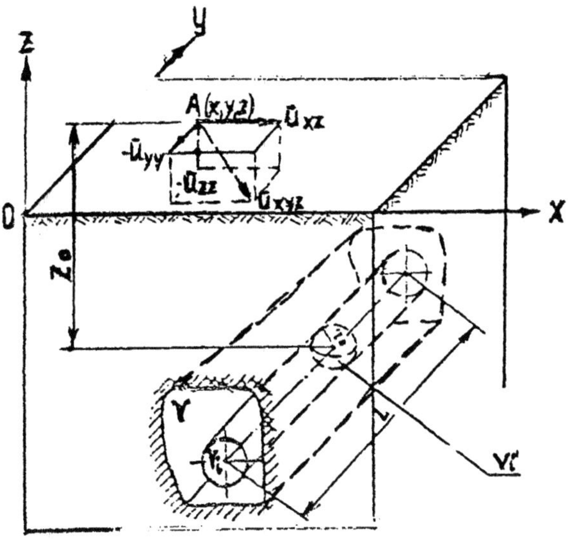

$$V = \iiint V_i \, \delta x \, \delta y \, \delta z$$

$$\tilde{U}_{xyz} = \Phi(u)\frac{V}{R^2} = K \cdot \Phi(u) \cdot r_0^3 \cdot \sum_{i=1}^{i=n} \frac{1}{(R_i)^2}$$

Fig. 2: Prognostication of parameters for process of rock shift

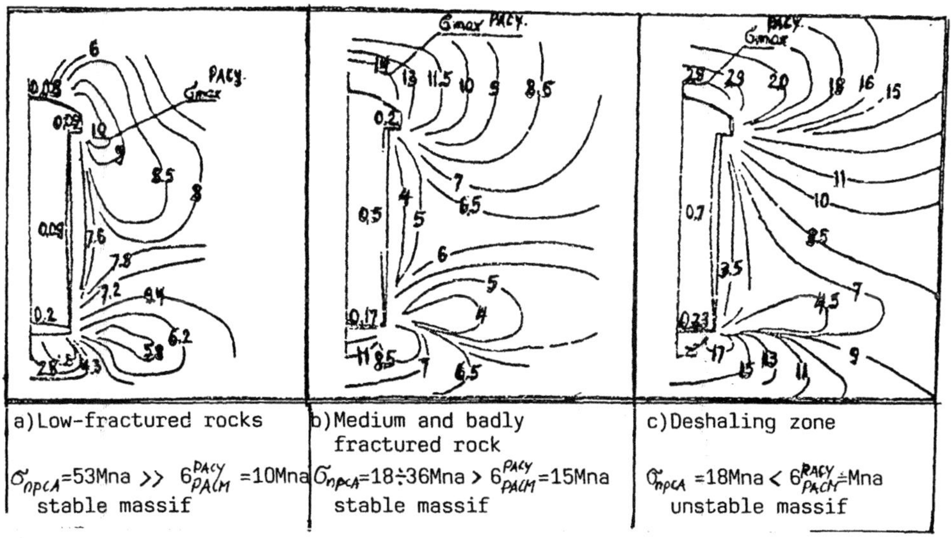

Fig. 3: Prediction of the time for rock destruction under the action of compression stresses

$$\sigma_s^i = \sum_{j}^{m} \left(A_{ss}^{ii} P_s^j + A_{sn}^{ij} P_n^j \right) + \sum_{j=M-1}^{N} \left(A_{ss}^{ii} D_s^j + A_{sn}^{ij} D_n^j \right)$$

$$\sigma_n^i = \sum_{j=1}^{M} \left(A_{ns}^{ij} P_s^j + A_{ns}^{ij} P_n^j \right) + \sum_{j=M+1}^{N} \left(A_{ns}^{ii} D_s^j + A_{nn}^{ij} D_n^j \right)$$

Calculation for time of destruction

$$t^0 = \frac{RT}{f_{k_1} f_{k_2} Y_i} \left[\left(U_0 - f_{k_1} f_{k_2} \gamma_0 \sigma_{0j} \right) + RTLn\left(\tau_0 f_{k_1} f_{k_2} \gamma Y_i \right) / RT \right]$$

where: R = Avogadro number
T = temperature
^0K = temperature
Y_i = rate of stress changes
U_0 = activation energy
γ = structural factor
f = correcting factors
τ_0 = 10-13

1.2 Prognosis for location of zones of probable massif destruction

After the construction of a massif structural and tectonic model and the calculation of the stress-strain state around the workings combining and analysis of the results of modelling of concentration of stresses is carried out, it gives mining and hydrogeological conditions. Correctness of used mathematical models has been checked by numerous field measurements of local field by method of unloading. Initial characteristics of rock (tension , elastic compression strength, elastic moduls, Poisson`s ratio, structural weakness coeffient, etc.) have been received in situ and in laboratory measurements by standard methods; relationship of natural stresses in tectonic blocks have been received by special methods.

To consider the conditions for emersion of water conducting channels, maximum values of stresses have been received when continuity break takes place:

ε_{cr} = 0.08 - 0.01 (charaterises presence of gas and water conducting channels)

ε_{cr} = 0.003 - 0.08 (newly appeared fractures do not form unique gas and water conducting fractures, and as a consequence on the way of underground waters movement additional resistance appears)

ε_{cr} = 0.002 - 0.003 (in this zone resistance accompanying mass transfer and through fractures do not form)

2. PROGNOSIS OF VALUES FOR WATER INFLOW TO UNDERGROUND STORAGES INDUCED BY FILTRATION OF UNDERGROUND WATERS THROUGH ZONES OF STRUCTURAL AND TECHNOGENIC DISTURBANCES

After consideration of the conditions for the disturbance of isolating properties of rock massif under the action of natural and technogenic factors, it is necessary to prognosticate volumes and rates of possible water inflows to underground working at the expense of infiltration of underground waters and conditions for emersion of hydraulic interrelation with surface waters.

For this purpose, results of numerous investigations in hard rocks were analysed using data from published sources dealing with object analogues. Relationships were received allowing to determine dependence of filtration coefficients on fracturing characteristics of hard rocks (see Figure 4). For this purpose the following was investigated:

- interrelations between filtration indices and fracture parameters in rock massif;
- interrelations between the conditions for formations of water conducting fractures in rocks and fractures in support;
- conformities with law when filtration pressures do not bring it to the formation of water conducting channels in massif and rocks as to function as an isolating screen. Using known relationships between permeability, filtration, fracturing factors, and hydraulic pressures (see Figure 5), and also analysis of interrelations between them, the condition has been received for the formation of water conducting channels.

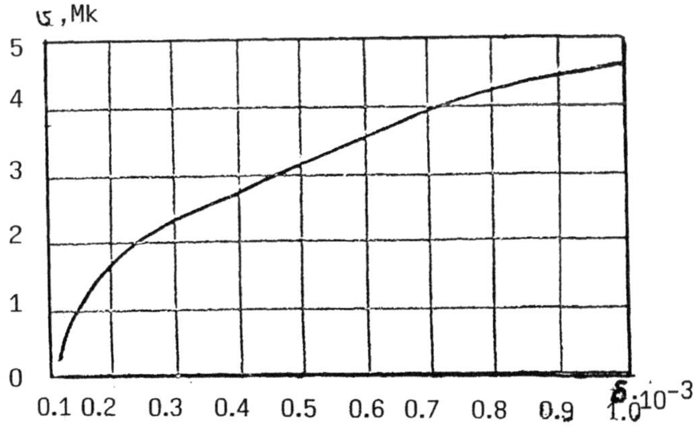

Fig. 4a: Dependence of the rate of soil destruction on the value of fracture opening

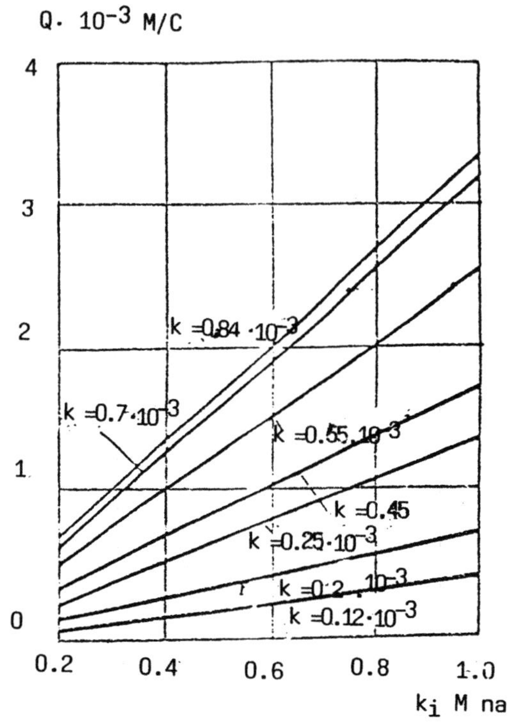

Fig. 4b: Dependence of the intensity of filtration through macrofractures on the value of underground water pressure

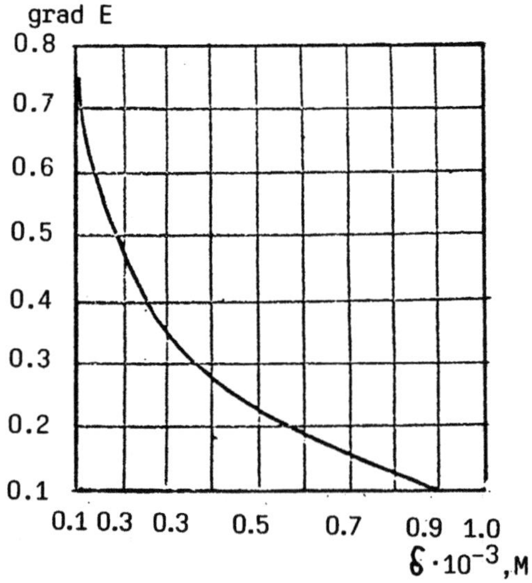

Fig. 5a: Dependence of gradient of underground water pressure through microfractures on the value of their opening

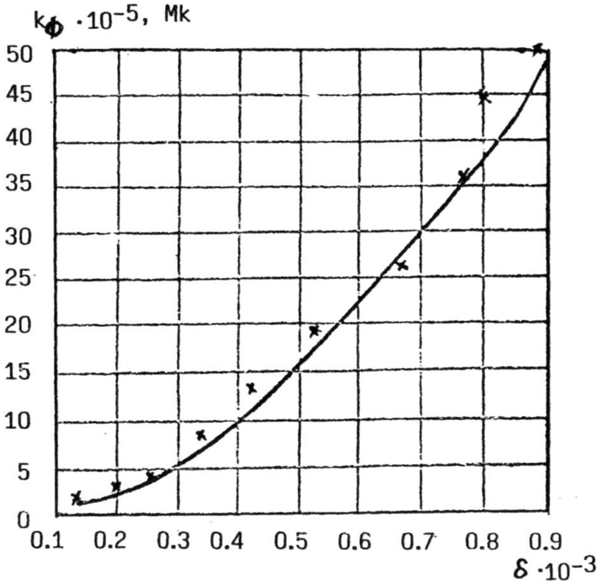

Fig. 5b: Dependence of filtration coefficient of rock on the value of fracture opening

X experimental results

— calculated curve

3.PROGNOSIS OF HYSROGEOLOGICAL CONDITIONS

To assess the halo of spreading of underground waters on the investigated site, possible ways have been explored for the movement of underground waters from the storage, dynamics of changing of fracture opening in rocks and in support, values of water inflows, maximum volumes and rates of filtration. It is necessary to prognosticate the various variants:

- on large zones of tectonic disturbunces,
- on zone of technogenic fracturing along the working,
- on zone of massif destruction and on rocks with maximum coefficient of structural weakening.

4. SAFETY ANALYSIS OF AN UNDERGROUND STORAGE FOR MATERIALS CONTAINING LONG-LIVED RADIONUCLIDES

Safety assessment during underground isolating of fissionable materials is based upon the investigation of possible items of emergency situations, analysis of scenarios of their development, successive descriptions of interrelated emergency sequences, quantitative substantiations of probabilities for their occurence.

For example, probablistic models of emergency situations have been considered for an underground storage for solidified radwaste containing complex compositions of radionuclides of which plutonium, cesium, and strontium are the most long-lived ones.

Probability of emergency situations includes:

- earthquakes,
- geodynamic movement,
- destruction of close-to-contour zones of openings,
- flodding of the storage.

All of them imply the probability of existence of parameters of models when an emergency situation will take place.

Total probability P_{Σ} for exceeding the values of allowable concentration of separate radionuclides on the boundary of sanitary protection zone is determined as a sum of corresponding probabilities for normal conditions of operation $P_{\Sigma,0}$ and every emergency situation $P_{\Sigma,i}$:

$$P_{\Sigma} = P_{\Sigma,0} + \sum_{i=1}^{4} P$$

4.1 Earthquakes

On the basis of mathematical modelling of conditions for fracturing in rocks and in support for given sizes of a working, depth of location, rock and concrete properties reverse task was decided, namely: what force of earthquakes create the conditions for fracturing in the considered multibarrier system? After that the probability for emersion of such an earthquake is determined for the region of the object location. To determine the repeatability o such seismic effects, statistical data of earthquake registration in the radius of 800 km from the supposed place of radioactive materials storage for the period from 1734 until today are listed in Figure 6.

The probability of emersion of such a situation for the required period of long-term storage will comprise

$$P_i = 5 \cdot 10^{-3}$$

To assess the probability of emergency situations sequence-flooding of working, the results have been used of numerous investigations in massifs of hard rocks having similar structural-tectonic conditions and rock properties. Maximum possible increase of the width of fracture opening in rock massif for considered conditions is fourfold, which corresponds to 16 times change of filtration coefficient, i.e. maximum possible range of change for filtration coefficient of rock massif will comprise 10.2 - 32.10 m/sec.

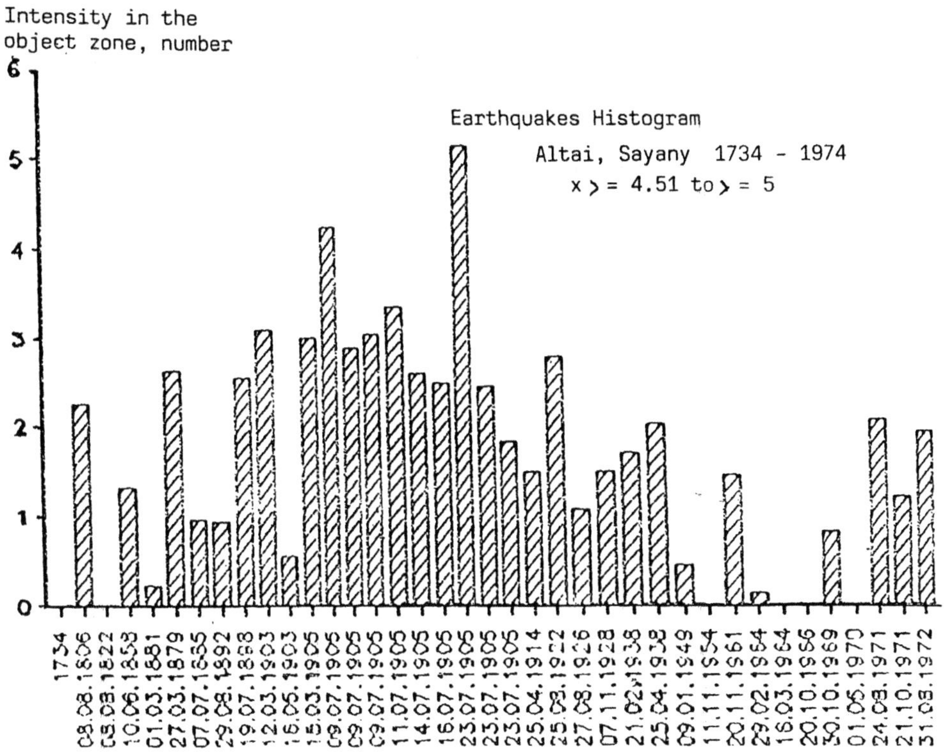

Fig. 6: Catalogue of intense earthquakes on the territory of the USSR from 1734 uo to 1975

4.2 Geodynamic shift

For the same mining and geological conditions on the basis of the results of mathematical modelling, critical values for shift amplitudes have been determined when fractures in the support can appear and their opening in zones of tectonic disturbances can increase which will create conditions for an emergency sequence: water influx into an underground working; the results are shown in Figure 7.

To assess possible amplitude of shift in an underground facility as criteria of modern geodynamic activity data were adopted of long-term observations over the change of the width of fracture opening in the support of mining workings crossed by tectonic disturbance. As a result of statistical processing of data received over 11 years on 17 boreholes prognosticated assessment of possible amplitude of shift for a required period of operation will comprise, as can be deduced from Figure 7, $= 1.17 \cdot 10^{-3}$ m. Probability for emergenc situation like "Geodynamic shift" is $P = 3.9 \cdot 10^{-2}$.

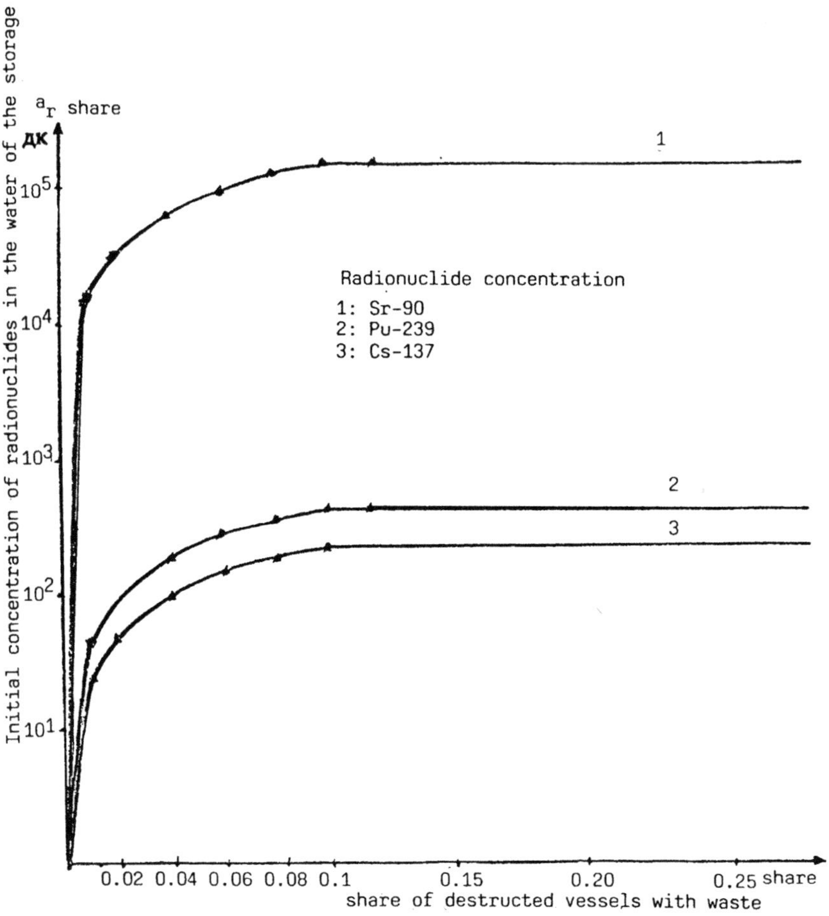

Fig. 7a: Dependence of radionuclide concentration in the water of the storage on stability of engineered barrier "metal"

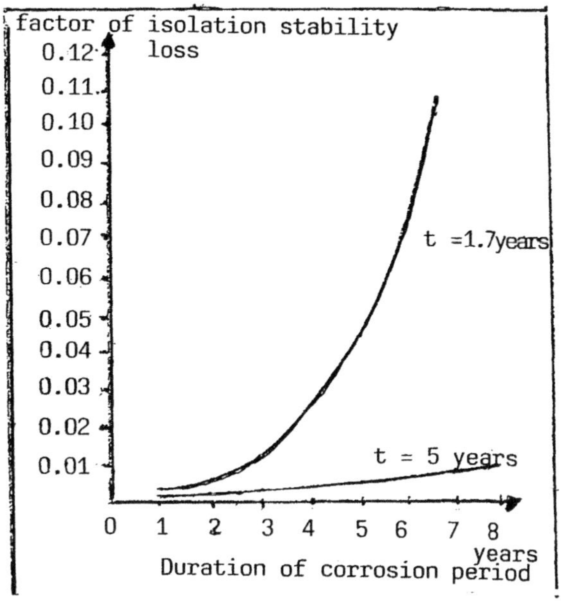

Fig. 7b: Characteristic of isolation stability of the engineered barrier "metal"

4.3 Working caving

Depending on the geological structure of contour zone of the working and the type of rock fracturing, stress-strain state of rocks have been modelled and time have been determined within which the working will keep its stability. Probability found is P = 0.29 3. To receive reliable assessments of possible water inflow into the working, values of real water inflows into existing underground workings under normal conditions of operation have been used.

Probability of workings flooding with water under conditions of non-regular situation after working caving is as follows: P = 0.45 3.1.

5. ASSESSMENT OF GRADE OF INFLUENCE OF ENGINEERED BARRIERS ON THE SAFETY OF UNDERGROUND ISOLATION OF LONG-LIVED RADIONUCLIDES IN GEOLOGICAL FORMATIONS

When maximum grade destruction of rock massif takes place and predicted value of water inflow into the storage are reached under the condition of situation: 4.1; 4.2; and 4.3, the grade has been assessed for keeping isolating stability by metallic container subjected to long-term effect of ground water. After isolating stability of a natural barrier (support and containers) have failed and leaching of radionuclides from matrix materials will begin. As a result of solidified blocks with waste, dependences of radionuclide concentrations in storage waters, the times have been determined when leaching with underground waters starts (see Figure 8).

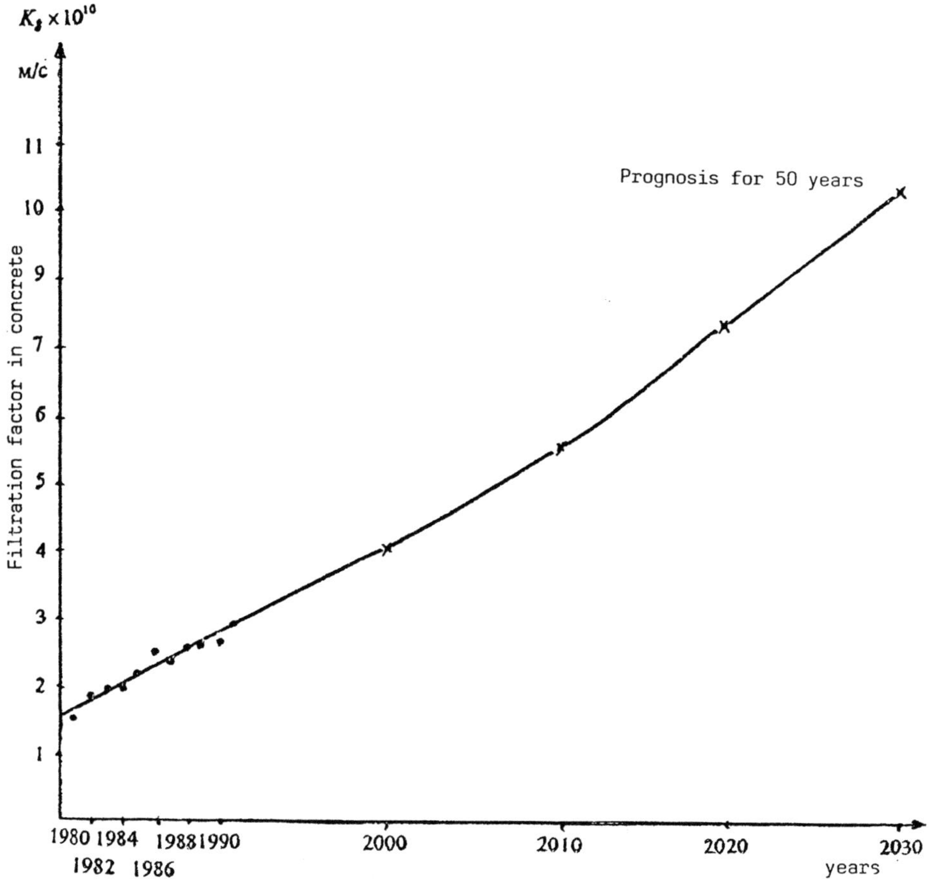

Fig. 8a: Prognosis for dynamics change of radionuclide concentration in underground waters during migration

- • results obtained from field measurements for fracture opening
- ✗ predicted dependence of the dynamics for filtration change

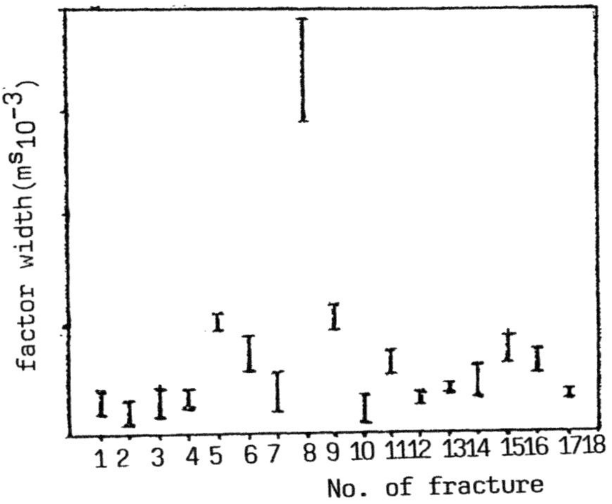

Fig. 8b: Amplitude of change for fractures within the object support 257 (arch) for 1980 - 1991

The volume of contaminated underground waters which can spread beyond the boundaries of the storage emergency situations is determined by durability of storage staying in flooded state and by filtration characteristics of rock massif.

Based on the analysis of possible ranges of values for massif filtration parameters, rate of underground waters movement, sorption properties of rocks, the range of radii for the halo of plutonium, cesium and strontium radionuclides spreading-out has been determined (see Figure 9).

CONCLUSIONS

1. Proposed methodical approach allows:
   - to conduct quantitative safety assessment of utilization of a concrete geological site for localization of given volume and composition of materials containing long-lived nuclides;
   - to assess the risk of long-term storage of fissionable materials on the surface and in the underground;
   - to determine most vulnerable points in a multibarrier system of isolation to adopt scientifically substantiated ecologically safe engineered decisions;
   - to find well grounded solutions on priority and details of investigations of rock massif state and economical expediency for the construction of an underground storage or repository;
   - to assess the risk for the public caused by location of an underground storage.
2. While analysing the stability of properties of multibarrier system for underground isolation of materials containing long-lived nuclides the main barrier is rock massif keeping its properties during geological periods.

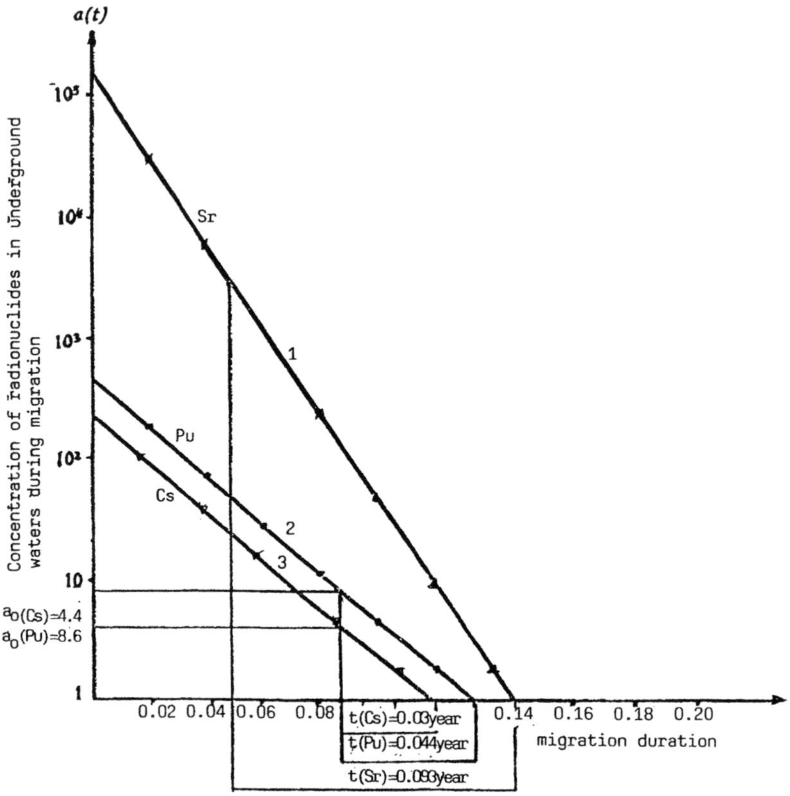

Fig. 9a: Scheme for determination of the boundary for sanitary-estrangement zone with account of real parameters of multibarrier isolation system

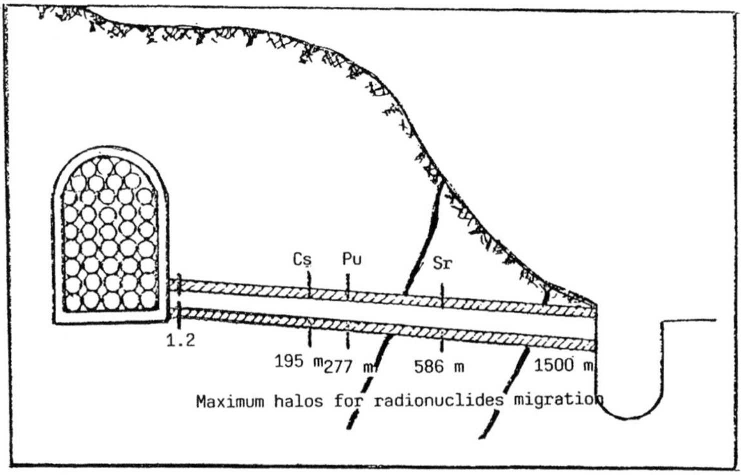

Fig. 9b: Maximum limits of halos for radionuclides spreading from the storage site

DISPOSITION OF PLUTONIUM IN DEEP BOREHOLES

WILLIAM G. HALSEY
LESLIE J. JARDINE
CARL E. WALTER
Lawrence Livermore National Laboratory
P.O. Box 808, Livermore, CA 94551 USA

ABSTRACT

Substantial inventories of excess plutonium are expected to result from dismantling U.S. and Russian nuclear weapons. Disposition of this material should be a high priority in both countries. Various disposition options are under consideration. One option is to place the plutonium either directly or in an immobilized form at the bottom of a deep borehole that is then sealed. Deep-borehole disposition involves placing plutonium several kilometers deep into old, stable, rock formations in which negligible free water is present. Containment assurance is based on the presence of ancient groundwater, indicating lack of migration and interaction with the biosphere. Recovery would be extremely difficult (costly) and impossible to accomplish clandestinely.

1. INTRODUCTION

As a result of recent changes throughout the world, a substantial inventory of excess separated plutonium is expected to result from dismantling U.S. nuclear weapons. The safe and secure management and eventual disposition of this plutonium, and of a similar inventory in Russia, are a high national priority.

Various options are under consideration for disposition of this material. These options can be broadly categorized in two groups: (1) use as an energy source followed by geological disposal and (2) nonuse followed by geological disposal. The use options have in common the generation of reactor spent fuel that would be disposed of in mined geologic repositories that are being developed for commercial reactor spent fuel and high-level waste (HLW). Excess weapon plutonium disposed of in this manner would produce a significant amount of energy and would subsequently result in spent fuel that, from a safeguarding point of view, would not differ significantly from commercial reactor spent fuel.

Plutonium disposition nonuse options can be divided into two types: (1) indirect and (2) direct. Indirect plutonium disposition options require physical immobilization of the plutonium in a particular solid material form. If appropriate amounts of HLW are included in the immobilized form, the material becomes similar to reactor spent fuel and

E.R. Merz and C.E. Walter (eds.), Disposal of Weapon Plutonium, 237-247.

subsequently can be placed in a mined geological repository. If HLW is not included in the immobilized form, disposition in a deep borehole is being studied. For direct plutonium disposition, little or no preparation of the plutonium and no addition of HLW are required. That is, transformation to a particular chemical or physical form may not be required of the end product, thus allowing plutonium metal, oxide, or other compounds to be dispositioned directly. Deep-borehole disposition is being studied both in the United States and in Russia. The U.S. National Academy of Sciences [1] recommended further assessment of the deep-borehole option because that method is potentially easier, quicker, and less expensive to implement.

Deep-borehole disposition involves placing excess plutonium several kilometers deep into old, stable rock formations with little free water present. The basis for long-term containment is the presence of ancient groundwater. Its presence indicates lack of migration and thus no expected interaction with the accessible environment until all the plutonium has substantially decayed (about a quarter-million years). Conceptual design studies serve as the basis for assessing the feasibility of deep-borehole disposition and allow systematic comparison to other options. Principal issues of concern include regulatory, statutory, and policy considerations and availability of sites with desirable characteristics. The technology for drilling deep holes, characterizing them, emplacing the plutonium, and finally sealing the boreholes appears to be well in hand.

1.1 SCREENING OF DISPOSITION OPTIONS

Evaluation criteria have been developed to enable a preliminary screening of the various options for further evaluation. These criteria include (1) resistance to theft or diversion by unauthorized parties; (2) resistance to retrieval, extraction, and reuse by the host nation; (3) technical viability; (4) compliance with environmental, safety, and health standards; (5) cost effectiveness; (6) timeliness; (7) fostering progress and cooperation with Russia; and (8) public and institutional acceptance. Disposition options, including the deep-borehole option, that survive this preliminary screening will be more fully analyzed in the future. Excess weapon plutonium exists in various forms and purity. Some of the purer forms are candidates for disposition in a use option such as mixed oxide (MOX) fuel for power reactors. However, the less pure and more dilute forms may require costly preprocessing, making them unsuitable for MOX fuel fabrication. The latter are excellent candidates for the deep-borehole disposition option.

1.2. PREVIOUS CONSIDERATIONS OF DEEP BOREHOLE DISPOSITION

Deep-borehole disposition has been considered in recent decades for disposal of both hazardous and radioactive wastes [2]. This method was investigated in the 1970s for disposal of HLW, including reactor spent fuel. Several characteristics of the deep-borehole option for disposal of reactor spent fuel led several nations to drop it in favor of a mined geologic repository. However, the same characteristics that were reasons for rejecting deep-borehole disposal of HLW and reactor spent fuel become assets in the context of deep-borehole disposal of excess weapon plutonium (see Table 1) [3].

TABLE 1. Deep-borehole issues for commercial reactor spent fuel and plutonium disposition applications.

Issue	Reactor spent fuel	Plutonium disposition
Retrievability	Difficult, not desirable	Difficult, desirable
Volume	Large volume, many holes	Small volume, few holes
Heat generation	Limiting	Comparatively low
Drilling technology	Limited then	Greatly improved now

Issues such as those in Table 1, and the desire to explore an option independent from the mined geological repository being considered for reactor spent fuel, are driving the current interest in investigating deep-borehole technology for disposition of excess weapon plutonium.

2. DESIGN CONFIGURATIONS

Preliminary design descriptions of the deep-borehole disposition option have been prepared to allow early comparison with other options [4,5]. These configurations are not site-specific but include descriptions of typical surface facilities, drilling and emplacement facilities, emplacement procedures, and as-emplaced disposition forms. Figure 1 shows a stepped borehole 4 to 6 km deep with plutonium emplacement below a 2-km isolation zone. Arbitrarily, it is assumed that the upper 2 km of deep boreholes would be suitably backfilled with nonfissile material and sealed. Figure 1 also illustrates emplacement of 2.5-cm-diameter ceramic pellets containing plutonium that are grouted into an uncased borehole 0.7 m in diameter at depths between 2 and 6 km.

2.1 DRILLING EXPERIENCE AND TECHNOLOGY

Several deep boreholes have been drilled in recent years [3]. Table 2 lists the characteristics of some of these boreholes. These examples indicate that drilling requirements for the deep-borehole disposition option fall within the range of successful experience.

The drilling cost of a deep borehole increases more than linearly with total depth. Drilling cost is not as strong a function of diameter as of depth. For example, with a 2-km isolation depth, one 8-km-deep borehole of a given diameter would accommodate as much plutonium as three 4-km boreholes of the same diameter and could cost about the same. The diameter of the less deep boreholes could probably also be greater, thus accommodating more plutonium. The number, diameter, and depth of deep boreholes required for cost-effective disposition of 50,000 kg of excess weapon plutonium has not yet been determined.

Table 3 summarizes the status of deep-borehole drilling technology applicable to plutonium disposition. Three classifications of the technology have been used to illustrate the status: (1) standard practice, (2) readily achievable, and (3) the limits of technology. The typical borehole diameter decreases as a function of depth in a stepwise manner. In general, the drilling process starts with a relatively large-diameter drill bit down to an interim depth. A casing (metal liner) is inserted into the hole. A cement

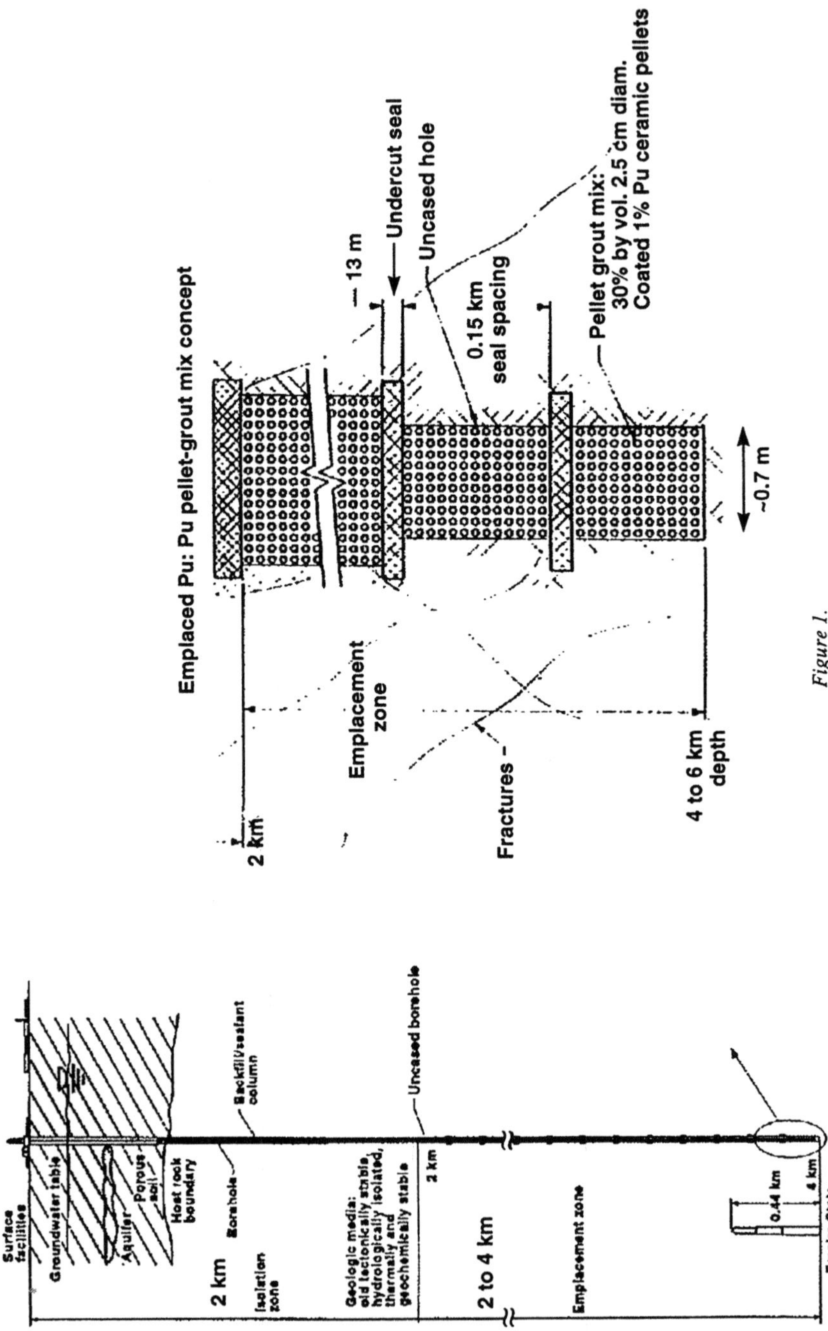

Figure 1.
Conceptual configuration of a deep borehole for disposition of weapon plutonium.

TABLE 2. Characteristics of actual (relevant) deep boreholes.

Country	Name of shaft or company	Total depth (TD) (km)	Diam. at TD, (m)	Rock type	Purpose	Start date
France	FRA-1 Sancerre-Couy	3.5	0.10	Partial crystalline	Scientific	~1987
Germany	KTB Hauptbohrung	7.5	0.31	Diverse crystalline	Scientific	1990
Germany	KTB Vorbohrung	4.0	0.15	Diverse crystalline	Scientific	1987
Russia	Kola	12.6	0.10	Crystalline	Scientific	1970
Sweden	Gravberg-1	6.6	0.10	Crystalline	Petroleum/ scientific	1986
U.S. (Nevada)	Mobil 1-A	6.0	0.15	Partial crystalline	Petroleum exploration	1979
U.S. (Texas)	Nellie-1	5.8	0.10	Partial crystalline	Petroleum exploration	1983
U.S. (New Mexico)	Fenton Hill	4.7	0.20	Crystalline overladen by volcanic	Scientific	1979
U.S. (Oklahoma)	Haraway 1-27	3.8	0.15	Crystalline exploration	Petroleum	1981
U.S. (Alaska)	DOE	1.9	1.40	Partial crystalline	Explosive test	1970

slurry is then pumped into the casing and up the annulus between the casing and the hole formation. The cemented casing serves several purposes. First, it prevents groundwater from entering in the upper portion of the hole. Second, at greater depth, it prevents brine from entering the borehole during drilling. Third, it prevents collapse of the upper regions of the hole, where more unstable geology may be found. Finally, it allows any fractured formations that may be encountered to be the sealed off. Once the casing is in place, a smaller drill bit is guided through the casing, and the hole is drilled deeper. This procedure can be repeated, resulting in several step-downs in casing size. At some depth, casing the hole is discontinued because it is no longer economical. Table 3 shows that 0.7-m-diameter, 4-km-deep boreholes are within what is considered to be the technology limit, and these parameters are the bases of current studies.

TABLE 3. Summary of deep-borehole drilling technology and experience applicable to plutonium disposition.

	Hole diameter at total depth (m)		
Hole depth (km)	Standard practice	Readily achievable	Technology limit
1.4[a]	0.5	0.8	1.3
3.0[b]	0.4	0.7	0.9
4.0[b]	0.3	0.6	0.7

[a] cased

[b] uncased

2.2. DESIRABLE SITE CHARACTERISTICS

Generic site characteristics that would be desirable for a deep borehole have been developed. The primary considerations concern the subsurface emplacement zone, for which desirable characteristics include (1) long history of tectonic stability, (2) limited open fracture/void space, (3) ancient and preferably connate water presence (if any), (4) sufficient formation thickness for cost-effective emplacement, (5) mechanical strength for hole integrity, (6) gradual thermal gradient, and (7) benign geochemistry. Rock types that have been considered include crystalline/metamorphic "basement" rocks (e.g., granite), tuffs, evaporites (e.g., salt), sedimentary rocks (e.g., shales), and mafic lavas (e.g., basalt). Additional site characteristics of importance include access to transportation, surface water for drilling, reasonable distance from population centers, and a lack of evident geological resources that could attract future exploitation.

The suitability of current technology for remotely determining downhole conditions and material properties is being evaluated. It is expected that an adequate understanding can be obtained by means of multiple characterization boreholes and extrapolation of existing techniques such as continually cored sample holes, detailed well logging, and cross-hole testing.

2.3. POTENTIAL DISPOSITION FORMS

A wide range of disposition forms has been considered. Preliminary design examples include plutonium metal, plutonium oxides, and plutonium immobilized in ceramics, glass, or metal alloys. It also may be possible to emplace plutonium-rich residues, scraps, and ash in deep boreholes. No conclusions have yet been reached on the preferred form. Direct disposition of plutonium is the simplest option if it proves to be viable. Indirect disposition of immobilized forms has potentially better isolation performance and criticality control. Radionuclides have not been included in the immobilized forms because they are considered unnecessary as a deterrent to diversion from deep-borehole disposition. If radionuclides were included, both the radiation and thermal output would needlessly complicate emplacement operations.

2.4. SURFACE FACILITIES

Preliminary surface-facility layouts have been prepared that satisfy functional requirements, including receiving plutonium from either rail or truck transport, processing as required, in-process storage, waste management, material security, and operating personnel. Facilities have been sized for disposition of 50,000 kg of plutonium over a 10-year operational period. It is assumed that these plutonium materials would require active security until the borehole is stemmed and sealed. It also is assumed that IAEA safeguards will be applied. Figure 2 is a typical site arrangement for such a facility [4,5].

2.5. DRILLING, EMPLACEMENT, STEMMING, AND SEALING

Technology of interest for drilling, emplacement, stemming, and sealing can be found in underground nuclear weapon testing technology, deep geotechnical research drilling

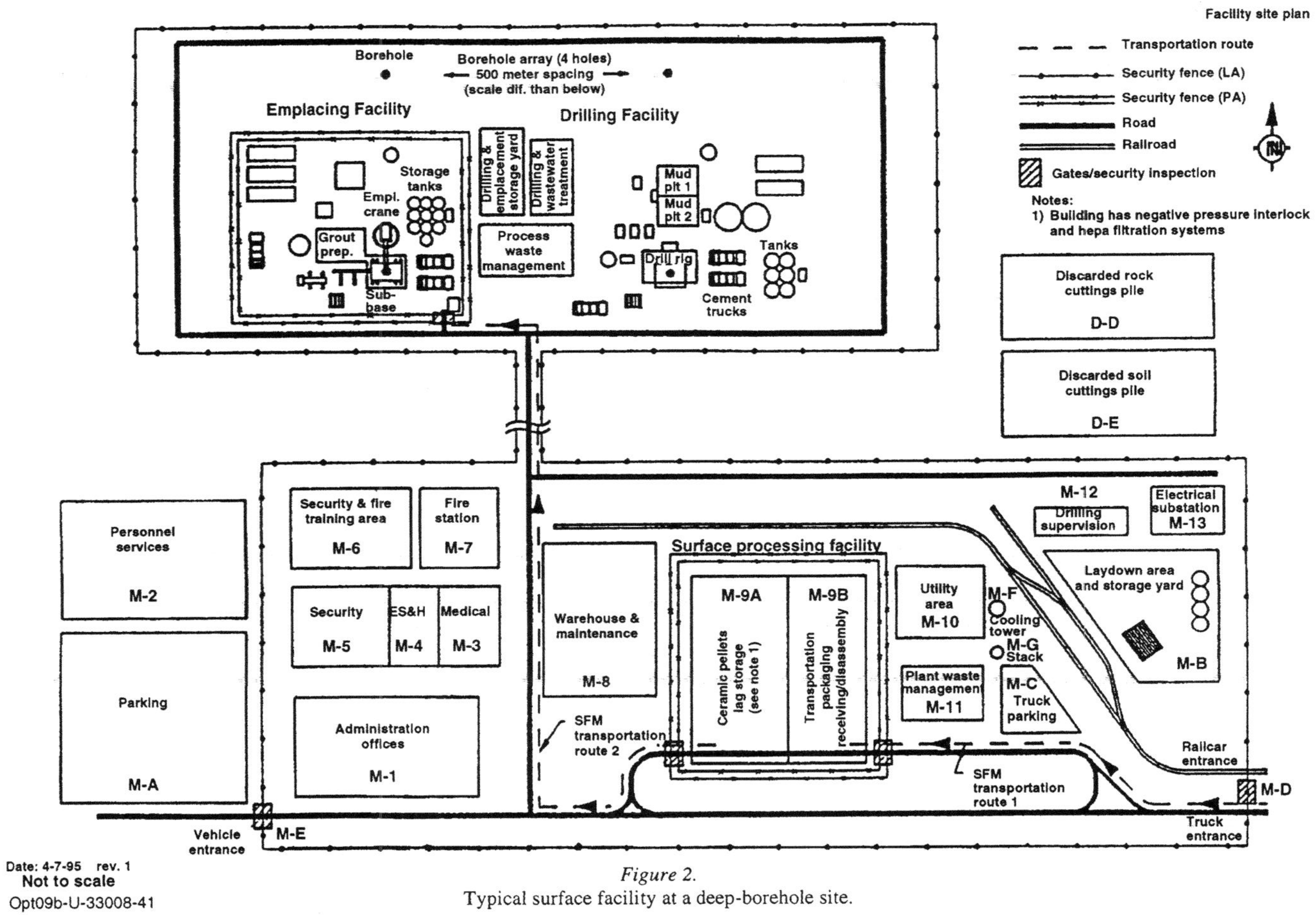

Figure 2.
Typical surface facility at a deep-borehole site.

programs, and the mineral exploration industry. Preliminary evaluations suggest that drilling boreholes in suitable rock to depths of 4 to 6 km with usable bottom diameters of 0.5 to 0.7 m can be achieved with reasonable time and cost. Emplacement of canister strings weighing thousands of kilograms can be accomplished with existing equipment. High-integrity seal technology is available from a variety of applications, but detailed designs for disposition have not yet been developed.

2.6. NUMBER AND SIZE OF DEEP BOREHOLES NEEDED

The number and size of deep boreholes required depend on the form and amount of plutonium to be dispositioned. In general, large-diameter holes are not required, and the appropriate size will likely not exceed ~0.7 m at the deepest emplacement point. In addition to other site and cost considerations, the total depth, surface isolation zone, and length of the emplacement zone depend on the in situ geology. Reasonable assumptions for total depth, surface isolation zone, and length of the emplacement zone are 4, 2, and 2 km, respectively. With these assumptions, the number of deep boreholes required to dispose of 50,000 kg of excess weapon plutonium can be determined for various forms of the material. Table 4 provides preliminary estimates.

TABLE 4. Number of deep boreholes required to dispose of 50,000 kg of Pu. Total depth = 4 km, hole diameter = 0.7 m.

Disposition option	Range of plutonium loading	Potential number of holes
Direct Pu oxide, metal, etc.	1–14 kg/m	2–25
Indirect, immobilized forms: Grouted ceramic pellets, glass and ceramic logs, and metal alloys	0.25–12 wt%	1–100

3. SAFETY CONSIDERATIONS

Preliminary safety considerations, both operational and postoperational, have been assessed. The unique considerations for deep-borehole disposition are the long-term safety of the emplaced material.

3.1. LONG-TERM PLUTONIUM ISOLATION

Permanent isolation of plutonium in deep boreholes relies primarily on the inherent characteristics of the geologic medium. The emplacement canisters are relied upon only for operations and are not expected to isolate the material from the host rock. Some immobilized disposition forms would retain integrity over the long term in the emplaced environment. In any case, if the site has suitable characteristics, by definition, no significant natural pathways exist for radionuclide migration. A remaining requirement is to show that insufficient potential exists for upward migration through the stemmed and sealed borehole. Also, it must be shown that vertical-flow pathways are not created in the host rock during drilling, operation, stemming, and sealing or during the postclosure thermal transient. The very high lithostatic load in both the host rock and the stemmed borehole at the emplacement depth are expected to inhibit opening of fractures and to limit voids significantly. Mobile water (if any) would have a high ionic strength, with a vertical density gradient sufficient to suppress thermal buoyancy. Periodic pressure or undercut seals would be used to interrupt new stress-relief fractures created parallel to the bore-hole. These are some of the key processes that would require verification in a characterization program.

3.2 CRITICALITY CONTROL

Emplacement of fissile material in a constrained volume, as in the case of the deep borehole, requires attention to the potential for critical configurations. Several approaches are being studied for operational and long-term (postclosure) criticality control. Individual emplacement packages can be designed to contain criticality-limited quantities of plutonium. Neutron absorbers can be incorporated into the disposition form and into the packing material surrounding the canisters. Plutonium can be physically dispersed throughout a matrix at concentrations that maintain subcriticality. Combinations of these approaches are included in current designs. In addition, preliminary studies are underway to consider the potential for reconcentration of plutonium in the geosystem and whether any adverse consequences would be expected from postclosure criticality.

4. REGULATORY AND LICENSING ISSUES

Perhaps the greatest current uncertainty in the feasibility of deep-borehole disposition is the possible imposition of regulatory and licensing requirements. Because concentrated fissile material in significant quantities previously has not been considered for direct disposition, current waste management regulations are not applicable. Existing or planned facilities may be useful as relative benchmarks to suggest regulatory stringency. Such facilities include the Greater Confinement Disposal Facility (GCDF) at the Nevada Test Site for disposal of small quantities (~6 kg of plutonium), the Waste Isolation Pilot Plant (WIPP) (near Carlsbad, New Mexico) being developed for disposal of larger quantities of defense transuranic wastes (<10 t of plutonium), and the High-Level Radioactive Waste repository (at Yucca Mountain, Nevada) designed to emplace much larger quantities of plutonium (~700 t) incorporated in reactor spent fuel. WIPP is a useful precedent for developing a custom regulatory environment to assess a deep-borehole facility because the hazard levels are similar and both are unique facilities. Some precedent exists for deep injection of hazardous material into waters that do not interact with accessible waters, and this approach is central to the feasibility of deep-borehole disposition. Current plans are to initiate dialogue with the Nuclear Regulatory Commission and the Environmental Protection Agency to consider possible regulation and licensing of such a facility.

5. CONCLUSION

Disposition of excess weapon plutonium in deep boreholes is in a preliminary scoping assessment stage to allow comparison with other disposition options. Initial studies of technical feasibility, potential safety issues, and potential cost and schedule are encouraging. Numerous issues require further assessment, with regulatory implementation being the greatest uncertainty. The possible occurrence of criticality in the long term and its effect, should that occur, require further study.

6. ACKNOWLEDGMENT

This work was performed under the auspices of the U.S. Department of Energy by the Lawrence Livermore National Laboratory under Contract No. W-7405-ENG-48.

7. REFERENCES

1. National Academy of Sciences (1994) Management and Disposition of Excess Weapons Plutonium, National Academy Press, Washington, DC.
2. O'Brien, M. T., Cohen, L. H., Narasimham, T. N., Simkin, T. L., and Wollenberg, H. A. (1979) The Very Deep Hole Concept: Evaluation of an Alternative for Nuclear Waste Disposal, Lawrence Berkeley Laboratory, Berkeley, Calif., LBL-7089.
3. Halsey, W. G. (1995) Regulatory Issues for Deep Borehole Plutonium Disposition, Lawrence Livermore National Laboratory, Livermore, Calif., UCRL-JC-120593.
4. Wijesinghe, A. M. (1995) Deep Borehole Disposal Facility PEIS Data Input Report for Immobilized Disposal, Lawrence Livermore National Laboratory, Livermore, Calif., UCRL-ID-119735.
5. Wijesinghe, A. M. (1995) Deep Borehole Disposal Facility PEIS Data Input Report for Direct Disposal, Lawrence Livermore National Laboratory, Livermore, Calif., UCRL-ID-119481.

DISCLAIMER

THE PROCESS OF THE EQUIPMENT AND ROOM SURFACE Pu-DECONTAMINATION AND MEANS TO PREVENT THEIR CONTAMINATION

Alexander B. Pavlov, Ye. I. Yulikov
All-Russian Research and Design
Institute for Enery Technology
VNIPIET, Savushkina str.,82
197228 St.Petersburg, Russia

The presentation is an attempt to give the general idea of the Russian practices to prevent plutonium contaminations in nuclear industry. The general impression of the measures is shown in Figure 1.

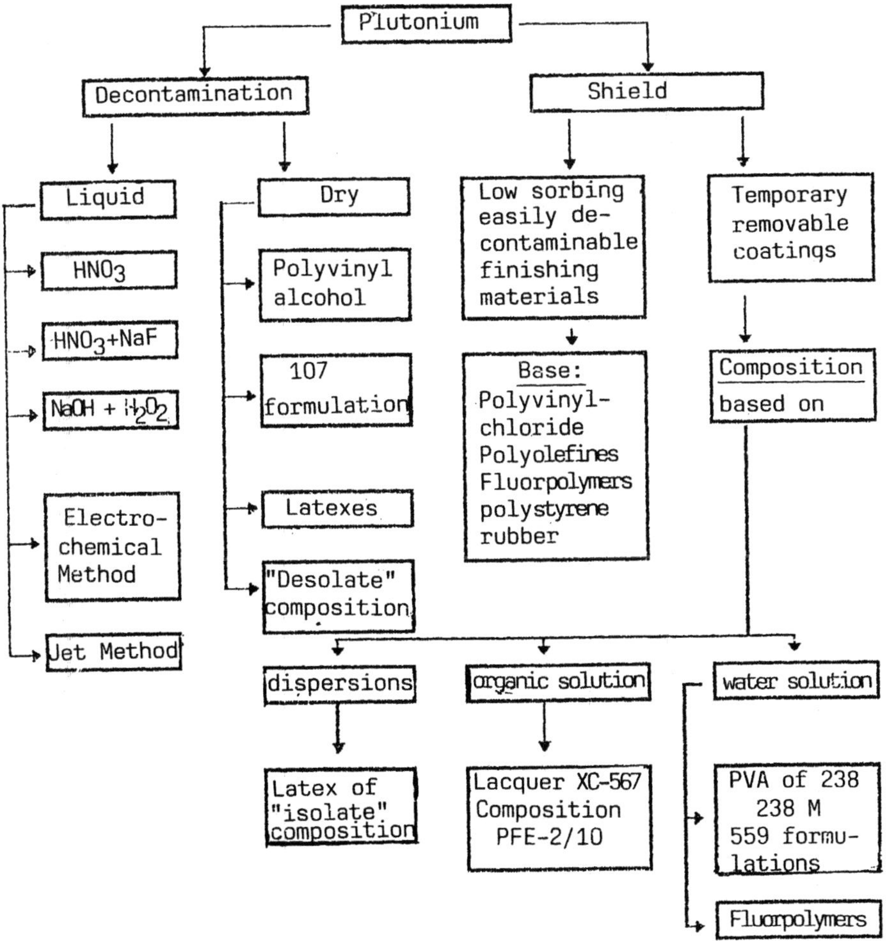

Fig. 1: Pu-Decontamination and Protective Measures

E.R. Merz and C.E. Walter (eds.), Disposal of Weapon Plutonium, 249-257.

The difficulties arising in plutonium-contaminated facilities purification are of general nature. They result from the chemical character of this element and its extremely high toxicity due to which the permissible levels of surface contamination according to the radiation safety standards are four times lower (5 alpha-particles/cm^2.min) than for most of alpha-radionuclides. This complicates the decontamination process.

The usual practice is to deal with the surfaces contaminated with plutonium from non-organic and organic media from air by solid and liquid aerosols, powder and dust plutonium contamination at nuclear power plants is not the determining factor and is to be considered as emergency, connected with the failure of fuel element claddings. Plutonium contaminations are most characteristic of radiochemical spent fuel reprocessing plants and research laboratories.

Thus, in chemical reprocessing plants the plutonium contamination is indicated already at the head end of the fuel reprocessing, that means during the fuel assembly cutting in the chambers of fuel element preparation in particular. By our estimates, the plutonium quantity in the chambers of fuel element preparation for cutting is in the order of 10 mg/m^2 of the surface. Further in the process, as plutonium is being extracted, the surface specific contamination increases coming to its maximum in the affinage and final product areas. So, at the start of the chemical metallurgical plant operation, when no special contamination protection measures were used, the contamination rate of the air in the maintenance area reached several hundred MPC (maximum permissible concentration).

It is known that the radioactive contamination of surfaces results from sorption processes (physical, chemosorption) and adhesion aggravated by diffusion, phase deposits as sediments of corrosion and process origin, radionuclide inclusion in oxide films, etc.

Because of the complex behavior of plutonium in aqueous acid solution, its sorption by an ion-exchange mechanism as true- and pseudocolloids of the polymerized plutonium is possible due to the formation of strong complexes with chemically active surfaces. The adsorption rate of colloids depends on the surface type: the glass adsorption is 1.6 mg/cm^2 /1/, that of steel is much higher. The polymeric materials depending on their nature can also adsorb plutonium nearly completely if it is carrier-free in solution.

The contaminant is fixed by the material still stronger if it diffuses into it forming an absorbed phase which cannot be removed without destruction of the surface layer of the material. As it is found out the polyvinylchloride plasticate of 57-40 formulation being contaminated with plutonium nitrate from the solution at pH 0.1 and 2.2 during 690 hours at a temperature equal to 42^oC sorbs 95% plutonium, while with the desalter ($NaNO_3$, 7 M/l) use, plutonium diffusion into the plasticate increases significantly, which can be approximately described by the Fick`s law with the diffusion coefficient of 8.3×10^{-14} cm^2/s.

In plasticate contamination by drops, plutonium diffusion occurs without desalter adding which is favored by salt and hydrogen ion concentration increase during the drops drying /2/.

The research workers who deal with plutonium pointed out the difficulty of this removing from polymeric materials, especially from rubber. Thus, plutonium is strongly sorbed by different surfaces and in significant quantities which requires special solutions and conditions for its decontamination.

The standard operation before decontamination of the tankage equipment is washing from plutonium and uranium usually carried out using 3 to 4 normal nitric acid solutions. After uranium and plutonium waste concentrations are received, decontamination from beta-gamma radionuclides is carried out using a bath redox composition. However, due to uranium and plutonium containing process deposits, concentration of these elements in the decontamination solutions is often higher than in the waste one, which is due to the insufficient dissolving power of nitric acid. Besides, due to the deposits, a large number of decontamination cycles (up to 10-15) using redox solutions is required which results in huge amounts of wastes. To eleminate the mentioned disadvantages of the standard decontamination system, the medium concentration nitric acid-based solutions with small content of fluoride ions were proposed with low corrosiveness to the equipment main structural materials (mainly stainless steel); these solutions permit efficient deposit removal together with plutonium and uranium for 1-2 cycles with the reduction of redox cycles from 10-15 to 1-3.

A special decontamination process is needed to wash the equipment in contact with the organic phase. This is due to the extractant films on the surfaces formation of organic deposits and interphase conglomerates. The deposits formed can pick up significant amounts of plutonium and uranium from process solutions. The organic deposits serve as screens preventing the action of decontamination solutions upon the surfaces, the decontamination factors being 6-60 times lower without their removal than in the case with the organics removed. The compulsory stage of the extraction equipment decontamination is, thus the organic deposits removal.

For this reason, alkali peroxide solutions were proposed and production-tested which showed high efficiency. Alongside with the efficient attack upon organic films, deposits and interphase conglomerates, the alkali peroxide solutions are useful for the extractant post-purification providing its more deep decontamination compared to usual alkali-soda solutions. After treatment with alkali peroxide solutions the extractors are emptied and decontaminated using standard washing solutions.

To increase the action of these solution, lower pressure jet devices (up to 1.2-1.6 MPa) are developed (see Figure 2). Extractor washing out the extractor deposits with their following decontamination. The mounting of the device should follow the positioning in its vertical position throughout passages of more than 40 mm diameter. The surface is treated using a jet of the solution along the spirals path in the design pitch by raising a nozzle with moving a rod and parallel rotation around the vertical axis of the device. The nozzle sizing angle is controlled by moving two terminal switches located in the terminal switches unit. The development stage provides the working design documentation for the model.

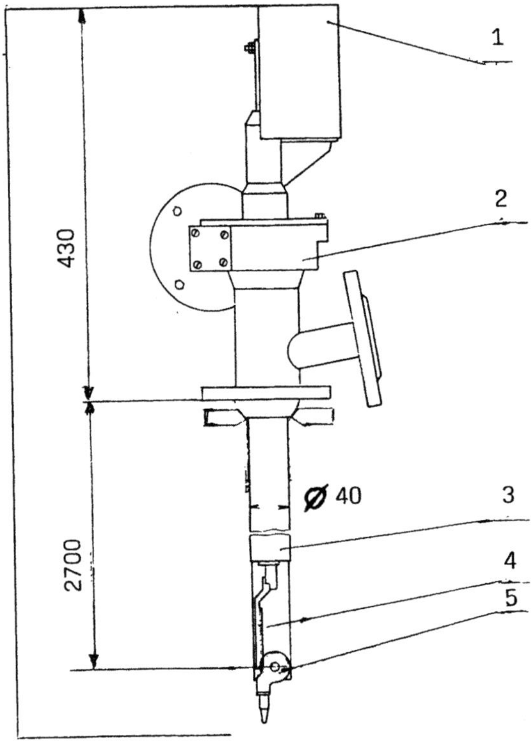

Fig. 2: Extractor washing device

1. Unit of the terminal switched
2. Reducing gear
3. Turning hull of device
4. Lath
5. Nozzle head

Technical characteristics:

Pressure of washing solution kg/cm^2, not more than	12
Temperature of the washing solution ^oC, not more than	90
Flowrate of washing solution m^3/h	2
Angle of the head turning, Degree	170
Frequency of the device rotation, min	14.7

Extractor washing devices are placed in extractors, mainly in the mixing chambers of the head-end extractors, as most liable to contamination by deposits through the penetrations for the instrumentation in pulsed mixer-settlers or for mixers in extractors with mechanical mixing of phases.

Thus, the use of the above operations makes it possible to decrease the volume of the washing and decontamination solutions with a factor of 2-3, to increase the plutonium and uranium outcome and to assess plutonium deposition on of the spent nuclear fuel dissolving equipment and extraction unit surfaces of 0.02 g/m^2 and 1.5 g/m^2, respectively.

In the finished product units plutonium dioxide as a rule, polmers are widely used as finishing protective and decontamination materials. The main requirement to the finishing materials is their easy decontaminability. This characteristics of the polymeric materials manufactured by the national industry and having a good raw material base, is reviewed in detail in the book by S.M.Gorodinsky and D.S.Goldstein /2/. These presentation deals only with those polymeric materials which are used as protective and decontaminating agents. The term "protective coatings" implies the materials which are intended either for fixation of radioactive contamination to prevent their resuspension in air and settling on clean surfaces, or for isolation of clean surfaces from radioactive contaminations. It is supposed that temporary and easy removable coatings will be used. Depending on the purpose the coatings are imparted with stronger protective and decontamination features. The original film forming compositions may be divided into three main groups:

- organic solvent-based compositions
- water soluble compositions
- water dispersal compositions.

An organic solvent-based film forming composition is a solution of monomers such as vinyl chloride, vinyl acetate, etc., or copolymers of different monomers with surface active material, complexing agent and other component addition. They are notable for short drying time, formation of elastic and very strong film and long lasting protection time. Of the compositions of this type recommended for nuclear power plants and radiochemical plants, XC-567 lacquer (vinyl chloride copolymer with vinyl acetate solution in the mixture of organic solvents with plasicizer addition) and PFA 2/10 composition (mehylol polyamide resin in ethyl alcohol) should be mentioned; the coating based on these compositions are water-acid solution and alkali resistant. The main disadvantage of these materials is their toxic and fire risk components content.

The coatings most widely used at nuclear power plants are those based on water soluble polymers, polyvinyl alcohol, in particular due to their low cost rise and fire-proofness as well as their ability to form the films in a wide range of mechanical, chemical and other specific properties. A large number of different formulations such as 107 /3/, 238, 238M, 559, etc., are worked out based on polyvinyl alcohol (PVA). Regular use of the protective coating of 238 formulation at chemico-metallurgical plants and at the final stages of the radiochemical plant resulted in 1-2 order decrease of the average annual plutonium concentration in the air of the operation areas in several years.

However, the use of the PVA-based coatings is limited due to some serious disadvantages:

- long time for drying (up to 24 hours)
- low water proofness
- insufficient strength, at least in some cases.

Lately, still more attention is paid to water dispersive film-forming agents, such as polyvinyl acetate dispersions (PVAD), water dispersion of vinyl acetate copolymer with ethylene (CVED), vinylchloride copolymer with vinyl acetate, etc. These filming agent-based coatings are recommended as fast hardening (2-4 hours of coating formation) protective compositions to isolate stainless-carbon steel and plasticate surfaces from radioactive contaminations. The main disadvantage of these dispersions is their shortage.

The disadvantage of PVA- and PVAD-based coatings are also their partial dissolution of previous layers and plutonium transfer, especially of its soluble forms, to the external surface of the coating, during fixation when a new layer is applied. So, after 5 layers of PVA-based coatings were applied to the surface contaminated with plutonium oxalate, the surface contamination of more than 1% compared to initial was indicated. The latex-based coatings are free of these disadvantages.

The optimum set of properties required from temporary removable protective coatings is characteristic of national synthetic latexes, butadiene-styrene latexes, in particular. The composition named "desolate" and "isolate" were worked out on the mentioned basis.

The specific feature of "desolate" is its ability to form a self-spalling,selfracking scales coating on the decontaminated surface, which is then removed using a sweeper or vacuum cleaner. These compositions were tested in the radiochemical laboratory of the Radium Institute in Gatchina. For the tests samples contaminated with the solution of 90 mCi/l total alpha-activity and pH 5.5 (contaminating isotopes being U-235 (1%), Pu-239 (18%), Am-241/Pu-238 (74%), Cm-244 (7%) alpha-contamination to 1200 particles/cm^2xmin) were taken. The tests were carried out on the bench in a box (alpha-contamination up to 1400 alpha-particles/cm^2xmin), contaminated with salt incrustations from transport basket decontamination. Alpha contaminations were removed from all three tested samples with the decontamination factor of 200-400 by 2-3 cycles of "desolate" application-removal, which gave 5-30 alpha-particles/cm^2xmin of residual alpha-contamination /4/.

The "isolate" composition is intended to isolate the radioactive contaminations. The isolating compositions are used in removal and shipping of contaminated equipment and protective chambers, in performing repair works connected with the opening of the zone 1 rooms, for protection of zone 2 room surfaces, for isolation of contaminations in air ducts to decrease the atmosphere releases, etc. The tests of the "isolate" compositions were also carried out in the radiochemical laboratory of the Radium Institute. The composition was applied to the plasticate floor of the maintenance area near the entrance to the protective chamber. If the coating contamination exceeded the limit, a second layer of the composition was applied. After repair work was completed, the coating was removed from the floor and the contamination level under the coating was measured. The results showed that the coating preserved its operation qualities during 70 days. Its appearance, strength, removability, elasticity,remained satirfactory. It also showed good protective and isolating properties: floor contamination after removal was lower than initial (1-5 alpha-particles/cm^2xmin, compared to 3-50 alpha-particles/cm^2xmin) /4/.

It should be mentioned that works in the development and utilization of isolating coating based on fluorine-organic compounds are carried out lately. Thanks to unique chemical and physico-mechanical characteristics of these compositions they may be considered promising in spite of their relatively high cost.

Of special attention is the problem of plutonium contaminated operation room floor decontamination besides difficulty of plutonium transfer from the surface into solution. There are problems connected with the secondary plutonium sorption on the surface, as well as with the possibility of contamination carrying by the washing solution to clean surfaces. As it was found out, the quantity of radionuclides retained on the surface after decontamination is proportional to that of the solution not removed from the surface. The surface washing by water is of low efficiency, as a rule, as with pH reduction, plutonium precipitates from the solution to the surface again.

Long-term investigations showed the following ways to solve the problem:

- frequent change of the washing solution (1 litre per m^2)
- use of solid sorbents and impregnated ion-exchange textile materials
- mechanization of the decontamination process with the solution recycling and its purification from suspensions.

It should be mentioned for the conclusion that decontamination from plutonium compounds and other radioactive contaminants is carried out best by combination of "liquid" and "dry" methods (see Fig.1). An example is the decontamination work conducted in two protective process chambers of the Scientific Institute of Nuclear Reactors. The chambers are intended for fuel element charging with mixed uranium-plutonium oxide fuel and fuel assembly comleting. The stainless steel lined benches of these chambers were first decontaminated using a film forming composition based on low molecular polyvinyl alcohol with the addition of plastisizer and nitric acid. The composition was applied by spillage and then spread. Two cycles were carried out.

As a result, the sustantial part of the contamination was removed (residual contamination was not higher than 4000 alpha-particles/cm^2 xmin), and the bench was prepared to final decontamination. The final decontamination was performed using an electrochemical unit with a remote low Ohm electrode. The setup in shown in Figure 3.

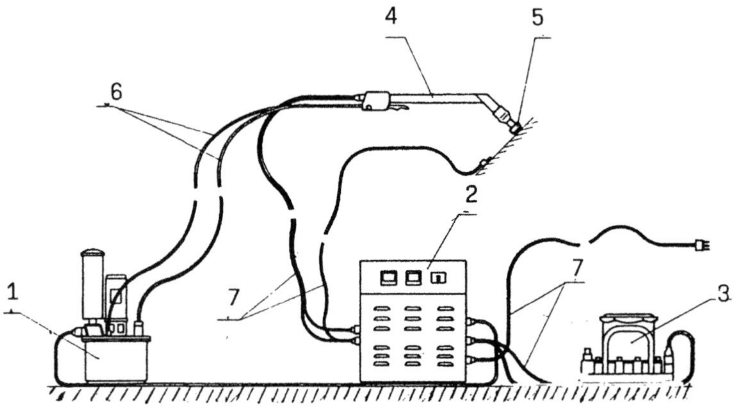

Fig. 3: Electrochemical washing device

1. Block for handing out and collecting of the electrolyte
2. Block of the apparatus operation
3. Block of the transformer
4. Electrode
5. Nozzle
6. Hose
7. Cable

Technical characteristics:

Average velocity of washing, m^2/min, not less than

smooth surfaces	0.1
figured surfaces	0.05

Useful volume of the block for handing out and collecting of the electrolyte, 1, not less than 10

Work time without the substitute of the electrolyte, h, not more than 6

Dimension, mm, not more than

- block for handing out and collecting of the electrolyte	530x380x340
- electrode	525x 90x100
- block of the transformer	780x460x350
- block of the apparatur operation	635x550x480

The electrochemical washing device is intended to the removal of the tenacious contaminations from the unpainted metallic surfaces of the various construction of equipment including parts of withered wells. The device can be employed in a broad range of environmental temperatures, e.g. from 0 to +25°C, and a relative huminity of the air up to 98% at a temperature of +20°C at all climatic conditions (rainfall, snowing, etc.). The washing is founded on using the electrochemical etching and polishing processes for the treatment of contaminated surface layers of metal with subsequent transference of the contaminants to the electrolyte. The development stage reached now the ability to carry out working documentation.

The unit consists of a power source which is a full-wave rectifier with 12 V output voltage and a brush-type electrode of 60 cm^2 working area. The electrolyte consisting of mineral and organic acids is delivered to the middle part of the electrode through the channel. The electrolyte flowrate is 150 ml/min with the treatment velocity of 30 dm^2/min. The electrochemical decontamination method under these conditions showed high efficiency and strongly fixed contamination removal from large pieces of equipment with well treated surfaces. The decontamination factors in the radiochemical laboratory of the Radium Institute in Gatchina achieved, were in the order of 7-40 alpha-particles/cm^2xmin, that means a decontamination factor between 50 and 235.

In conclusion, it should be mentioned that irrespective of the attained success in the development of ways and means of the decontamination and surface protection from plutonium, investigation to improve the process efficiency must be continued with parallel efforts towards the decrease of liquid radwaste arisings. Introduction of new plant designs offer the opportunity to develop new approaches which allow to significantly improve the radiation situation and thus, to decrease harmful releases into the environment both in plant operation and decommissioning.

BIBLIOGRAPHY

/1/ Wick, O.J.(ed.), Plutonium Handbook, Moscow, Atomisdat (1971)

/2/ Gorodinsky, S.M., Goldstein, D.S., Decontamination of Polymeric Materials, Moscow, Energoisdat (1981)

/3/ Below, A.I., Galkin, G.A., Mamaev, L.A., The Development of the easily removable decontamination coatings based on the film-forming emulsions. Optimization of methods and means to prevent radioactive contaminations and decontamination of the equipment of NPP, research centres and nuclear fuel reprocessing plants. Council of mutual Economic Assistance Standing Commission for Peaceful Utilization of Atomic Energy. Presentation at the Experts Meeting of CMEA countries, October 20-24(1975)68-72

/4/ Pavlov, A.B., Petrova, L.A. et al., Investigation of Possibility to Receive the Decontamination Compounds based on Industrial Latexes forming Scaly Selfspalling Coating when Drying. Report on the scientific research work: Development of water diluted latex composition forming easily removable protective coatings. VNIPIET, St.Petersburg (1990) 90-00605

TECHNICAL BASIS FOR Pu SAFEGUARDS PRACTICES, HANDLING AND DISPOSAL, PERLA and TAME EXPERIENCES

H. Dworschak, S. Guardini, B.A. Hunt
Safety Technology Institute of the European Commission,
Joint Research Centre, Ispra Establishment, I-21020 Ispra

ABSTRACT

The practices applied in safeguarding Special Nuclear Materials (SNM) for civil use are outlined together with the basic techniques aiming at an efficient and effective control by the entitled inspectorates. The role of the Joint Research Centre (JRC) Ispra is described to support these inspectorates in developing and implementing cost-effective, reliable and user-friendly instrumentation and in improving the performance of the inspections and verifications in field by making available appropriate training facilities and courses.

1. INTRODUCTION

The chemical and physical properties of Pu are at the origin of the world-wide exceptional attention devoted to this material. The outstanding feature for the broad public lies in its application for producing detrimental weapons, whilst at a political level the major concern is the risk of proliferation of such weapons. The challenge for the technical and scientific world is to identify means for its appropriate control if not irreversible elimination. The much bigger apprehension provoked by Pu as compared to the other nuclear weapons material U-235, is due to the fact that the latter can be "poisoned" with natural U to such a degree of enrichment to become non applicable as a nuclear explosive. The required technological skill and effort for its re-enrichment are considered to represent a sufficiently high threshold to prevent easy proliferation risks.
For Pu such "inherent" poisoning possibilities do not exist; it must indeed be borne in mind that the distinction made between "weapons-grade" and (power) "reactor-grade" Pu is merely arbitrary and does not mean that the latter can not be used for warheads production. This feature together with the chemistry of Pu that would allow its separation and isolation from any kind of other elements and compounds by rather straightforward chemical process technologies is said to represent the difference in concern about the potential misuse of these elements for weapons production.
The possibilities of how Pu can be safeguarded shall be outlined in the following.

E.R. Merz and C.E. Walter (eds.), Disposal of Weapon Plutonium, 259-267.

2. HOW TO SAFEGUARD WEAPONS Pu

The Special Nuclear Materials Control & Accountancy (SNMC&A) system internationally and regionally applied to safeguard fissile and fertile materials used for civil or as often said peaceful purposes is based on well established nuclear material balances, which are periodically verified through Physical Inventories Taking (PIT) by plant operators and Physical Inventory Verification (PIV) by inspectors; continuity-of-knowledge of these balances by appropriate Containment/Surveillance (C/S) measures during the periods between verification is maintained by recording whether the material be in process or in storage. Inspectors from the national and international inspectorates must have unlimited access to nuclear materials (obviously in compliance with radiological protection rules) and to the relevant accountability and transition records.

Typical methods for the control and verification activities are

- Non-Destructive Assay (NDA) and Destructive Assay (DA) techniques
- hold-up determination techniques in bulk handling facilities by volume-concentration measurements
- Containment/Surveillance (C/S) methods by sealing and camera monitoring/image processing techniques
- information processing and data evaluation by statistical techniques

This brief description already indicates very clearly that the safeguards regime could be rather intrusive and includes the potential risk of creating some problem even for the commercial non-military nuclear industry. It is therefore obvious that particularly for SNM of military origin special measures will have to be applied in the initial stages of dismantling operations in order to maintain the secrecy of weapons related technology. as long as by the composition and configuration of the material sensitive information could be disclosed,

Solutions proposed[1] to solve this issue are for example:

- a two tiered verification scheme with national and international inspectors, in which the latter would be kept away from sensitive matters or
- the creation of an international inspection staffing restricted to members from nuclear weapon states.

Two aspects merit particular attention in this respect. The difference in nuclear disarmament verification from non-proliferation verification goals is that the former aims at the reduction of warheads from tens of thousands to some thousands, whilst the latter are designed and implemented to prevent diversion or undeclared production of fissile material even for only a single weapon. All the material identified and released for dismantling will have to be dealt with according to the goals for non-proliferation. This holds of course not only for the warheads themselves, but equally also for all scraps and waste containing appreciable quantities of fissile material.

Inventory differences (ID) associated especially with the latter materials often only poorly measured in weapons facilities[2] , need to be resolved by both administrative provisions and thorough re-measurement campaigns before transition to international safeguards, since for the latter the figures settled in the transition records will represent the basis ("initial inventory"), on which all further verifications and eventual discrepancies will refer to.

Despite the above mentioned differences in verification goals it appears appropriate that technologies and procedures for international safeguards (and physical protection) be used in support of weapons reduction initiatives already in their initial phase. To this extent people in charge of this initial yet still sensitive activity need to be acquainted with the safeguards strategies and the tools for their implementation. In case the knowledge of these strategies and tools is not available within the competent organisation, adequate training should be provided to the responsible inspectors teams and plant operators.

3.. SAFEGUARDS TECHNIQUES[3]

3.1. Nuclear Materials Accountancy (NMA)

The starting point of any safeguards system is the existence of a NMA system at each facility. It must include essentially the material description and information on its quantity, (origin), number of items (where applicable) and location. Fully developed computerised means are available for the management and statistical evaluation of the NMA data. The adaptation of these techniques to storage or chemical processing plants of the National Defence Fuel Cycle (NDFC) seems rather straightforward. Any NMA will depend on the knowledge available on the nuclear material characteristics and content, including the uncertainty for each of the batches.

3.2 Analytical Techniques and Volume Determinations

In bulk handling facilities, *e.g.* reprocessing or fuel fabrication plants DA techniques are applied by taking samples at key measurement points, usually the input and output tanks of a facility. The precision of the obtained data depends not only on the analytical methods applied, but even much more on the determination of the liquid volumes and on the homogeneity of the sampled solutions. Special procedures for tank volumes calibration and homogenisation techniques as well as volumes and density measurement determinations have been established over the past decades and can be routinely applied.

3.3 Non Destructive Assay

A vast experience has been gained in the use of NDA techniques for safeguards purposes on U of all kind of enrichment and on Pu of different isotopic compositions. The measurements are performed routinely on material of large variety of chemical and physical nature. The techniques for Pu measurement rely on neutron coincidence counting, calorimetry and high resolution gamma-ray spectroscopy for isotopic composition determination. On well defined products the measurement errors might be as low as fractions of a percent. On scrap and waste materials the uncertainty may reach more than 20 %. Detector heads in combination with Time Correlation Analysers (TCA) based on passive neutron counting are available for packages (drums) of up to a few hundred litres volume[4].

3.4 Sealing/Identification Techniques

Sealing/identification techniques appear particularly attractive for sensitive weapons materials, which can be put in "neutral" containers for transfer operations and prolonged interim storage periods. The most used type is still the copper wire closed with a unique marked cup which is identified by photography. A recent development made by the JRC, is based on ultrasonic techniques, applicable also to containers stored in pools, *e.g.* on so-called

Multi Element Bottles (MEB) for spent fuel elements storage. Their application to PuO_2 transport containers is under study at present[5].

3.5 Optical Surveillance

Remote optical surveillance by video cameras represents an efficient complementary tool for the control of itemised material in storage areas. For the processing of round-the-clock pictures taking computerised review systems are now becoming available which condense the records to phases only when changes in the pictures happen.

Integration of such Containment/Surveillance (C/S) systems with automated NDA instrumentation[6] would allow to perform an efficient control of nuclear materials even in sensitive facilities in a widely "unattended" manner, *i.e.* without the need of inspectors being present all the time and hence much less intrusive.

3.6 Training

An increasing number of techniques has been applied over the years for safeguarding nuclear material in the various parts of the civil fuel cycle. There is a steady evolution of these techniques, their efficient application requires to transfer continuously technology from development laboratories to the final users, which includes also appropriate instruction and training of inspectors.

JRC dedicates important resources to the training of safeguards inspectors for the application of NDA and volume measurement techniques, statistical methods and inspection procedures (*e.g.* Physical Inventory Verification). One to two week courses are given on basic principles of the techniques and procedures as well as on their practical application in realistic conditions. For this purpose special facilities as PERLA and TAME are available, in which "near-real field" conditions are established. A description of these facilities is given in the last part of this paper.

4. JRC SAFEGUARDS TRAINING FACILITIES

4.1 Training and Technology Transfer

Training programmes are oriented towards providing a capacity to design a verification sampling plan, execute the measurements and carry out an integrated data evaluation of both accounting and verification data. In future years emphasis will be shifted towards ensuring that verification activities are carried out in the most cost efficient manner possible, employing new techniques such as remote monitoring and integrated decision making processes.

With these assumptions several years ago the Joint Research Centre started a programme for establishing test and training facilities for safeguards[7], designed to improve the transfer of technology and bridge the gap between laboratory developments and the application of measurement techniques in an industrial environment.

The following laboratories are now being operated:

a. PERLA : PERformance and Training LAboratory for Non Destructive Assay[8].

b. TAME : TAnk MEasurement Laboratory[9].

c. LaSCo : LAboratory for Surveillance and Containment[10].

4.2 The PERLA facility

With the scope of improving and implementing NDA methods in field, the JRC has decided to set up a specialised laboratory called PERLA: PERformance Laboratory, which has the particular characteristics of being equipped with a large inventory of bulk material standards, representative of fabrication plant process materials and products:

- Massive (about 3 kg) UO2/PuO2/MOX cans.
- HEU (MTR) fuel assemblies.
- LEU mock-up fuel assembly.

This gives developers the opportunity to test their models on real-scale samples and inspectors the opportunity to calibrate their instruments with near-field items. The PERLA standards are characterised with high quality control procedures. Several analytical laboratories participated in batch sampling and analysis exercises, employing certified international reference materials, such that the mass and isotope contents of the standards are very accurately known and traceable back to international, primary reference materials. This aspect is a considerable contribution in assuring traceability to field inspector verification measurements when using calibration curves established with the use of PERLA standards and ensures the link between NDA field measurements with international reference systems.

PERLA has in recent years developed a training programme (Table 1) that follows the

TABLE 1 - PERLA training menu.

- Basic Physics for Non-Destructive Assay of Nuclear Material
- Introduction to Applied Statistics for Safeguards Purposes
- Gamma Spectronomy for Enrichment Determination of Uranium
- Determination of the Isotopic Abundance of Plutonium
- Passive Neutron Measurements
- Active Neutron Measurements: AWCC
- Active Neutron Measurements: PHONID
- Advanced Applied Statistics for Safeguard Purposes
- Physical Inventory Verification (PIV) for a HEU Facility
- Physical Inventory Verification for a Pu (MOX) Facility
- Physical Inventory Verification for a LEU Facility

above concepts. There are three successive levels of training:

- Level 1 : Inspectors are trained in Basic Physics for NDA and Introductory Statistics for Safeguards Purposes.
- Level 2 : Training on NDA instruments e.g. active/passive neutron devices, gamma spectrometry, etc.
- Level 3 : Integrated physical inventory verification training courses, specifically tailored for fabrication plants, in which inspectors are taught to

combine safeguards requirements (e.g. detection probability, statistical sampling, measurement, data analysis, decision making processes) to arrive at a final safeguards statement.

Regular feedback from trainers and trainees ensures a continuous tuning of those aspects eventually not consistent with the user's needs.In parallel with the evolution of the training course contents, the technology underlying the training is progressing quite rapidly as well. New technologies have become available or are under development thanks to software/hardware improvements or are under development, that will represent powerful training tools: interactive multimedia systems are amongst them. JRC is developing a multimedia training package for Euratom inspectors that focuses specifically the training aspects of new and integrated NDA-C/S.

4.3 TAME Laboratory

The TAnk MEasurement Laboratory (TAME) is now a fully operational facility dedicated to training and studying the problems related to the measurement of volume and mass of liquids in a nuclear facility. The laboratory has been installed on a 8m by 8m area reaching to a height of 15m and is equipped with the following :

- an annular input accountancy vessel of 12.5 m^3 resting on 3 load cells
- an annular output vessel of 250 dm^3
- an output 'D' shaped vessel of 400 dm^3 suspended from a weighing balance
- three intermediate feed tanks of combined volume 14 m^3
- two mono-tubes of 4 metres height having diameters of 200 and 400 mm respectively
- a dosing station for calibration
- an off-gas system for maintaining and controlling tank under pressure
- recycling and homogenisation systems of all tanks
- ejector air-lift systems for remote sampling and liquid transfer
- instrumentation for measuring temperature, pressure, humidity density, air and liquid flow
- series of differential pressure measurements for determining volume and mass of tank contents, (Ruska, Diptron, industrial type electromanometres etc.,)
- a dedicated process control and data acquisition system.

TAME represents most aspects of a reprocessing facility for what concerns volume and mass measurements of vessel contents, including sampling and homogenisation techniques and capabilities, various means of solutions transfer, for example pump, air-lift ejector, gravity etc. The laboratory has been set-up to provide a training facility for inspectors and operators with an objective to cover as many as closely related phenomena as found in a reprocessing facility. In this respect TAME is generally dedicated to training on :

- tank calibrations of different shapes and forms
- tank calibrations and re-calibration techniques
- continuous monitoring of vessel contents over short and long time frames
- physical effects and influences associated with temperature and other varying operating conditions
- data acquisition and elaboration.

5. PLUTONIUM SAFEGUARDS PRACTICES

Solid and liquid Pu bearing bulk material is assayed for safeguards and NMC&A purposes by Destructive (DA) and/or Non-Destructive (NDA) techniques, illustrated in the following.

5.1 Plutonium in solid form

For monitoring Pu in solid form preferentially NDA methods are used. such as passive neutron techniques in combination with high resolution gamma spectrometry. The passive neutron technique is based on the correlation of spontaneous fission neutron rates with the Pu-240 equivalent weight. This weight represents the sum of even Pu mass isotopes undergoing spontaneous fission:

$$\text{Pu-240 equ.} = 2.431 \times (\text{Pu238}) + 1 \times (\text{Pu240}) + 1.667 \times (\text{Pu242})$$

In order to establish this weight factor for mass determination the isotope vector of the Pu to be analysed has to be known. For this determination high resolution gamma spectrometry (HRGS) is the reference method. The spectra are evaluated by thoroughly tested and verified computer codes like MGA[11] for safeguards accountancy purposes by the EURATOM and the IAEA inspectors.

Passive neutron techniques have been significantly imp[roved during the last decade by the introduction of more sophisticated algorithms, which take into account higher neutron multiplicities as in the case of the Time Correlation Analysis (TCA)[12]. The TCA technique in particular is acquiring now an important role in assaying poorly defined Pu bearing material (inhomogeneous and/or mixed with impurities) as scrap material and α -waste.

For monitoring Pu in storage facilities with low material movement requirements calorimetry represents a valid alternative or complementary independent method to passive neutron counting

Achievable accuracies by NDA methods depend evidently on the materials nature[13]:

- Conventional shift register (SR) techniques combined with HRGS can easily give 1 % accuracies, when applied to well characterised homogeneous material like metal, pure PuO_2 or MOX powder, pellets, pins *etc.*
- Multiplicity counters and TCA instruments are able to measure Pu in scrap or waste barrels, *i.e.* in undefined physical and chemical form within an error range of 10 to 20 %.
- Calorimetry[14], especially when coupled with passive neutrons counting is a very precise and accurate assaying method, provided the isotopic composition is known to better than 1 %. The limitation for NDA assaying techniques are indeed represented mainly by the uncertainties due to HRGS rather than to passive neutrons or calorimetric measurements, especially for high-burnup or recycled Pu.
- For Pu with a low burnup and hence at very low Pu-240 equivalent values passive neutron methods might suffer from the low coincidence count rates. Active neutrons methods like Sb-Be based instruments interrogating the fissile isotopes content might give in this case better and quicker results.

5.2 Pu in liquid form

When Pu is recovered from spent fuel or when weapons Pu has to be processed (conversion from metal to other chemical forms), it has to be assayed very accurately in liquid solutions

in the in- and output accountancy tanks, which represent the "key measurement points" of any processing facility.
The classical method is the sample taking from the previously carefully homogenised solution and the determination of the Pu concentration by destructive analytical methods (DA) with the total mass determination in combination with the measurement of the volume and density of the solution in the accountancy tanks. The mass/volume determination accuracies represent a growing concern for safeguards in facilities with high throughput and consequently large tank volumes involved, which has attracted attention for special training facilities like TAME dealt with above.
For the measurement of the Pu (and U) concentration in solution a very accurate NDA method is represented by x-ray absorption edge densitometry[15] or K-edge. A very high accuracy and precision of the order of a few tenths of percent can be achieved provided vials of a few cm thick with a very exact liquid pathlength are applied.

6. CONCLUSIONS

Some aspects linked to the technical feasibility of safeguarding Pu coming from the dismantling of warheads are discussed and the major concern this material generates with respect to U-235 is briefly addressed.
The first requirement to efficiently control Pu under a non-military regime is to assess accurately the "initial inventory" at the border between military and civil safeguarding responsibility, wherever this transition will be established by political agreements.
Destructive and non-destructive analytical techniques are available, the performance of which is sufficiently developed to accurately determine the initial inventories.
Conventional HRGS and neutron interrogation techniques are applicable to the weapons stockpiles guaranteeing less than 1 % error for well defined pure material and a few percent for scrap material generated by the weapons construction and dismantling processes. Passive neutron counting may suffer from the low Pu-240 content; active neutron techniques and calorimetry represent useful complementary techniques.
Containment and surveillance techniques will be essential tools for effectively safeguarding the important stockpiles of nuclear weapons material that will have to be kept in appropriate stores for rather long time periods independently of the solutions chosen eventually for their disposal..
Integrated C/S and NDA techniques can be designed to allow an efficient control of such stockpiles over long periods in a cost effective unattended mode. Attractive new data processing and decision making tools (*e.g.* neural networks and fuzzy logics) are becoming available that will help developing integrated verification approaches permitting the inspectors to perform their tasks mainly in the remote mode.
The personnel in charge of controlling nuclear material, also originated from weapons dismantling, needs to be specifically trained in NMC&A techniques and methodologies. Special training facilities and programmes implemented for this purpose at the JRC Ispra of the EC are described.

REFERENCES

1. T.E. Shea, International Verification of Nuclear Disarmament, *34th annual meeting proceedings of INMM, Scottsdale, July 1993, vol. XXII, p 808;*
2. D.W. Crawford, O.B. Johnson, Nuclear Safeguards Challenges in a Changing Department of Energy, *15th annual ESARDA symposium on safeguards and nuclear material management proceedings, Rome, May 1993, EUR 15214 EN, p 41;*
3. M. Cuypers et al., Verification Technology in an Evolving Nuclear Environment, *15th annual ESARDA symposium on safeguards and nuclear material management proceedings, Rome, May 1993, EUR 15214 EN, p 59;*
4. L. Bondar, W. Hage, F.W. Ledebrink, B. Pedersen, M.T. Swinhoe, Triple Neutron Correlation for MOX Waste Assay, *Proc. ANS annual meeting (June 1993), San Diego (CA).*
5. B.C. D'Agraives, J. Toornvliet, E. Mascetti, J. Clegg, *Second Exercise of the long term Experiment with Ispra Sealing Bolts at the BNFL Sellafield Facility, Proc. 11th ESARDA annual symposium on safeguards and nuclear materials management, Luxembourg 1989.*
6. D. Van der Eecken, M.T. Swinhoe, M.T. Franklin, F. Sorel et al., An Unattended Measurement System for Safeguarding fresh LEU Fuel Assemblies at the Output of a Fabrication Plant, *Proc. 17th ESARDA annual symposium on Safeguards and nuclear Materials management, Aachen (D), May 1995.*
7. M. Cuypers, H. Dworschak, C. Foggi, S. Guardini, B.A. Hunt, D. Landat, Test and Training Facilities at JRC Ispra for Safeguards Inspectors, *Proc. 32nd INMM symposium, New Orleans (LA), July 28-31 1991*
8. S. Guardini, W. Kloeckner, The Role of PERLA in EURATOM Safeguards, *Proc. 17th ESARDA annual symposium on safeguards and nuclear materials management, Aachen (D), May 1995.*
9. B.A. Hunt et al., Volume and Mass Measurement Techniques in Large and Small Tanks for R&D and Training at the JRC Ispra Establishment, *IAEA symposium on international safeguards, Vienna, March 1994, IAEA-SM 333/70*
10. M. Cuypers, D. Landat, "LaSCo", a Performance Assessment and Training Laboratory for C/S, *Proc. 13th ESARDA annual symposium on safeguards and nuclear materials management, Avignon (F), May 1991.*
11. R. Gunnink, et al., MGA, a Gamma Spectrum Analysis Code for Determining Plutonium Isotopic Abundances, *UCRL-LR-103220, LLNL, Livermore (CA), 1990.*
12. S. Guardini editor, Proc. ESARDA International Workshop on Passive Neutron Counting, PERLA, JRC Ispra (I), *April 1993, EUR 15102 EN, 1993.*
13. G. Bigman, Performance Values of Non Destructive Assay Techniques Applied to Safeguards, *Proc. 4th Int. Conf. on Facility Operation Safeguards Interface, Albuquerque 1991.*
14. S. Guardini, W.W. Strohm editors, Proc. 1st Int. Workshop on Plutonium and Tritium Calorimetry, PERLA, JRC Ispra (I), March 1992, *EUR 14307 EN, 1992.*
15. H. Ottmar et al., Striving for Improved Accountancy in Reprocessing Input Measurements by x-ray Densitometry, *Proc. 15th ESARDA annual symposium on safeguards and nuclear materials management, Rome (I), May 1993.*

LONG-TERM SAFETY ASPECTS OF GEOLOGICAL DISPOSAL OF LARGE QUANTITIES OF PLUTONIUM

B. BALTES
Gesellschaft für Anlagen- und Reaktorsicherheit (GRS)
Schwertnergasse 1
50667 KÖLN

1 Fundamentals of the Disposal of Radioactive Waste

1.1 Safety Objectives of Disposal

On an international level the fundamentals of radioactive waste disposal are of a general nature. Safety objectives were defined which have general validity for the disposal of waste. They account for the protection of life, objects and the environment by requesting a long-term, effective isolation of the waste from the biosphere. Furthermore, provisions have to be made that the isolation remains effective during very long periods which account for the half-lives of long-living radionuclides. The fundamentals of the disposal of radioactive waste, which form the foundation of rules and regulations, include the following:

1. the protection of human life from ionizing radiation,
2. a reduction of the radiation exposure of future generations to a minimum, and
3. environmental protection.

These general fundamentals were specified in rules and regulations, e.g. accounting for the request for negligible or no cost for future generations with respect to present disposal practices. There is consensus in the international community for fundamental demands to ensure the protection of the living space of man and animals, the protection of economical resources and the optimization of this protection.

1.2 Rules and Regulations

A demand that has to be met before a repository may be established is the quantification of radiological protection regulations. The International Commission on Radiological Protection (ICRP) during the past decades has developed a system of recommendations to protect human life from ionizing radiation /5; 4/. The recommendations of both the ICRP and the IAEA imply use of probabilistic safety assessments of the post-closure phase of a repository to account for data uncertainties and probabilities of events. In the

E.R. Merz and C.E. Walter (eds.), Disposal of Weapon Plutonium, 269-273.

ICRP this approach led to the development of risk-oriented limiting values for individuals. The ICRP recommends the following:

"The risk in a year from probabilistic events should be restricted to less than 10^{-5} health effects to members of the critical group".

The national institutions develop their own rules, regulations and criteria. However, they take credit from the recommendations of the ICRP, IAEA and NEA. In some countries the risk concept forms the framework for post-closure assessments. In other countries a combined risk and dose concept was developed or, alternatively, the dose concept only.

2 Proof of Long-Term Safety of High Level Waste Disposal

2.1 Site Selection

The disposal site represents the basis for assuring the repository safety. It has to assure compliance with the above mentioned safety objectives during all stages of repository operation and post-closure period. The site selection criteria which were developed within an international framework were not primarily designed for finding the most ideal of all sites. The criteria rather serve for finding a site, which, in combination with a suitable concept of natural and technical barriers, complies with safety requirements for a repository /2; 3/. For an assessment of the natural barrier function of a particular site general criteria for the geological, hydrological and geochemical conditions at the site were developed /3/. In a qualifying assessment of a site, criteria for a selection have to met such as a sufficient thickness and extent of the host rock, stability of the host rock towards erosion, location of a site in a stable region, limited groundwater supply in the host rock and limited groundwater movement at the site, favorable groundwater chemistry, high retardation characteristics of the aquifers.

In addition to the above, the following assessment criteria of a site are looked at: The time span covered in a safety assessment, the long-term site evolution, assessment of potential human activities and the sociological environment at the Earth's surface of the site.

2.2 Repository Design

For the implementation of above mentioned criteria for safe disposal the efforts of the international community of nuclear energy users focuses on the isolation of the waste (HLW and spent fuel) in deep geological settings. All repository designs using geological disposal are based on the co-called "multi-barrier system" (consisting of conditioned waste, waste container, backfill, seals and the geosphere). Once the choice of the host rock type (e.g. granite, salt, clay) has been made, a repository concept guided by technical aspects on safety may be developed. The repository design will be conceptually adjusted to the site conditions. The conditioned waste (e.g. glas container, spent fuel) will be packed in canisters which are designed to function as an inpermeable barrier during time spans ranging from several hundreds to several ten thousands of years. The canisters will be stored in boreholes or in drifts, and the remaining void space around the canisters and in the drifts and shafts will be filled and sealed with backfill materials (e.g. bentonite). The backfill material has the function to minimize a potential groundwater supply to the canisters. Furthermore, once the integrity of the canisters has failed, the backfill material should delay the travel times of the radionuclides dissolved in groundwater from the repository to the biosphere.

The design criteria with respect to a spatial arrangement and loading density of a repository, which contains heat-producing waste and fissionable radionuclides, are governed by the requirements that the host rock temperature may not exceed a certain value and that an accumulation of critical masses of fissile material has to be avoided. The maximum temperatures that may not be exceeded by the host rock are 200 °C for salt rock and 100 °C for granite and clay. Additional limiting values e.g. for the nuclide inventory of the canisters will be established based on the outcome of safety assessments of the post-closure period for the repository in question.

Based on a repository concept, repository site, compliance criteria and the repository design, waste specifications will be formulated. There is international consensus on the general waste specifications. Individual waste specifications will be developed in accordance with a particular repository concept.

2.3 Long-term Safety Assessments

It is generally accepted within the international community that a release of radionuclides from a repository into the biosphere cannot be avoided in the long term. However, the mulit-barrier system will retard and reduce any release such that above mentioned safety objectives will be complied with. Evidence of the safety of a repository during the post-closure phase will be provided by carrying out numerical safety analyses and by other means. Such safety analyses have to be carried out while considering the specific repository concept and the site. Safety analyses include modelling and analysis of the radionuclide inventory, the site evolution, the system of technical and natural barriers and the potential transport of radionuclides through the barrier system to the biosphere. The radiological impact resulting from a nuclide release into the biosphere will be compared with given exposure limits. The repository system has to be laid out in a way that guarantees compliance with exposure limits.

3 Safety Aspects of Plutonium Disposal

3.1 Conceptual considerations

The long-term safety assessments which were carried out within the international community are based either on the disposal of vitrified HLW, or on the direct storage of spent fuel. The fraction of Pu present in spent fuel amounts to approximately 10.5 kg/t U, whereas in HLW the Pu fraction is by more than two orders of magnitude below the one of spent fuel. Seven years after unloading from a nuclear reactor, a MOX element will contain ~48 Kg Pu/tU, which corresponds to ~19 Kg Pu-239/tU. Presently no safety assessments are known to the author regarding disposal of large quantities of Pu in the form of vitirfied waste. The following considerations are based on the assumption that 2.5 Kg Pu-239 are present in a vitrified waste canister.
The following major aspects have to be taken into consideration when storing HLW containing a significant fraction of Pu-239 and spent fuel:

Criticality:
Any chances for re-criticality during the post-closure phase caused by dissolution of fuel elements or vitrified waste and subsequent concentration of fissile material have to be entirely avoided. This has a bearing on the repository concept, its layout and loading requirements. Consequently there is an upper limit value for the presence of Pu-239 in vitrified waste.

Heat Production:
After unloading from a reactor the heat production from a MOX fuel element exceeds the one of a UO_2 fuel element by a factor of three. Compliance of the maximum temperature requirement for the respective host rock will be guaranteed by the layout of the repository and the packing of the waste containers. As a result of the temperature requirement for e.g. granite (100 °C), containers which are used for the disposal of MOX can be loaded with only one fourth of the quantity of elements when compared to containers which are used for UO_2 fuel elements.
Additional loading of Pu-239 in a vitrified HLW waste canister results in an increased heat production of ~1.9 W/Kg Pu-239. This is a comparatively minor contribution which probably does not lead to a loading restriction of Pu-239 for waste containers.

Long-term Safety:
During recent years several long-term, site-specific and generic safety assessments for different types of host rocks and repository concepts involving HLW and spent fuel were carried out and published. These assessments demonstrate the long-term safety, i.e. the compliance of geological disposal with safety objectives. Moreover, radionuclides were identified which control the potential radiological impact because of their relatively high release rate into the biosphere and a high biological effectiveness. Those are e.g. the following fission and activation products: Cl-36, Se-79, C-14, Nb-94, Cs-135 and J-129, and actinides such as Pa-231, Pu-239, Pu-242, Th-230, Ra-226 and

Np-237. Particular attention is required for the behavior of the actinide decay series and their evolution during the transport through the barrier system. Although almost no Pa-231 is associated with the disposal of spent fuel, it will build up in the geosphere as a result of Pu-239 and U-235 migration and decay, respectively. The same is true also for Ra-226. In case of HLW disposal, the evolution of the actinides is similar, although radiologically it takes place on a lower level because of a smaller inventory when compared to the disposal of spent fuel. The decay series N+3 dominates the radiological impact due to the presence of Pa-231 for time periods after 10^6 years. Evaluations of long-term safety assessments show that disposal of UO_2 fuel elements can be carried out safely. However, when disposing MOX fuel elements, special attention is required for the enhanced Pu contents of Pu-242 and Pu-239 and their decay products. Studies of the disposal of UO_2 fuel element allow the deduction to be made that safe disposal of MOX fuel elements is feasable.
Studies of the disposal of vitrified HLW in canisters point to the possibility, that an enhanced Pu-239 level in HLW from the perspective of long-term safety can be handled if the near field barriers are effectiv. Long-term pilot analyses show that after degradation of the glass the solubility limits of Pu and U will be reached in the near-field, resulting in a constant nuclide release into the geosphere regardless of the inventory. The following parameters have substantial bearing on the results of the analyses and therefore have to be verified: the solubility limits of the groundwater for the radionuclides as a function of the chemical environment of the respective barrier, the nuclide retention in the near-field and the far-field, the groundwater supply and the type of scenario modelled in the biosphere. Special attention is required to the prevention of re-criticality during this process. However, when compared to safety assessments, the initial analyses allow the conclusion to be drawn that final disposal of vitrified HLW/Pu mixtures in canisters will be feasable from a safety view, provided that a suitable disposal site and disposal concept are chosen. However, site-specific and concept-specific assessments will be required to corroborate above findings.

Literature

1. Objectives, standards and criteria for radioactive waste disposal in the European Community, EUR 12570, 1989

2. Safety principles and technical criteria for the underground disposal of high level radioactive waste, Safety Series No. 99, IAEA, WIEN, 1989

3. Siting of geological disposal facilities, Safety Series N0. 111-G-4.1, IAEA, Wien, 1994

4. Radiation protection principles for the disposal of solid radioactive waste, ICRP Publication 46, Pergamon Press, Juli 1985

5. Recommendations of the international commission on radiological protection, Annals of the ICRP, ICRP Publication 60, Pergamon Press, Juli 1990

MPC&A FOR PLUTONIUM DISPOSITION IN THE RUSSIAN FEDERATION

W. G. SUTCLIFFE
Center for Security and Technology Studies
Lawrence Livermore National Laboratory, L-019
P.O. Box 808
Livermore, CA 94551 USA

1. Introduction

The issue of what to do with excess fissile materials from dismantled nuclear weapons has been discussed for a number of years [1]. The options or alternatives commanding the most attention were identified by the American National Academy of Sciences [2]. For plutonium these options are: (1) the fabrication and use of mixed-oxide (MOX) reactor fuel followed by the disposal of the spent fuel, or (2) vitrification (immobilization) of plutonium combined with highly radioactive material followed by direct disposal. The Academy report also identified the alternative of disposal in a deep borehole as requiring further study before being eliminated or accepted. The report emphasized security of nuclear materials as a principal factor in considering management and disposition decisions. Security of materials is particularly important in the near term—now—long before ultimate disposition can be accomplished.

The MOX option was the subject of a NATO workshop held at Obninsk, Russia in October 1994 [3]. Hence this paper does not deal with the MOX alternative in detail. It deals with the following: materials protection, control, and accounting (MPC&A) for immobilization and disposal; the immobilization vs MOX alternatives; the security of disposed plutonium; the need to demonstrate MPC&A for plutonium disposition; and, finally, a recommended investment to quickly and inexpensively improve the protection of fissile materials in Russia. It is the author's view that near-term management is of overriding importance. That is, with respect to the ultimate disposition of excess nuclear materials, how we get there is more important than where we are going.

2. MPC&A for Immobilization and Disposal

Materials protection, control, and accounting (MPC&A) measures for plutonium are necessary whether the plutonium is embedded in reactor fuel or in glass, ceramic, or other material. Existing MPC&A practices for mixed oxide (MOX) reactor fuel containing plutonium can provide a basis for the establishment of MPC&A for direct immobilization without any unusual difficulties.

E.R. Merz and C.E. Walter (eds.), Disposal of Weapon Plutonium, 275-283.

In the MOX option, the plutonium passes through a number of stages or processes. These include recovery of the plutonium from weapons components, conversion to oxide, fabrication into reactor fuel, exposure in a reactor, storage of spent fuel, and geologic disposal of spent fuel (or reprocessing and disposal of high-level waste). For direct immobilization and disposal the sequence is similar, except that the processes that correspond to fuel fabrication and reactor exposure are combined in the immobilization step, in which waste forms (e.g., glass logs) are formed with embedded fission products or high-level waste. See Fig. 1. For both the MOX and immobilization options, a number of transportation and interim storage (staging) steps will be required.

MPC&A is most crucial before the plutonium is combined with radioactive materials, especially when the plutonium is in bulk form. Standard MPC&A methods used in Europe and Japan to protect and safeguard the transportation and storage of plutonium, MOX fuel fabrication facilities, and reprocessing facilities could be adapted for use in a facility that immobilizes plutonium mixed with highly radioactive material.

Inventorying acts with other measures to provide for the security of the plutonium. It also provides important assurance that criticality-based limits for the immobilization processing and repository emplacement are met. For security, however, it is only necessary to ensure that all items (containers) are accounted for and have not been tampered with. Directly inventorying the amount of plutonium immobilized with fission products or high-level waste is not necessary. Methods used for the protection of plutonium without fission products could be adapted for this purpose. Tags on the containers can facilitate inventorying, which could be done remotely, possibly with the use of robots. Seals or continuous surveillance could ensure that no material has been removed. Visual surveillance using television and/or a combination of weight, motion, and radiation sensors could be used to detect removal of material or tampering with containers.

Periodic inventories of containers would be required in any storage facility and in a mined geologic repository until the repository was sealed. Inventorying would likely be supplemented by containment and surveillance methods such as portal-perimeter monitoring. Sensors might include motion and seismic sensors on the perimeter and gamma and neutron radiation detectors at the portals. Inventorying plutonium disposed of in a deep borehole is not practical.

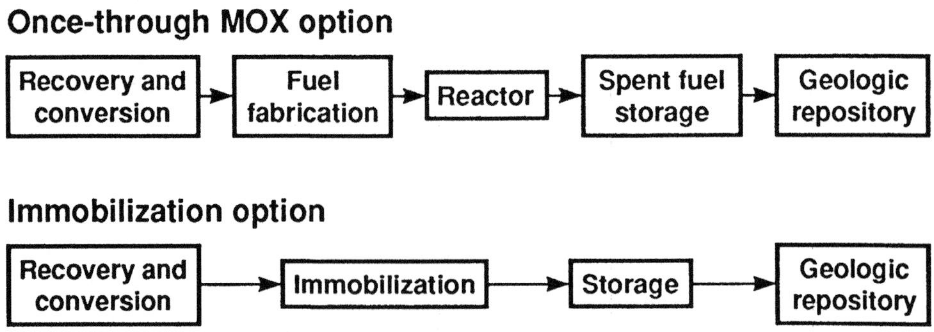

Figure 1. Plutonium disposition processes.

3. Immobilization vs MOX

A number of criteria have been put forth for the decision of whether to choose MOX or immobilization for the disposition of excess weapons plutonium [2, 4]. In particular, considerable attention has been given to economic criteria such as net cost [5]. However, the National Academy of Sciences' committee noted that economics should not be the primary criterion driving the decision. This is because the national security benefit from the safe, secure disposition of plutonium is great, and the cost for such disposition is considerably smaller than the budget for national security. The disposition cost (for a nominal 50 tonnes of plutonium) is expected to be a few billion dollars over several years, which is of the order of 1% of what the United States spends for national security each year.

As discussed in Sec. 2, required MPC&A would likely be comparable for MOX and immobilization options. Immobilization could have an advantage of less storage (staging) and transportation because of fewer distinct operations. That is, the addition of radioactive material may be combined with immobilization (vitrification) in one facility. The disadvantage is that there is no operational experience for MPC&A measures for these new processes.

In Russia it is argued that not only is MOX a more economical option, it utilizes a valued resource. Rather than discuss these somewhat contentious arguments, we will argue for another important criterion that discriminates between the choices in Russia.

Perhaps the most important criteria for disposition are those collectively called proliferation resistance. They deal with making materials less attractive and/or less available for use in weapons. One measure of the exposure of the material to possible theft or reuse is the time until the start of implementation of the disposition decision. It has been said that weapons plutonium (and unsafeguarded civil plutonium) represents a "clear and present danger." Starting (or even finishing) the disposition of excess material will not completely remove this danger, because Russia (and the United States) will retain significant stockpiles of nuclear weapons and materials for some time to come, and more material can be expected to be declared excess as time goes on. However, there is great value in making the decision and starting the disposition program as soon as possible. Such action will generate a programmatic momentum toward controlling, processing, and disposing of excess nuclear materials, and as such could bolster progress toward a more comprehensive regime to protect and safeguard all nuclear materials. In any case, an early decision, accompanied by the identification of the initial "excess" material (e.g., the first 10 tonnes) would focus attention and provide a specific target for assistance.

It is important that this programmatic momentum be established while conditions in Russia and the United States are favorable. Economic and political conditions in Russia make the future somewhat uncertain. Impatience in the United States could eventually reduce support for the disposition program, particularly in Russia. The United States has declared that excess material will be placed under IAEA safeguards; a similar declaration from Russia would be welcomed.

The reasons that a MOX decision could be implemented sooner than an immobilization and disposal decision in Russia are twofold. The first is institutional. In

Russia, the majority opinion, especially espoused by MINATOM, is that MOX is the preferred option. This is undoubtedly connected to the desire to benefit from previous expenditures by reaping the intrinsic energy value in plutonium. (In the United States there is no such clear majority of opinion.) The second reason is that vitrification of plutonium would require more R&D than the MOX option; and there may be less infrastructure to support vitrification or immobilization than MOX. In particular, it has been reported that currently there may not be enough high-level waste available to provide an adequate radiation barrier if all excess plutonium were to be vitrified [6].

Another factor may be that an immobilization option in Russia is more closely tied to the development of a geologic repository than is a MOX option; and the technical, institutional, and political processes required to develop a repository in Russia are not fully developed. Throughout the world, long-term surface storage of spent fuel is, by necessity, becoming more accepted; and dry storage for aged (cooled) spent fuel is being increasingly utilized. Even if a once-through MOX cycle is adopted, Russia may decide not to dispose of spent fuel in a repository, in order to preserve a future reprocessing option.

MINATOM prefers the option to use weapons plutonium in fast BN-type liquid metal reactors, arguing that this technology results in reduced amounts and toxicity of waste. However, it appears that the Russians cannot afford to build BN reactors, and the rest of the world is unlikely to fund such development. Burning MOX in light water reactors, on the other hand, appears to be acceptable, as indicated at the NATO meeting last October in Obninsk [3]. The West may provide some funding in this case, which is consistent with Russia's desire to exploit plutonium's intrinsic energy value.

Whether plutonium is slated for MOX or direct disposal, it will have to be recovered from weapons components and (most likely) converted into oxide. The requirements for the oxide, however, might differ depending on whether it will be an ingredient of MOX or will be immobilized in a glass log or other matrix. Although MOX might be able to tolerate the expected impurities (Am and Ga) the grain size and structure could be crucial. Thus the decision on whether or not to pursue the MOX option should be made early so that planning and implementing the metal-to-oxide conversion process can be initiated as early as possible.

Although there are a number of differences between the MOX and immobilization options (for example, MOX use changes the isotopic composition of the plutonium), none of these differences appears compelling, except that the MOX option could start the disposition process significantly sooner. Doing that could be critically important in establishing a programmatic momentum towards a regime of enhanced protection and safeguards for nuclear materials.

4. Security of Disposed Plutonium

As with plutonium embedded in spent fuel (which is a percent or less for LWR spent fuel, a few percent for MOX spent fuel), the security of disposed plutonium rests on various characteristics (barriers) of the disposal system. The most important barrier may be the radiation barrier, achieved by the addition of radioactive materials that effectively preclude human contact with the plutonium. The addition of high-level waste or cesium-

137 to plutonium immobilized in glass or other material has been proposed to make it as inaccessible as plutonium in spent fuel. This concept of inaccessibility was termed the "spent fuel standard" by the U.S. National Academy of Sciences in their report on the management and disposition of plutonium [4]. If the radiation barrier is absent, other barriers will have to be considerably augmented to achieve the same level of protection.

The "self protecting" standard for spent fuel is often taken to be 100 rad/hr (1 gray/hr) at 1 m. Depending on the physical security measures, this may not provide sufficient protection for either spent fuel or immobilized plutonium. It takes an exposure of about 450 rad for a lethal dose (LD50, a 50% chance of death). In the absence of guards or other measures, one can imagine a scenario in which spent fuel or immobilized plutonium is not self-protecting. The dose rate 1 m from the ends of a spent fuel assembly are reduced by a factor of 5 to 10 from those 1 m from the midpoint. The dose rate at 5 m from the midpoint is reduced by over an order of magnitude from that at 1 m [7]. Similar reductions with distance and angle can be expected for glass logs containing fission products [8]. To be truly self-protecting (i.e., secure in the absence of other barriers), dose rates should be more than 1000 rad/hr. Fission products, such as ^{137}Cs, which provide the barrier have a half life of about 30 years, after which other barriers will have to be augmented. It is possible that the plutonium could be sealed in an geologic repository in that time, replacing the radiation barrier by lack of accessibility, i.e., an isolation barrier.

Chemical dilution alone is not a practical barrier, because chemical separation methods for plutonium are well known [9] and practiced in many places around the world. Before (and after) immobilization, plutonium must be diluted to less than about 20% to preclude use as a weapon. Coupled with the radiation barrier, dilution is more effective because chemical separation becomes more difficult (dangerous), costly, and visible.

The immobilization material will likely contain a low percentage of the plutonium because of performance and criticality considerations in a repository or borehole. A low percentage or density of plutonium also means that more packages would have to be stolen or diverted to obtain a given amount of plutonium.

Physical size and weight can be a significant impediment to the theft or diversion of packages containing immobilized plutonium. Fabrication, transportation, and criticality considerations will dictate a maximum size and weight. The minimum size and weight should be set so that equipment such as cranes and forklifts are necessary to move the packages. That is, they should not be transportable by a few, or even several, individuals. Glass logs being considered are about 0.6 m in diameter, 3 m long and weigh about 2 tonnes. Although longer, spent fuel assemblies are smaller and weigh considerably less (300 to 600 kg). When coupled with the radiation barrier and with physical protection measures such as guards and fences, size and weight are important impediments.

The security of plutonium in a geologic repository or borehole depends on effective, and often synergistic, performance of the barriers designed into the disposal system. If immobilized plutonium is not provided with a radiation barrier, other barriers will have to be significantly enhanced. Currently, theft, rather than covert diversion by the host country, is the primary threat. As well as considering the on-site guards as a response to this threat, the design and evaluation of a security system should include consideration of available backup forces, such as off-duty guards, local and regional police, and nearby military forces. If these "extra" forces are not available, then additional guards and physical protection measures and devices will be required on the site to reduce the risk of theft and provide sufficient security. Analyses (including time and motion studies and conflict simulation) of threat scenarios can help determine the appropriate trade-offs and level of security.

5. Demonstrating MPC&A for Plutonium Disposition in Russia

In a previous paper [10], the need for "monitoring" was discussed as referring to activities designed to assure the international community that nuclear materials are secure and not subject to unauthorized use. There is still widespread apprehension that some materials in Russia are vulnerable to theft and black market trafficking. Officials in Russia have taken this vulnerability seriously and have moved to increase security [11]. Through various programs, the United States and other countries are working with Russia to enhance the security of nuclear materials [12]. The primary assistance has been in the MPC&A area [13]. Although the work on enhancing MPC&A in Russia has not been secret, it has not been transparent enough to satisfy the world community that Russia has indeed made a long-term commitment to bring MPC&A in Russia up to world standards, such as those promulgated by the IAEA or EURATOM. Plutonium disposition offers an excellent opportunity for Russia to implement and demonstrate world-class MPC&A. In particular, any new or modified fuel fabrication, vitrification, or other facilities should incorporate modern MPC&A consistent with IAEA standards. As a step in this direction, and as a demonstration of transparency, Russia could request help from IAEA in planning MPC&A measures for plutonium disposition facilities.

6. Enhancement and Demonstration of MPC&A

The enhancement and demonstration of MPC&A need not, and should not, wait until plutonium disposition decisions are made and implemented. In particular, the enhancement and demonstration of the protection of nuclear materials could be accomplished quickly at a very reasonable cost. Russian Interior Minister Viktor Yerin recently reported that 80% of Russian nuclear sites lacked basic equipment at their gates for detecting radioactive substances [11]. Recently, a United States–Russian laboratory-to-laboratory effort has installed a portal radiation monitor at a facility of the Russian Institute of Physics and Power Engineering (IPPE) in Obninsk [14]. Installation of such monitors is also planned for some facilities at Arzamas-16, Chelyabinsk-70, and Tomsk-7. However, at these and other sites in Russia, there may be as many as 100 facilities, all together, that store or handle nuclear materials [15]. The number of required portal monitors could be higher. The cost for the type of portal monitor installed at the IPPE facility is about $23,000 [16]. One hundred such portal monitors would cost $2.3M. Cost savings and increased security could be realized if the materials to be protected were consolidated at fewer facilities.

Although the cost for walk-through portal monitoring is not prohibitive, it will take some time to produce and install portal radiation monitors at all the necessary facilities [17]. In the meantime, it would be possible to institute portal monitoring quickly and cheaply by using hand-held monitors. The TSA Systems PRM-470 monitor sells for about $1625 [18], and the company's affiliate has already translated the users' manual into Russian. This type of monitoring could be augmented by the use of metal detectors, especially at facilities that handle plutonium and highly enriched uranium in metal form. Metorex Inc. sells walk-through metal detectors for about $4600 installed [19]. Thus it would cost about $6225 to equip each portal with a hand-held radiation monitor and a walk-through metal detector, and about $622,500 for 100 portals. The cost could be reduced to about $1825 per portal ($182,500 for 100 portals) if hand-held metal detectors were used. While much faster and less expensive to initiate, a hand-held monitoring program would require the extensive use of monitoring personnel and would depend for effectiveness on the reliability on these people.

Finally, if Russia were to commit to a program to install portal monitoring at all facilities handling nuclear materials, one would expect that countries worried about the theft of nuclear materials in Russia would offer financial support. In any case, it would be important for Russia to make such a program transparent, so that the world community could have confidence that all nuclear materials in Russia are secure. This could be accomplished by allowing inspections of portal monitoring by individual countries or international groups.

7. Summary, Conclusions, and a *Recommendation*

1. MPC&A for disposition of excess weapons plutonium is essential, and it is feasible whether the disposition option is MOX reactor fuel or immobilization and direct disposal.
2. Ensuring that all plutonium containers are accounted for and have not been tampered with is easier and more effective than verifying the precise inventory of plutonium in containers.
3. The security of immobilized plutonium depends on a number of barriers built in to the disposal system that may work together synergistically to prevent theft and reuse.
4. The decision, in Russia, between MOX use and immobilization and direct disposal will not likely be driven by MPC&A considerations.
5. Early implementation of disposition decisions can establish a programmatic momentum toward protection of nuclear materials while conditions in the United States and Russia are favorable.
6. In Russia, a MOX option could be started sooner than an immobilization option.
7. Demonstration of MPC&A is essential, and the use of new facilities offers an excellent opportunity to incorporate transparency.
8. *The protection of nuclear materials in Russia could be enhanced and demonstrated, quickly and inexpensively, by the use of hand-held radiation monitors at nuclear facility portals.*

Acknowledgments

The author thanks colleagues at Livermore, Los Alamos, and Savannah River who provided information for this paper, and Allan D. Frymire of TSA Systems Ltd. and Carol Shaltis of Metorex Inc. for providing information on radiation and metal detectors, respectively. In addition the comments and suggestions provided by P. W. Murphy and P. L. Chrzanowski of LLNL, D. F. Bunch of Management Strategies, Inc., and Oleg Bukharin of the Center for Energy and Environmental Studies of Princeton University are gratefully acknowledged.

Work performed under the auspices of the U.S. Department of Energy by Lawrence Livermore National Laboratory under Contract W-7405-Eng-48.

References

1. W. G. Sutcliffe, Ed., *Fissile Materials from Nuclear Arms Reductions: A Question of Disposition*, Lawrence Livermore National Laboratory, Livermore, CA, CONF-910208 (February 1991).
2. Committee on International Security and Arms Control, National Academy of Sciences, *Management and Disposition of Excess Weapons Plutonium*, National Academy Press, Washington, DC (1994).
3. E. R. Merz, C. E. Walter, and G. M. Pshakin, Eds., *Mixed Oxide Fuel (MOX) Exploitation and Destruction in Power Reactors*, NATO ASI Series, Kluwer Academic Publishers, Dordrecht (1995).
4. *Long-Term Storage and Disposition of Weapons-Usable Fissile Materials Programmatic Environmental Impact Statement, Implementation Plan*, U.S. Department of Energy, Office of Fissile Materials Disposition (March 1995).
5. Specific cost estimates for the reactor and vitrification options can be found in *Management and Disposition of Excess Weapons Plutonium: Report of the Panel on Reactor-Related Options*, National Academy Press, Washington, DC (July 1995). The U.S. Department of Energy has also concluded a study of reactor options that analyzes costs: *Plutonium Disposition Study, Phase II* (to be published).
6. Foreign Broadcast Information Service, JPRS Report, Arms Control & Proliferation Issues, Special Edition: Plutonium in Russia: Ecology, Economic, Policy, JPRS-TAC-95-005-L, Sec. 4.2, p. 15 (2 February 1995).
7. W. R. Lloyd, M. K. Sheaffer, W. G. Sutcliffe, *Dose Rate Estimates from Irradiated Light-Water-Reactor Fuel Assemblies in Air*, Lawrence Livermore National Laboratory, Livermore, CA, UCRL-ID-115199 (1994).
8. L. J. Jardine, G. A. Armantrout, and L. F. Collins, *Utilization of Cs137 to Generate a Radiation Barrier for Weapons Grade Plutonium Immobilized in Borosilicate Glass Canisters*, Lawrence Livermore National Laboratory, Livermore, CA, UCRL-JC-118741, Rev. 1 (1995).
9. O. J. Wick, Ed., *Plutonium Handbook, A Guide to the Technology, Vols. I & II*, American Nuclear Society, LaGrange Park, IL (1980).
10. W. G. Sutcliffe, *A Perspective on Safeguarding and Monitoring Excess Military Plutonium*, NATO Advanced Research Workshop: "Mixed Oxide Fuel (MOX) Exploitation and Destruction in Power Reactors," Obninsk, Russia, October 17, 1994.
11. Margaret Shapiro, *Russia Orders Tightened Security to Protect Nuclear Materials*, The Washington Post, February 24, 1995.

12. Oleg Bukharin, *Cooperation between U.S. and MINATOM in the Area of Fissile Material Safeguards*, Presentation at a meeting of the President's Committee of Advisers on Science and Technology, Washington, DC, January 27, 1995.
13. Walter Y. Kato, "Nuclear Material Protection, Control, and Accounting (MPC&A) in the Russian Federation," Workshop on Plutonium Disposition sponsored by the German-American Academic Council, Bonn, March, 23, 1995.
14. Gennady Pshakin, Igor Matveenko (IPPE, Obninsk, Russia), Leslie Fishbone (BNL), Roger Vogel (SNL), Jack Blasy (LLNL), John Phillips (LANL), Bradley Weil (ORNL), Ken Byers (PNL), "MPC&A Lab-to-Lab Program Activities at IPPE," submitted for Institute of Nuclear Materials Management Annual Meeting, Palm Desert, July 1995.
15. *Nuclear Successor States of the Soviet Union*, The Carnegie Endowment for International Peace and the Monterey Institute of International Studies, Number 2, December 1994.
16. The monitor sells for $20,000; it is estimated to cost about $3000 to install in Russia. Allan D. Frymire, TSA Systems Ltd., Longmont, CO, personal communication (1995).
17. TSA Systems Ltd. has an affiliate, Polimaster, in Minsk that can produce portal monitors. Bob R. Miller, TSA Systems Ltd., Longmont, CO, personal communication (1995).
18. Allan D. Frymire, TSA Systems Ltd., Longmont, CO, personal communication (1995).
19. Carol Shaltis, Metorex Inc., Langhorne, PA, personal communication (1995). (Hand-held metal detectors cost about $200.)

6. SUMMARY OF WORKSHOP PRESENTATIONS

General

The third workshop explored the technology status for immobilizing excess weapon plutonium in glass or ceramic matrices and subsequent perpetual storage in a mined geologic repository or a deep borehole. There do not appear to be any technological show-stoppers. Clearly, Russia`s official position continues to view disposition of plutonium in this manner to be without merit. Although Prof. Victor Murogov (MINATOM) did not attend the meeting, his representative, Vladimir Kagramanian from the Institute of Physics and Power Engineering (IPPE), as well as Prof. Aleksander Karelin from the V.G.Khlopin Radium Institute emphasized their position that excess weapon plutonium is a valuable resource that should be utilized beneficially, and not "just buried".

Country: Russia
Presenter: Aleksander I. Kárelin
Title: (1) Weapons Plutonium Management: Status, Approaches, Prospects

A number of factors must be taken into account in the management of weapon plutonium and uranium such as the quantities in both military and civilian sectors, possibility of reuse in weapons, accounting reliability, Physical protection, safety, the political situation, technology, and economic conditions.

More than 50 t of plutonium and several hundred tons of HEU (highly enriched uranium) are to be released from military use. The optimal concept for management of this material should satisfy the main criteria: safety, ecological protection (workers, public, environment) and low cost. Scientists and specialists, not politicians, should choose the concept. Published contraversial positions (New York Times, 19 Aug.1994) on concept approach does not promote a solution.

The cited article asserts that although Russia preserves plutonium the U.S. wants to destroy plutonium by vitrification and mixing it with HLW. Such articles are deceptive and mislead the public as fissile material may only be destroyed by nuclear transformation. It makes no sense to separate isotopically impure plutonium from HLW (as at the BNFL plant in Sellafield) and store it as a material of value and on the other hand mix high-purity military plutonium with HLW.

Russian scientists and specialists steadfastly hold that military plutonium and uranium have a high energy potential and should be used for civilian electric power. Former Russian producers of military plutonium refused to participate in this workshop, whose theme is immobilization of plutonium by vitrification and mixing with HLW. Russian scientific investigations are directed toward efficient dissolution of plutonium metal and oxide, new glass and ceramic matrices for high-background plutonium that is unsuitable for further use, and methods of equipment decontamination.

E.R. Merz and C.E. Walter (eds.), Disposal of Weapon Plutonium, 285-303.

Although there is contraversy at the political level, it appears that scientific approaches to fissile material management are coming closer among the affected countries. For example, Russian scientists agree with most of the recommendations of the recent U.S. NAS study. A point of contention is the NAS recommendation to stop production of military plutonium. Because fissile material cannot be rigorously distinguished by its end use, this recommendation is inconsistent with continued development of nuclear power. Further steps should be taken for decreasing the long-term risk of potential diversion and application of plutonium to military use.

The cost of carrying out the recommendations of the U.S. NAS, except as noted above, will be recompensed by beneficial use of the fissile materials for electric power generation. The main result from appropriate management of weapons and excess military fissile materials is national and international security.

Country: Sweden
Presenter: Johan Swahn
Title (2) Retrievability and Safeguard Concerns Regarding Plutonium in Geologic Repositories

Three options for managing surplus separated plutonium are (1) converting it to the "spent fuel standard", (2) storing temporarily pending its use as an energy source, and (3) making it irretrievable. Considerations pertinent to these options are examined. Regardless of whether plutonium is used as MOX fuel or immobilized with HLW, it has the same proliferation risk as spent fuel with attendant long-term safeguard requirements. Going beyond the "spent fuel standard" is reasonable only if all spent fuel is similarly treated.

If breeder reactors are predominately used, there will not be long-term plutonium waste to safeguard. Difficult proliferation issues arise if breeder reactors are used, however, because separated high-military-quality plutonium would be in the system. These issues could prevent the breeder reactor economy from being realized.

Proliferation concerns and the cost of long-term safeguards can be avoided through use of two available methods for irretrievable plutonium disposition: transmutation and deep boreholes. In the latter case, interim retrievability could be provided until the time that the deep boreholes were sealed.

Country: Germany
Presenter: Erich R. Merz
Title (3) Multibarriers: Their Efficacy Against Intentional Intrusion into a Geologic Repository to Retrieve Disposed Plutonium

The primary goal of any plutonium disposition option is to exclude or at least to minimize the possibility that discarded ex-weapon plutonium could be reused as strategic arms. Mined repositories in deep underground geologic formations have been identified as the preferred disposal method for spent reactor fuel, solidified high-level waste, and transuranic elements. In the case of containment of only radioactive wastes, preventing the return of burried radioactive material into the biosphere is the prime protection objective. For the sake of secure and ultimate plutonium disposition on the other hand, another safety criterion is of decisive importance, namely the efficacy of the various barriers to illicit retrieval of the disposed plutonium.

Without relying on futuristic technology, disposition of plutonium can be accomplished by geological disposal, whether in the form of spent fuel or glass blocks. Even so, 30 - 60 years may be required to dismantle weapons, decide on a course of action, and select of a geologic repository. Multi-barriers are provided at four levels: material form (glass, ceramic), canister (stainless steel), backfill material (bentonite clay) in a geologically stable formation, and overburden (underground depth). Stability of this arrangement is defended by calculations for 10^4 years. Intentional human intrusion, that cannot be ruled out, is seen as the greatest risk, but one that is mitigated by arrangements that are conducive to international inspection. Intrusion would not then go undetected.

Country: Russia
Presenter: Vladimir S. Kagramanian
Title (4) Prospects for Using Plutonium as a Fuel in Russia

Russian experience with fast nuclear reactor technology is reviewed. Use of plutonium in thermal reactors is being studied. MINATOM policy with regard to management of both commercial and weapon plutonium is based on maximum use of Russia`s plutonium experience, importance of plutonium diversion resistance, ecological and economic acceptability, and development of a long-term closed fuel-cycle-option. In this context, disposition of weapon (and commercial) plutonium is inseparably linked with completion of the BN-800 fast reactors using traditional MOX fuel. The BN-800 design has received all applicable certifications and represents a new generation of world-level nuclear power plant safety. Use of MOX-fueled LWRs is unacceptable in Russia for several reasons, including: probable low throughput requiring construction of 15 to 20 reactors, high radiotoxicity of the spent fuel produced, and economic improvements made on BN-800 would not be realized in new LWRs.

Plutonium disposition in Russia depends on development of a nuclear power center at PO Mayak (including the RT-1 reprocessing facility, Complex-300 fuel fabrication facility, and three BN-800s). Other reactor design studies and fuel cycle strategy development, however, are also ongoing.

Nevertheless, investigations are underway to evaluate plutonium utilization options, possibly calling upon foreign technology including those on LWR technology in the framework of defense nuclear centers concepts.

Country: Belgium
Presenter: Etienne Trauwaert
Title (5) Plutonium Handling and Vitrification: Main Process Steps and Their Cost Evaluation

Two proven ways are available for disposition of weapon fissile material. One way is to make reactor fuel, the other is to condition it into a high-level-waste package. The latter is likely to be more expensive, however, other arguments may override cost. In any case, it is likely that at least small batches of contaminated plutonium may be vitrified. Based on waste vitrification experience, at the PAMELA plant in Mol, vitrification would cost $ 280/kg glass containing from 10 to 20% waste oxides (but see next paragraph). If the oxides are all PuO_2, at the lower concentration the product would cost approx. $ 3200/kg Pu. A cost estimate of $ 200/kg glass (or approx. $ 2300/kg Pu) is also stated for interim storage and final disposal of the glass product in deep geologic clay layers.

In this process liquid waste would be mixed with plutonium nitrate. After denitrating, the concentrate would be mixed with borosilicate glass frit and fed into a joule-heated ceramic melter containing already-molten glass at 1425 K. The amount of plutonium that can be incorporated in the glass is subject to limits imposed by consideration of radiation damage and criticality. These limits are believed to be 1% and 1.6% (by weight) in glass, respectively. (The cost estimates in the paper are based on a 10% oxide fraction, an apparent conflict).

Country: Russia
Presenter: Lev B. Shpunt
Title (6) Dissolution of Materials Containing Plutonium Metal and its Oxide Forms

The general approach to dissolution of metals and oxides of varying purity containing plutonium is discussed. Typically, metal is oxidized and then dissolved in a mixture of nitric and hydrofluoric acids. Metal can also be anodically dissolved using metal or liquid contact. Four processes are used for dissolution of plutonium oxides. KRI is proposing (and studying) a new concept for anodic dissolution of metals or oxides that combines the metal and liquid contact approaches. Individually, these approaches suffer from many disadvantages. Used in combination, however, these approaches avoid their individual difficulties, and 0.999 of the plutonium present is dissolved. By constructing equipment with titanium (and platinum coated anodes), corrosion rates are acceptably low. The approach is considered advantageous for a wide range of plutonium-bearing materials.

Typically, at metal plutonium chargings of 1 kg, a current density of 1 - 2.5 A/cm^2, a temperature between 30 and 65^oC, and HNO_3-concentration between 4 and 7.5 M, the measured dissolution rate amounts to 200 g/h.

Country: U.S.A.
Presenter: Rodney C. Ewing
Title (7) Crystalline Ceramics: Waste Forms for the Disposal of Weapons Plutonium

There is every reason to expect that waste performance can be much improved over what is now accepted for borosilicate glass. Ceramic phases, like the proposed ones, may represent an interesting alternative if applied in conjunction with the deep borehole concept. However, prudence of such a concept requires further intensive research and development with regard to several unsolved aspects.

Immobilization of excess weapons plutonium in the highly durable crystalline ceramics: apatite, pyrochlore (zirconolite), monazite, and zircon is reviewed. Zircon is chosen as the preferred host for this application. Long-term durability must extend for 500,000 y as far as plutonium is concerned. Criticality of U-235, the decay product of Pu-239, must be considered indefinitely beyond that period. As these ceramics occur in nature, observations of past durability can be made. Zircon sometimes contains up to 0.5% U or Th. Samples have been determined to be 4 billion years old. Improved performance over that of borosilicate glass is expected.

Country: Russia
Presenter: Evgeniy B. Anderson
Title (8) The Crystalline Host-Phases for Immobilization of Weapons Plutonium and Waste Actinides

It is proposed to immobilize plutonium and the actinides in polymineral ceramics based on the crystalline host phases: zircon and cubic zirconium dioxide. Plutonium and actinides form a broad range of substitutional solutions in these hosts. Polymineral ceramics have been synthesized in the laboratory by hot pressing. An original method resulting from the study of Chernobyl "lavas" of high uranium-zircon, was applied to simulate (with cerium dioxide) considerable amounts of plutonium and actinides in a zircon lattice. Both zircon and zirconium dioxide are exclusively /sic/ durable materials with natural analogs. Zirconium is easily obtained from fuel reprocessing waste.

The polymineral ceramic would be synthesized from ultra-fine oxide powders of plutonium, actinides, zirconium, and silicon produced from a plasma. Powders produced in this way are partially or fully amorphous and chemically reactive. Zirconium metal powder is added to the oxide powders to allow induction heating and to avoid the need for precise stochiometry. The zirconium powder promotes incorporation of plutonium and actinides into the host lattice, according to a model deciphered during the Chernobyl "Lava" study. Excess zirconium is subsequently oxidized.

Hot pressing of the powders is carried out at temperatures below 1725 K and pressures below 4500 MPa. All experiments have been successful and experimental work is continuing. The method of producing zircon doped with isomorphous admixtures has been patented. Other applications, e.g. strong ceramics, are expected.

Country: Russia
Presenter: Boris A. Nadykto
Title (9) On the Long-Term Storage of Weapon Plutonium in Form of Criticality-Free Ceramics Containing Lithium

For safe, long-term storage of excess fissile materials it is suggested that they be placed in a form where neither criticality nor reactions with oxygen can occur. Oxides of plutonium or uranium would be mixed with oxides of lithium or boron, and followed by cold-pressing and sintering. Constructing a nuclear weapon from material in this form would require considerable knowledge, equipment, effort, time, and expense. Calculations indicate that with lithium and boron enriched to 90% in their respective absorber isotopes, 50 wt% oxide mixtures would not be critical at any size. The (α,n)-reaction with Li-7 would provide a (protective) radiation source.

The sintered material would be placed in a leak-tight container and suspended in a 300-m deep borehole having a diameter of 1 - 2 m. The container could be raised to the surface as needed for monitoring and inspection. The hole could be water-filled if necessary to remove heat from the canister.

Country: Germany
Presenter: Hj. Matzke
Title (10) Incorporation of Plutonium and other Actinides in Borosilicate Glass and in Waste Ceramics

Properties of borosilicate glasses and ceramics containing plutonium are reviewed. Cm-244 (approx.18 y half life) is used to simulate long-time radiation damage effects and helium accumulation resulting from alpha decay. Solubility of plutonium in borosilicate glasses varies with glass type and composition. At 5% PuO_2, only 4.6% is retained in the vitreous phase, with the second phase being crystalline PuO_2. Addition of HLW lowers the solubility of PuO_2. PuO_2 solubility is further lowered (to 1.2%) by addition of rare earths. No segregation of Cm is noted in ceramic materials containing approx. 3% Cm. A sample of titanate ceramic, type SYNROC B was made containing 10% NpO_2. (Upper limits of Cm or Np solubilities are not given).

In glasses, radiation damage tends to lower leach rates, hardness, and Young`s moduls. Fracture toughness is increased. A large amount of helium is formed that produces swelling if the glass is annealed at a sufficiently high temperature. In ceramics, significant swelling is observed even without annealing. Near the damage level corresponding to saturation in swelling, a phase change from crystalline to amorphous is observed, resulting in a higher leach rate. Ceramics also exhibit a lower hardness and increasing fracture toughness with incresing damage level. Because annealing of damage occurs at a higher temperature in ceramics, radiation effects are more important in ceramics than glasses. Addition of fission products to either ceramics or glasses is not expected to result in added damage effects.

There is a large data base that can be used to suggest a plutonium-bearing matrix for long-term disposition. More research seems necessary to achieve one that is fully optimized.

Country: Japan
Presenter: Hirotaka Furuya
Title (11) Feasibility of Rock-Like Fuel and Glass Waste Form for Disposal of Weapons Plutonium

Results from continuing studies of a "rock-like" fuel (non-fertile PuO_2-bearing fuel) for disposition of weapon plutonium and borosilicate glass containing HLW and PuO_2 are presented. Radiation damage studies on a glass sample doped with plutonium (approx. 1%) and curium (~3%) oxides (containing Pu-238 and Cm-244) are reported.

Large ($>$0.8) fractions of plutonium destruction (especially Pu-239) can be achieved in a PWR, according to early studies. The phases that are present and the leaching characteristics of the rock-like structure of the spent fuel have been investigated. To compensate for unfavorable reactivity conditions (ß, Doppler, etc.), additions of depleted uranium or thorium oxides to the fuel material are being considered as well as partial reactor loadings with standard (LEU) fuel. (These actions would reduce the net plutonium destruction fraction).

Conditions of helium release from borosilicate glass are modeled, and have been determined from a self-damaged (alpha decay) sample. The model includes trapping of bubbles and re-solution from bubbles into the glass in addition to simple diffusion. Bubble behavior is believed to strongly influence the volume and mechanical properties of the glass. A volume change of 0.4 to 0.6% is observed at 1.7×10^{25} α-decay/m^3. Hardness and Young`s moduls decrease with dose and fracture toughness increases with dose to saturation values reached at $1.0 \times 10^{24}$ $\alpha$-decay/$m^3$. Hardness and Young`s moduls changes are generally recovered during annealing above 675 K for 10 hours, whereas fracture toughness remains high.

Country: Russia
Presenter: Yuri I. Matyunin
Title (12) Possible Ways of the Weapon Plutonium Processing to the Steady Glass Materials Suitable for Further Safe Disposal

Disposition of weapon plutonium by utilization as fuel in a reactor and subsequent disposal as spent fuel is considered to be the preferable option. Direct loading of plutonium in phosphate glass has also been investigated. Phosphate glass is preferred to borosilicate glass because higher plutonium content is possible, PuO_2 is less likely to disperse as a separate phase, and the technological process is more suitable. With glass containing high-level waste, (HLW) 3 to 5% plutonium can be uniformly distributed in the glass.

Without HLW, up to 20% PuO_2 can be incorporated in the phosphate glass. A uniform distribution of up to 3.4% PuO_2 results if the molten temperature (1275 K) is held $<$8 hours. Settling to the bottom of small glass cylinder samples is observed beyond these conditions. Lowering the alumina content of the glass lowers both these limits if there is no HLW present.

Two methods of processing plutonium-glass products are proposed. After dismantlement, the weapon plutonium is oxidized to a chemically active state and either converted to a nitrate (wet process) or directly vitrified by melting with fluxed additives (dry process). Use of HCl or HF is not recommended because their use increases the amount of plutonium-containing waste. Various phosphate glass compositions and methods of synthesis can be devised depending on the desired plutonium content and inclusion of an HLW component.

Additional research and development required for full production of the proposed glass products includes: (1) optimization of plutonium oxidation methods and parameters, (2) specification of glass matrix composition to satisfy product requirements (plutonium content, chemical and radiation stability, mechanical strength, etc.), (3) investigation of mineral-like matrices for plutonium, (4) development of necessary manufacturing equipment.

Country: Germany
Presenter: Friedrich W. Ledebrink
Title (13) Management of Operational Waste from MOX Fuel Plants

Experience with plutonium carry-over in waste from MOX fuel fabrication at the Siemen`s Hanau plant is reviewed. The MOX fuel fabricated to date has contained 8500 kg of reactor-grade plutonium. The waste generated during fuel fabrication has been conditioned and placed in 2900 drums (0.2 m^3/drum).

Based on this experience, the waste stream from an automated plant should contain 0.2% of the plutonium processed. If MOX fuel containing 120 Mg of weapon plutonium were fabricated, a carry-over of 240 kg of plutonium in the waste generating 20,400 waste drums could be expected. The waste volume that would result is based on a conservatively imposed criticality limit of 0.011 kg Pu (assumed to be 93.5% Pu-239) per drum at the Konrad repository. A cementation process to encapsulate these wastes is being proposed in order to establish a higher criticality limit of 0.12 kg Pu per drum, thus reducing the waste volume for geologic disposal (on this basis) by an order of magnitude.

Plutonium content of individual waste packages (20 cm diam by 35 cm high) is determined from passive neutron coincidence measurements with a Los Alamos-developed device (DRC-100). By making measurements on small packages, as low as 30 mg of plutonium is detected. Total individual measurement error is between 20 and 30%.

Conditioning processes to achieve the final waste form are described. Solid waste is sorted as organic (60%) and inorganic (40%) and the organic waste is shredded to avoid floating to the top in the subsequent cementation process. The final waste form contains three types of waste uniformly mixed with cement: liquid (resulting from wet chemical processes and analyses), solid organic, and solid inorganic. The mixture is cast into special 0.2 m^3 drums fitted with high efficiency filtered vents to prevent pressure buildup in the drum. The process of cement solidification avoids high pressure and temperature, providing a high degree of safety. The cementation facility has proven to be robust and trouble-free.

Ten waste drums are subsequently packed in a rectangular container and the intermediate spaces are filled with cement for final placement in the repository. The resulting repository volume per drum is thus increased by a factor of 2.7 and the 20,400 drums (see above) would occupy about 11,000 m^3 of repository volume. By raising the criticality limit and more aggressive evaporation of the liquid waste, the solid organic waste would become the limiting consideration in establishing repository volume. A factor of three volume reduction could be realized.

Country: U.S.A.
Presenter: George G. Wicks
Title (14) SRS Vitrification Studies in Support of the U.S. Program for Disposition of Excess Plutonium

Vitrification is considered to be one of the two leading options for disposition of weapon plutonium. Advantages of the vitrification option include potential for early implementation, acceptability of a wide range of plutonium-containing materials, technology maturity, and relevant waste glass performance. The leach rate of actinides has been shown to be 10 to 100 times lower than for other glass additives. Vitrification as a means of managing high-level waste (HLW) has been developed and demonstrated in France, Belgium, and England, as well as in Russia and the U.S. Immobilization of HLW in borosilicate glass is the only internationally accepted waste form. The potential for use of existing "waste glass" facilities reduces the need to build and later decommission new facilities.

In early work (1979) 7 wt% PuO_2 was immobilized in borosilicate glass at the Savannah River Site. Recent laboratory tests have achieved 15 wt% PuO_2 in a commercial borosilicate glass. The maximum plutonium content will likely be established by criticality rather than by solubility limits. Criticality can be mitigated by neutron absorbers such as boron and lithium, normally inherent in HLW forms, or intentionally-added erbium and gadolinium.

Proliferation resistance can be achieved by direct vitrification in which plutonium and HLW (or Cs-137) are simultaneously incorporated in the glass melt, or in a staged approach (called can-in-canister). Metallic plutonium would be converted to a suitable process feed material by oxidation or dissolution in acid. The vitrification process can use either liquid or oxide feed. In the staged approach, a plutonium-only glass is produced as a relatively small interim ingot (called a can). Subsequently, these cans could be transported to another location where HLW is available and remelted during vitrification of the HLW or placed in a canister into which molten glass containing HLW would be poured to fill the interstices. (hence, can-in-canister). In either the direct or staged approach, the same final product form could result.

The final form and composition of the plutonium bearing glass would be tailored for optimum technical performance in meeting political (proliferation) considerations. Many existing facilities in the U.S. could be used, and these are described. Combination of the plutonium disposition mission with an on-going mission for disposal of defense waste has the potential for minimizing the cost of plutonium disposition.

Country: Belgium
Presenter: Maurits D. Demonie
Title (15) On the Feasibility of Vitrifying High-Level Liquid Waste Containing High Amounts of Plutonium

Preliminary indications are that vitrification of a mixture of HLW and plutonium is feasible in the Pamela plant at Belgoprocess. A summary of the experience in waste vitrification at the Pamela plant is presented. The vitrification process was applied to immobilization of the liquid HLW produced during reprocessing of spent fuel from LWRs fueled with LEU and from research reactors fueled with HEU. A single step process is used with a joule-heated ceramic melter. In two campaigns spanning 6 years, 78 t of glass were produced from reprocessing 181 t of LEU spent fuel, and 412 t of glass were produced from reprocessing 31 t of HEU spent fuel. The total glass produced contained less than 180 kg uranium and less than 14 kg plutonium.

The Pamela system was analyzed for plutonium vitrification. The criticality limit on melter feed concentration is 6 kg/m^3. It is concluded that Pamela, after specific process development steps with a few facility modifications, could produce glass containing 1.6 wt% plutonium, below the observed concentration (2 wt%) for complete dissolution in borosilicate glass. The melter would hold a maximum of 12.8 kg plutonium in 800 kg glass. A standard 400-kg glass product canister of 0.15 m^3 would contain 6.4 kg plutonium. The Pamela plant would be able to process 1 t/y of plutonium in this manner.

Country: Russia
Presenter: Albert S. Aloy
Title (16) Vitrification of Weapons Plutonium: A Hypothetical Option or a Real Way to Solve the Problem?

The basic criterion for selecting the technology for disposition of weapon plutonium is safety in its broadest sense - safety independent of geographical borders or geological time sacales. A number of factors argue against vitrification as the technology for disposition of weapon plutonium. To the present, development of HLW vitrification technology has concentrated on minimizing the content of actinides in the glass, as the actinides are too valuable to be considered waste. From a storage standpoint, high fired PuO_2 has greater integrity as a solid form than plutonium contained in glass because of the much lower solubility of PuO_2 in water than the leaching rate of plutonium from glass. These observations are counter to the U.S. National Academy of Sciences study that concludes that vitrification is an alternative that deserves attention for plutonium disposition.

Attention is given to three scientific aspects of vitrification: criticality, radiation stability, and homogeneity of composition.

From a criticality standpoint, the theoretical maximum concentration of plutonium in borosilicate glass is 8 to 10 wt%. European standards, however, currently limit plutonium content in glass canisters to 0.11 kg. Even at a low concentration of 1-wt% plutonium in the glass canister, the plutonium content would be 4 kg, forty-fold over the limit. This "disconnect" in current European regulations would have to be resolved. Criticality is further facilitated by the high leachability of boron and the ingrowth of U-235 and Am-241.

Concern has arisen that low energy photons contribute to the formation of oxygen microbubbles in borosilicate glass. Oxygen generation (in addition to helium) is accompanied by increasing internal pressure, change in chemical composition, and increased surface area. Although these effects do not require radical changes for HLW management, incorporation of significant amounts of actinides in the glass will undoubtedly affect radiation stability of the glass, as well as the leachability of the α-decay products.

Homogeneity of HLW-containing borosilicate glass is affected by immiscibility, segregation, and formation of insoluble precipitates during vitrification. These phenomena have a profound effect on the vitirification process as well as the properties of the glass product. There are essentially no fundamental spectroscopic analyses of actinide bearing glasses. Such analyses would eliminate discrepancies observed in the solubility limits of actinides as reported by various investigators.

Development of HLW vitrification technology to an industrial level has consumed a considerable amount of money for R&D and time. Two continuous processes have been developed: the rotary calciner/metallic melter used in France and UK; and the liquid/slurry-fed ceramic melter used in Belgium, Japan, Russia, Germany, and U.S. For the low actinide content of HLW, criticality has not been an issue in these processes.

It is not likely that at high plutonium concentration these processes would be acceptable from the standpoint of criticality and material homogeneity. Capacity of existing plants would need to be increased to process 100 t plutonium in 15 to 20 years.

Considering the experience with vitrification of HLW, the unknowns associated with significantly higher plutonium content, and the large energy potential of plutonium, the implication is clear that, at least in Russia, weapon plutonium should be used as fuel in reactors.

Country: U.K.
Presenter: Erich Merz for J.A.C. Marples
Title (17) Vitrification of Plutonium for Disposal

Immobilization of plutonium by vitrification is feasible. Coprocessing with HLW makes recovery of the plutonium difficult. Probably the best method for incorporating plutonium in glass is to dissolve the metal in nitric acid containing HF and hydrazine and use this solution as feed material to the vitrification plant. Electrolytic dissolution is a promising alternative.

Borosilicate glass (used or planned to be used exclusively in the West) has a solubility limit of 5 wt% PuO_2, while phosphate glass (used in Russia) has a higher solubility limit. Criticality considerations make borosilicate glass preferable. For added criticality safety, gadolinium or other rare earth oxides could be included in the glass.

The leach rate of plutonium from the glass depends on the solubility of plutonium in the repository ground water. If the water has a pH $>$ 9, the solubility of plutonium, and hence the leach rate are very low. Heat emission from plutonium is low (in the absence of HLW), thus plutonium-containing glass can be disposed of without a waiting period. Radiation damage data from accelerated testing provides an indication of glass performance after approx. 2000 years. At this level no obvious ill effects have been observed.

Recovery of plutonium from standard size (0.4 -1.4 t) glass containers would require large-scale facilities. If the glass also contains HLW, the facilities would need to be shielded. Added complications in separating pure plutonium would result from the presence of silica.

Country: France
Presenter: Antoine Jouan
Title (18) French Appraisal of the Vitrification Option: Plutonium - or How to Get Rid of it?

Plutonium vitrification would be the first time man has ever considered burying an energy source rather than extracting an energy source from the earth. The French position is to utilize the energy that is contained in weapon plutonium.

Notwithstanding, various technologies are available to vitrify liquid or solid PuO_2 or plutonium metal. Many years of development would be needed to achieve an industrial process. A dry process has significant advantages and reduce criticality concerns. A wet process simplifies metering the feed. Vitrification of plutonium with fission products is feasible, but will not be a simple process.

The continuous vitrification process is used in France and U.K. to convert liquid HLW from Purex plants to borosilicate glass. The HLW nitrate solution is fed into a heated calciner where salts are concentrated and the nitrates are converted to oxides. The calcine and glass additives are then introduced to an induction-heated metal crucible. A batch vitrification process was used in France 30 years ago. In the batch process, evaporation, calcining, and vitrification are all accomplished in the induction-heated melter.

About 5200 borosilicate glass canisters, each weighing about 400 kg, have been produced in France in three facilities since 1978. France is now developing a new glass melting technique that uses an induction-heated cold crucible. This technique allows higher melt temperature ($>$1775 K), thus capable of producing more refractory glass or vitrocrystalline formulas providing better containment.

Country: U.S.A.
Presenter: Leonard W. Gray
Title (19) Safety Aspects with Regard to Plutonium Vitrification

One disposition option for excess weapon plutonium is immobilization. A team has been assembled to assess the technical feasibility of immobilization and to describe the required infrastructure. To be economically viable, it is assumed that the concentration of plutonium must be $>$ 0.4 wt%. Above 10.0 wt% plutonium, the immobilization form is considered to be too attractive a diversion target.

Immobilization must satisfy the radiological, physical, and chemical aspects of the "spent fuel standard". The literature discusses 72 uniquely-named waste forms that on review represent only 45 unique forms. Decision analysis techniques were used to screen these 45 waste forms in two stages. In the first stage, a coarse screen eliminated all but 16 forms that were, in retrospect, clearly inappropriate for immobilization. A more formal comparison of the 16 remaining waste forms was made in the second stage using multiattribute utility analysis techniques. A clear ranking was obtained with borosilicate glass, Synroc (a crystalline ceramic), and phosphate glass being the leading contenders with value rankings compared to a maximum (highest rank) of 1.0, of 0.89, 0.66, and 0.55, respectively. Only borosilicate glass is considered further in the paper.

Criticality safety is a major concern in designing the vitrification process. It is required that no single failure could produce a critical accident. Specific criticality analysis of the vitrification process has yet to be done. The issues involved drive the design of the process. In the initial process designs, the primary method of achieving criticality control is by conservatively limiting plutonium mass and its possible geometry. For example, a two-step process is envisioned with the first step a small batch process producing plutonium-borosilicate glass. After homogenization (dilution) the potential for criticality is greatly reduced and a larger melter containing higher amounts of plutonium can be used in the second step. Additional neutron absorbers can be incorporated during the second step. As process designs become more definitive more detailed criticality analysis may indicate that a cheaper one-step process is safe.

Criticality safety in the repository also needs to be addressed. The immobilization matrix, plutonium concentration, and neutron absorbers are criticality factors that are directly established in the vitrification process design. Additional criticality factors are established by the repository design. Both designs need to be performed in a system context.

The primary safety problem is criticality control during form fabrication. More detailed calculations will be performed to determine whether mass limits can be increased. Requiring both mass and geometry control dictates a two-step vitrification process. Raising the mass limit sufficiently would allow a lower-cost one-step process to be used instead.

Country: Russia
Presenter: Tatyana A. Gupalo
Title (20) Disposition of Radionuclides in Deep Geologic Formations Either in Containers or Vitrified Form

It is universally recognized that the only technical approach for management of radioactive material not destined for further utilization (waste) is to solidify the material and isolate it in the bowels of the earth. Recently, special attention has been given to plutonium. Two differing points of view regard plutonium to be waste or a national resource. In either case, the possibility of safely isolating plutonium for long-term storage, subsequent utilization, and disposal must be scientifically considered.

Underground storage of HLW relies on multibarriers for isolation. The role of engineered barriers, however, is limited to the near term. Natural barriers (natural rock formations) provide the ultimate isolation in the long term. One looks therefore for blocks in the earth`s crust with low tectonic activity, having structural stability, low permeability, and low water content.

To determine the suitability of a specific geological formation it is necessary to (1) assess the properties of the rock such as structure, physical processes, and changes in the massif under natural forces by modeling over the required period of time, (2) investigate scenarios associated with possible emergency events and determine their probability, and (3) conduct a probability safety analysis for both normal and emergency events. The main physical processes that need to be considered for both the near and far fields are geomechanic, thermal and mass transfer, and hydrogeologic. A data base on earthquake magnitudes is available dating back to 1743.

It is concluded that a quantitative safety assessment can be made for a specific repository configuration and geological site.

Country: U.S.A
Presenter: Leslie J. Jardine
Title (21) Disposition of Plutonium in Deep Boreholes

Deep-borehole disposition involves placing plutonium several kilometers deep into old, stable, rock formations that have negligible free water present. Containment assurance is based on the presence of ancient ground water indicating lack of migration and communication with the biosphere. Recovery would be extremely costly and impossible to accomplish clandestinely. A number of relevant deep boreholes have been drilled, and the technology is considered to be mature. A number of disposition forms could be used - none has been selected. The number of deep boreholes required could be as low as one or two. Hole diameters at the material emplacement depth need not be over approx. 0.7 m. A hole depth of 4 km with the bottom 2 km containing plutonium material seems reasonable. Major issues involve long-term criticality, licensing, and public acceptance. Regulatory implementation is the greatest uncertainty.

Country: Russia
Presenter: Alexander B. Pavlov
Title (22) Technology for Decontamination and Prevention of Contamination of Equipment Surfaces and Rooms by Plutonium

An overview of practices in Russia to prevent plutonium contamination in the nuclear industry is presented. Difficulties with plutonium arise from its chemical nature and its extremely high toxicity (four times higher than most other alpha-decay radionuclides). Plutonium contamination at nuclear power plants results generally only in connection with fuel element cladding failure, and is not a routine concern. Contamination is a continuing concern in research laboraties and in spent fuel reprocessing plants.

Contamination of surfaces in a reprocessing plant begins at the head end during initial cutting of the fuel elements at a level of 10 mg/m^2 and increases to a maximum in the final product areas. Contamination of surfaces results from absorption and diffusion processes. Absorption rates depend on the surface characteristics. Special solutions and conditions are required for decontamination. Initially washing with a medium concentration nitric acid solution that contains a small amount of fluoride is used. This solution permits efficient deposit renewal and reduces the waste that is generated. Specially designed jet sprays are used to reduce the amount of washing solutions.

Numerous advances have been made in techniques for protection of surfaces from plutonium and decontamination of surfaces from plutonium. Means to increase decontamination effectiveness and decrease liquid waste generation will continue to be investigated, however. Significant improvements, especially in new plants, are possible to decrease releases to the environment during operation and in decommissioning.

Country: Italy
Presenter: Heinz Dworschak
Title (23) Technical Basis for Plutonium Safeguards Practices, Handling and Disposal, PERLA and TAME Experience

The practices and basic techniques applied in safeguarding special nuclear material that is in civilian use aim at efficient and effective control by entitled inspectorates. The Joint Research Center supports these inspectorates with instrumentation development and implementation as well as training. Worldwide attention is now focused on safeguarding plutonium.

The distinction between reactor and weapon plutonium is arbitrary, as either can be used to produce a nuclear weapon. Also, straightforward chemical processes can be used to isolate plutonium from all other elements. Nevertheless, plutonium has been efficiently controlled for decades under IAEA and Euratom systems of material control and accountancy that are based on well-established material balances periodically verified by physical inventories. The inspectorates must have unlimited access to the nuclear materials. Thus, safeguarding plutonium derived from military sources will be intrusive. Solutions for accomplishing the safeguarding and preserving military sensitive information have been proposed. In this regard, the difference in the goals of nuclear disarmament verification and non-proliferation verification must be noted.

Safeguard techniques are based on the existence of a material accountancy system. Computerized means of managing the data including statistical analysis are available. Destructive and non-destructive assays are routinely conducted to make volume and density measurements. Techniques for plutonium determinations rely on neutron coincidence counting, calorimetry, and high-resolution gamma ray spectroscopy. Sealing/identification techniques are effective. Ultrasonic techniques are being investigated for surveillance of PuO_2 in transport containers. Optical surveillance used with data reduction techniques is effective.

Training programs and facilities are in place that provide the capacity to design verification plans. The training provides a means of transfering technology between the laboratory and the industrial measuring environment. Plutonium safeguard practices vary depending on whether the plutonium is liquid or solid, its isotopic content, and whether it is relatively pure or scrap in waste barrels.

The accuracy of measurements varies from less than 1% to as high as 20%. The first requirement for safeguarding excess weapon plutonium in a civilian environment is to accurately assess its initial inventory as it crosses the military/civilian border. Well-defined pure plutonium can be assayed within 1%, while scrap can be assayed within a few percent.

Country: Germany
Presenter: Bruno Baltes
Title (24) Long-Term Safety Aspects of Geologic Disposal of Large Quantities of Plutonium

Fundamental requirements for disposal of radioactive waste (HLW or spent fuel) include reduction of ionizing radiation exposure to humans now and in the future, and environmental protection. These requirements are the basis for rules and regulations promulgated by the ICRP and the IAEA. Probability safety assessments are implied for the post-closure phase of a repository to account for data uncertainty and event probabilities. The ICRP recommends that the risk of health effects to members of the critical group be less than 10^{-5} per year.

The disposal site represents the basis for assuring repository safety during all stages of repository operation and the post-closure period. All geologic repository designs rely on a multi-barrier system consisting of conditioned waste, waste container, backfill, seals, and the geosphere. The containers are designed to function as an impermeable barrier for several hundred to several ten-thousand years. Spatial arrangement and loading density of material that generates heat is governed by host-rock temperature limits (475 K for salt rock and 375 K for granite and clay). Also, formation of critical masses has to be avoided. It is generally accepted that release of radionuclides to the biosphere cannot be avoided in the long term. However, the multi-barrier system must reduce the release sufficiently to satisfy the health criterion.

Long-term safety assessments have been carried out for disposal of vitrified HLW or LEU spent fuel. Safety assessments have not been carried out for MOX spent fuel or vitrified waste with significant amounts of plutonium, but deduction from the LEU spent-fuel assessments lead to the conclusion that safe disposal of MOX spent fuel is feasible. Criticality considerations will probably limit the concentration of plutonium in vitrified waste, whereas heat generation considerations will not likely be limiting. The heat rate from MOX spent fuel can contain only one-fourth (sic) the quantity of LEU spent fuel elements. The migration of Pa-231, a long-term decay product from Pu-239 and U-235, needs to be assessed as part of post-closure performance.

Country: U.S.A.
Presenter: Carl E. Walter for William G. Sutcliffe
Title (25) MPC&A for Plutonium Disposition in the Russian Federation

Material protection, control, and accounting (MPC&A) is feasible for immobilization and disposal of excess weapon plutonium. Security of immobilized plutonium depends on a number of barriers. It is not likely that MPC&A would be a determining factor in choosing between MOX spent fuel or immobilization as a means of plutonium disposition. Demonstration of MPC&A is important - new facilities offer an excellent opportunity to incorporate transparency. The need to enhance and demonstrate the protection of nuclear materials exists, and could be pursued at a low cost.

Country: Russia
Presenter: Valery F. Menschikov
Title (26) Ecological Safety Aspects with Regard to Plutonium Disposal

(Paper not available)

The speaker confined his statements to an overview of radioactive contaminations existent in the neighbourhood of Russian nuclear weapon complexes. He did not touch how to tackle the environmental challenges arising due to weapon disarmament requirements. His contribution was an impeachment concerning the misdeeds of the nuclear community.

Indeed, there are heavily contaminated areas located at or near facilities that reprocessed fuel from weapon material production reactors as well as from nuclear weapon fabrication plants. In Russia, the corresponding sites are Majak and Chelyabinsk, Tomsk-7, Krasnoyarsk-26, and Novgorod/Arzamas-16. Most recently, world attention has been focused on Russian contamination of Artic and Pacific oceans and seas by discharge of radioactive wastes from operation of nuclear reactors on submarines and icebreakers. Although the original estimate of total activity discharged was about 2.5 million curies, the current activity remaining is estimated to be about 140,000 curies. However, this amount still constitutes less than 0.5% of the current radioactivity in Lake Karachai at the Mayak site alone.

The largest environmental releases of radioactivity by far are derived from reprocessing at the sites Chelyabinsk, Tomsk-7, and Krasnoyarsk-26. Early radioactive waste management of reprocessing wastes included discharging of wastes directly to rivers, lakes and reservoirs. The underground disposal of radioactive wastes via well injection, starting in the 1960ties, has introduced radioactivity into the environment in the order of estimated 1.7 billion curies. This practice continues today.

Certainly, the determination of the potential impacts to the environment and human health from the discharges to surface water systems and underground formations is believed to be one of the critical needs of waste management studies in Russia. However, polemic utterances are a meagre help.

REFERENCES

1. "Managing the Plutonium Surplus: Applications and Technical Options", edited by Richard L. Garwin, Michael Grubb, and Emma Mantanle, NATO Advanced Science Institutes Series 1: Disarmament Technologies, Vol. 1 (1994)

2. "Mixed Oxide Fuel (MOX) Exploitation and Destruction in Power Reactors", edited by Erich R. Merz, Carl E. Walter, and Gennady M. Pshakin, NATO Advanced Science Institutes Series 1: Disarmament Technologies, Vol. 2 (1994)

GLOSSARY

BNFL	British Nuclear Fuels, Limited
HEU	High-enriched Uranium
HLW	High-level Waste
IAEA	International Atomic Energy Agency
ICRP	International Commission on Radiological Protection
KRI	V.G.Khlopin Radium Institute
LEU	Low-enriched Uranium
LWR	Light Water Reactor
MOX	Mixed Oxide
NAS	National Academy of Sciences
PWR	Pressurized Water Reactor

7. NATO WORKSHOP SUMMARY EVALUATION

The third workshop explored the technology status for immobilizing excess weapon plutonium in glass or ceramic matrices and their subsequent perpetual storage in a mined geologic repository or a deep borehole. There do not appear to be any technological show-stoppers. Additional thought is required to bring to closure some unresolved issues. Some research and development remains to be done.

7.1 Strategy

A very firm position is taken in Russia and France that the energy potential in excess weapon plutonium must be realized beneficially and not squandered in a geological repository. The point was made that it makes no sense on one hand to separate isotopically impure plutonium from HLW (as at the BNFL plant in Sellafield) and store it as "material of value" and on the other hand mix high-purity military plutonium with HLW. Although there is controversy at the political level, it appears that technical approaches to fissile material disposition are coming closer among affected countries. There was general consensus that the disposition method should be chosen on a technical basis. Also there appears to be a strongly held conviction on the part of most speakers at the workshop, whether they are addressing disposition strategy or not, that disposition without energy utilization is not a reasonable solution.

Russia`s policy with regard to management of both commercial and weapon plutonium is based on maximum use of their experience, importance of plutonium diversion resistance, ecological and economic acceptability, and development of a long-term closed fuel-cycle-option. In this context, disposition of weapon (and commercial) plutonium is inseparably linked with the completion of the BN-800 fast reactors using traditional MOX fuel. The BN-800 design has received all applicable certifications and represents a new generation of world-level nuclear power plant safety. Use of MOX-fueled LWRs in Russia, on the other hand, would be a poorer choice for several reasons, including probable low throughput requiring construction of 15 to 20 reactors, high radiotoxicity of the spent fuel produced (true anywhere), and higher reactor cost.

Despite this general belief, it is recognized that some of the excess plutonium may be of such a condition that permanent disposal in a geological repository is appropriate. Russian scientific investigations are directed toward efficient dissolution of plutonium metal and oxide, new glass and ceramic matricies for high-background plutonium that is unsuitable for further use, and methods of equipment decontamination.

The three options for managing surplus plutonium (converting it to the "spent fuel standard", storing temporarily pending its use as an energy source, and making it irretrievable) were discussed. Regardless of whether plutonium is used as MOX fuel or immobilized with HLW, it has the same proliferation risk as spent fuel with attendant long-term safeguard requirements. Going beyond the "spent fuel standard" is reasonable only if all spent fuel is similarly treated.

E.R. Merz and C.E. Walter (eds.), Disposal of Weapon Plutonium, 305-316.

If breeder reactors are predominantly used, there would not be long-term plutonium waste safeguards issues, but there would be difficult proliferation issues if separated military-quality plutonium was present in the system. The editors note that new pyrochemical processes avoid the existence of separated plutonium.

Proliferation concerns and the cost of long-term safeguards can be avoided through use of two available methods for irretrievable plutonium disposition: transmutation and deep boreholes. In the latter case, interim retrievability could be provided until the time that the deep boreholes were sealed.

Disposition of plutonium can be accomplished by geologic disposal without relying on futuristic technology, whether in the form of spent fuel or glass. Even so, a long period of time (30 - 60 years) may be required to dismantle weapons, decide on a course of action, and select a geologic repository. Multi-barriers are provided at four levels: material form (glass), canister (stainless steel), backfill material (bentonite clay) in a geologically stable formation, and overburden (underground depth). Stability of this arrangement can be defended by calculations for 10^4y. Intentional human intrusion, that cannot be ruled out, is seen as the greatest risk, but one that is mitigated by arrangements that are conducive to international inspection. Intrusion would not then go undetected.

7.2 Product Form

A thorough study resulted in a quantitative ranking of product forms ranked borosilicate glass highest, followed by Synroc (e.g. zircon) and phosphate glass in that order. In this section the information presented at the workshop on product forms other than borosilicate glass is summarized and evaluated. The editors note that there are differences of opinion on the best product form for the immobilization of plutonium. These differences should be resolved. The potential spin-offs of new materials, (e.g. tough ceramics) for other applications should not be overlooked.

Immobilization of weapon plutonium in the highly durable crystalline ceramics: apatite, pyrochlore (zirconolite), monazite, and zircon has been reviewed, and among these, zircon is preferred. Zircon advocates predict improved performance over that of borosilicate glass. Long-term durability of the disposition system must extend for $5 \cdot 10^5$y. Criticality of U-235, the decay product of Pu-239, must be considered indefinitely beyond that period. As zircon is a naturally occuring mineral, observations of past durability, over a period of $4 \cdot 10^9$y have been made.

Plutonium and the other actinides form a broad range of substitutional solutions in polymineral ceramics: zircon and cubic zirconium dioxide. Both zircon and zirconium dioxide are said to be exclusively durable materials with natural analogs. Polymineral ceramics have been synthesized in the laboratory by hot pressing.

An original method resulting from the study of Chernobyl "lavas" of high uranium-zircon, was applied to simulate (with cerium dioxide) considerable amounts of plutonium and actinides in a zircon lattice. The polymineral ceramic is synthesized from ultra-fine oxide powders of plutonium, actinides, zirconium, and silicon produced from a plasma. Powders produced in this way are partially or fully amorphous and chemically reactive. Zirconium metal powder is added to the oxide powders to allow induction heating and to avoid the need for precise stochiometry. The zirconium powder promotes incorporation of plutonium and actinides into the host lattice, according to a model deciphered during the Chernobyl "lava" study. Hot pressing of the powders is carried out at temperatures below 1725 K and pressures below 4500 MPa. All experiments have been successful and experimental work is continuing.

It was suggested that excess fissile materials be placed in a form where neither criticality nor reactions with oxygen can occur. Oxides of plutonium or uranium would be mixed with oxides of lithium or boron, cold-pressed, and sintered. Subsequent use for nuclear weapons would encounter significant difficulties. Calculations indicate that with lithium and boron enriched to 90% in their respective absorber isotopes, 50 wt.% oxide mixtures would not be critical at any size. The (α,n) reaction with Li-7 would provide a protective radiation source. The sintered material would be placed in a leak-tight container and suspended in a 300-m deep hole having a diameter of 1-2 m. The container could be raised to the surface as needed for monitoring and inspection. The hole could be water-filled if necessary to remove heat from the canister.

A new concept for anodic dissolution of metals or oxides that combines the two approaches of metal and liquid contact (generally used separately) is being studied. Used in combination, these approaches are said to avoid their individual difficulties, and 0.999 of the plutonium is dissolved. The approach is considered advantageous for a wide range of plutonium-bearing materials.

"Rock-like" fuel (non-fertile PuO_2-bearing fuel) for disposition of weapon plutonium is being studied. According to early studies, large fractions of plutonium destruction (especially Pu-239) can be achieved in a PWR. The phases that are present and the leaching characteristics of the rock-like structure of the spent fuel have been investigated. Ways to compensate for unfavorable reactivity conditions are being considered, such as additions of depleted uranium or thorium oxides to the fuel material as well as partial reactor loadings with standard (LEU) fuel. The editors note that these actions would reduce the net plutonium destruction fraction.

Waste from MOX fuel (containing 8500 kg of reactor-grade plutonium) fabrication at the Siemens Hanau plant has been conditioned and placed in 2900 drums (0.2 m^3/drum) fitted with high-efficiency filtered vents to prevent pressure buildup. The process of cement solidification avoids high pressure and temperature, thus providing a high degree of safety. The cementation facility has proven to be robust and trouble-free.

Based on this experience, the waste stream from an automated plant should contain 0.2% of the plutonium processed. This waste occupies a large volume because of the conservatively imposed criticality limit of 0.011 kg Pu (assumed to be 93.5% Pu-239) per drum at the Konrad repository. A cementation process to encapsulate these wastes is being proposed in order to establish a higher criticality limit of 0.12 kg Pu per drum, thus reducing the waste volume for geologic disposal (on this basis) by an order of magnitude. Plutonium content of individual drums is determined from passive neutron coincidence measurements. Ten waste drums are subsequently packed in a rectangular container and the intermediate spaces are filled with cement for final placement in the repository. The resulting repository volume per drum is thus increased by a factor of 2.7. As low as 30 mg of plutonium can be detected with a total individual measurement error of 20-30%.

In ceramics, significant swelling is observed even without annealing. Near the damage level corresponding to saturation in swelling, a phase change from crystalline to amorphous is observed, resulting in a higher leach rate. Ceramics also exhibit a lower hardness and increasing fracture toughness with increasing damage level. Because annealing of damage occurs at a higher temperature in ceramics, radiation effects are more important in ceramics than in glass. Addition of fission products to either ceramics or glass is not expected to result in added damage effects.

7.3 Vitrification Technology

Vitrification is considered to be the leading non-utilization option for disposition of weapon plutonium. Advantages of the vitrification option include potential for early implementation, acceptability of a wide range of plutonium-containing materials, technology maturity, and relevant waste glass performance. The leach rate of actinides has been shown to be 10 to 100 times lower than for other glass additives. Vitrification as a means of managing high-level waste (HLW) has been developed and demonstrated in France, Belgium, Germany, and England, as well as in Russia and the U.S. Immobilization of HLW in borosilicate glass is the only internationally accepted waste form. The potential for use of existing "waste glass" facilities reduces the need to build and later decommission new facilities.

It is likely that at least small batches of contaminated plutonium may be vitrified. Based on waste vitrification experience, at the Pamela plant, vitrification would cost $ 280/kg glass containing 10-20 wt.% waste oxides. An additional cost of $ 200/kg glass is estimated for interim storage and final disposal of the glass product in deep geologic clay layers.

In the Pamela process liquid waste would be mixed with plutonium nitrate. After denitrating, the concentrate would be mixed with borosilicate glass frit and fed into a joule-heated ceramic melter containing already-molten glass at 1425 K. The amount of plutonium that can be incorporated in the glass is subject to limits imposed by consideration of radiation damage and criticality. These limits are believed to be 1 wt.% and 1.6 wt.% in glass, respectively. On this basis the editors conclude that disposing of 50 Mg of weapon plutonium by vitrification would cost $ 2.7 B.

In Japan, radiation damage studies have been performed on glass samples doped with plutonium (~1%) and curium (~3%) oxides (containing Pu-238 and Cm-244). Conditions of helium release from borosilicate glass have been modeled and determined from a self-damaged alpha-decay sample. The model includes trapping of bubbles and re-solution from bubbles into the glass in addition to simple diffusion. Bubble behavior is believed to strongly influence the volume and mechanical properties of the glass. Hardness and Young's modulus decrease with dose and fracture toughness increases with dose to saturation values. These results are substantiated by work in Germany.

German results indicate that solubility of plutonium in borosilicate glasses varies with glass type and composition. At 5% PuO_2, only 4.6% is retained in the vitreous phase, with the second phase being crystalline PuO_2. Addition of HLW lowers the solubility of PuO_2. PuO_2 solubility is further lowered (to 1.2%) by addition of rare earths. In glasses, radiation damage tends to lower leach rates, hardness, and Young's modulus. Fracture toughness is increased. A large amount of helium is formed that produces swelling if the glass is annealed at a sufficiently high temperature.

Direct loading of plutonium in phosphate glass has been investigated. Some workers prefer phosphate glass to borosilicate glass because higher plutonium content is possible, PuO_2 is less likely to disperse as a separate phase, and the technological process is more suitable. This statement is questionable. With glass containing high-level waste (HLW), 3 to 5% plutonium can be uniformly distributed in phosphate glass.

Without HLW, up to 20% PuO_2 can be incorporated in the phosphate glass. A uniform distribution of up to 3.4% PuO_2 results if the molten temperature is held $<$ 8h. Settling to the bottom of small glass cylinder samples is observed beyond these conditions. Lowering the alumina content of the glass lowers both these limits if there is no HLW present.

Two methods of processing plutonium/glass products are proposed. After dismantlement, the weapon plutonium is oxidized to a chemically active state and either converted to a nitrate (wet process) or directly vitrified by melting with fluxed additives (dry process). Use of HCl or HF is not recommended because their use increases the amount of plutonium-containing waste. Various phosphate glass compositions and methods of synthesis can be devised depending on the desired plutonium content and inclusion of an HLW component.

Additional research and development required for full production of the proposed glass products includes: (1) optimization of plutonium oxidation methods and parameters, (2) specification of the glass matrix composition to satisfy product requirements (plutonium content, chemical and radiation stability, mechanical strength, etc.), (3) investigation of mineral-like matrices for plutonium, (4) development of necessary manufacturing equipment.

In the U.S. in 1979 PuO_2 (at 7 wt.%) was immobilized in borosilicate glass. Recent laboratory tests have achieved 15 wt.% PuO_2 in a commercial borosilicate glass. The maximum plutonium content will likely be established by criticality rather than solubility limits. Criticality can be mitigated by neutron absorbers such as boron and lithium, normally inherent in HLW forms, or intentionally-added erbium and gadolinium.

Proliferation resistance can be achieved by direct vitrification in which plutonium and HLW (or Cs-137) are simultaneously incorporated in the glass melt, or in a staged approach (called can-in-canister). Metallic plutonium would be converted to a suitable process feed material by oxidation or dissolution in nitric acid. The vitrification process can use either liquid or oxide feed. In the staged approach, a plutonium-only glass is produced as a relatively small interim ingot (called a can). Subsequently, these cans could be transported to another location where HLW is available and remelted during vitrification of the HLW or placed in a canister into which molten glass containing HLW would be poured to fill the interstices; (hence, can-in-canister). The same final product form could result using either the direct or staged approach.

The final form and composition of the plutonium bearing glass would be tailored for optimum technical performance in meeting proliferation requirements. Many existing facilities in the U.S. could be used. Combination of the plutonium disposition mission with an on-going mission for disposal of defense waste has the potential for minimizing the cost of plutonium disposition.

Preliminary indications are that vitrification of a mixture of HLW and plutonium is feasible in the Pamela plant at Belgoprocess. A single step process is used with a joule-heated ceramic melter. The criticality limit on melter feed concentration is 6 kg/m^3. It is concluded that Pamela, after specific process development steps and with a few facility modifications, could produce glass containing 1.6 wt.% plutonium, below the observed concentration (2 wt.%) for complete dissolution in borosilicate glass. The melter would hold a maximum of 12.8 kg plutonium in 800 kg glass. The Pamela plant would be able to process plutonium at the rate of 1 t/y in this manner.

In Russia, development of HLW vitrification technology has concentrated on minimizing the content of actinides in the glass, as the actinides are too valuable to be considered waste. From a storage standpoint, high fired PuO_2 has greater integrity as a solid form than plutonium contained in glass because of the much lower solubility of PuO_2 in water than the leaching rate of plutonium from glass. These observations are counter to the U.S. National Academy of Sciences study that concludes that vitrification is an alternative that deserves attention for plutonium disposition.

From a criticality standpoint, the theoretical maximum concentration of plutonium in borosilicate glass is 8 to 10 wt.%. European standards and current practice, however, currently limit plutonium content in glass to significantly lower concentrations. Long-term criticality is further facilitated by the high leachability of boron and the ingrowth of U-235 and Am-241.

Concern has arisen that low-energy photons contribute to the formation of oxygen microbubbles in borosilicate glass. Oxygen generation (in addition to helium) is accompanied by increasing internal pressure, change in chemical composition, and increased surface area. Although these effects do not require radical changes for HLW management, incorporation of significant amounts of actinides in the glass will undoubtedly affect radiation stability of the glass, as well as the leachability of the α-decay products.

Homogeneity of HLW-containing borosilicate glass is affected by immiscibility, segregation, and formation of insoluble precipitates during vitrification. These phenomena have a profound effect on the vitrification process as well as the properties of the glass product. There are essentially no fundamental spectroscopic analyses of actinide bearing glasses. Such analyses would eliminate discrepancies observed in the solubility limits of actinides as reported by various investigators.

Research and development of HLW vitrification technology to bring it to an industrial level has consumed considerable funds and time. Two continuous processes have been developed: the rotary calciner/metallic melter used in France and the U.K., and the liquid/slurry-fed ceramic melter used in Belgium, Germany, Japan, Russia, and the U.S. For the low actinide content of HLW, criticality has not been an issue in these processes. It is not likely that at high plutonium concentration these processes would be acceptable from the standpoint of criticality and material homogeneity.

In the U.K. the best method for incorporating plutonium in glass is considered to be to dissolve the metal in nitric acid containing HF and hydrazine and use this solution as feed material to the vitrification plant. Electrolytic dissolution is a promising alternative. Coprocessing with HLW makes recovery of the plutonium difficult.

Borosilicate glass (widely used or planned to be used) has a solubility limit of 5 wt.% PuO_2, while phosphate glass (used in Russia) has a higher solubility limit. Criticality considerations make borosilicate glass preferable. For added criticality safety, gadolinium or other rare earth oxides could be included in the glass.

The leach rate of plutonium from the glass depends on the solubility of plutonium in the repository ground water. If the water has a pH $>$9, the solubility of plutonium, and hence the leach rate are very low. Heat emission from plutonium is low in the absence of HLW, thus plutonium-containing glass can be disposed of without a waiting period.

Radiation damage data from accelerated testing provides an indication of glass performance after ~2000 y. At this level, no obvious ill effects have been observed.

Recovery of plutonium from standard size (400 to 1400 kg) glass containers would require large-scale facilities. If the glass also contains HLW, the facilities would need to be shielded. Added complications in separating pure plutonium would result from the presence of silica.

Various technologies are available to vitrify liquid or solid PuO_2 or plutonium metal. Some consider that a long period of development ("years") would be needed to achieve an industrial process. A dry process has significant advantages and reduces criticality concerns. A wet process simplifies metering the feed. Vitrification of plutonium with fission products is feasible, but will not be a simple process.

The continuous vitrification process is used in France and the U.K. to convert liquid HLW from Purex plants to borosilicate glass. The HLW nitrate solution is fed into a heated calciner where salts are evaporated and the nitrates are converted to oxides. The calcine and glass additives are then introduced to an induction-heated metal crucible.

About 5200 borosilicate glass canisters, each weighing ~400 kg, have been produced in France in three facilities since 1978. France is now developing a new glass melting technique that uses an induction-heated cold crucible. This technique allows higher melt temperature (>1775 K), thus capable of producing more refractory glass or vitrocrystalline formulas providing better containment.

7.4 Geological Disposal

It is universally recognized that the only technical approach for management of radioactive material not desired for further utilization (waste) is to solidify the material and isolate it deep underground. The possibility of safety isolating plutonium for storage, subsequent utilization, and disposal in a similar manner to waste must be scientifically evaluated.

Underground storage of HLW relies on multibarriers for isolation. The role of engineered barriers, however, is limited to the near term. Natural barriers (natural rock formations) provide ultimate isolation in the long term. One looks, therefore, for blocks in the earth`s crust with low tectonic activity, having structural stability, low permeability, and low water content.

To determine the suitability of a specific geological formation it is necessary to (1) assess the properties of the rock, such as structure, physical processes, and changes in the massif under natural forces by modeling over the required period of time, (2) investigate scenarios associated with possible emergency events and determine their probability, and (3) conduct a probability safety analysis for both normal and emergency events. The main physical processes that need to be considered for both the near and far fields are geomechanic, thermal and mass transfer, and hydrogeologic. A data base on earthquake magnitudes is available dating back to 1743. It is concluded that a quantitative safety assessment can be made for a specific repository configuration and geological site.

Deep-borehole disposition involves placing plutonium several kilometers deep into old, stable, rock formations that have negligible free water present. Containment assurance is based on the presence of ancient ground water indicating lack of migration and communication with the biosphere. Recovery would be extremely costly and impossible to accomplish clandestinely. A number of relevant deep boreholes have been drilled, and the technology is considered to be mature. A number of disposition forms could be used, however, none has been selected as yet. The number of deep boreholes required could be as low as one or two. Hole diameters at the material emplacement depth need not be over~0.7 m. A hole depth of 4 km with the bottom 2 km containing plutonium material seems reasonable. Major issues involve long-term criticality, licensing, and public acceptance. Regulatory implementation is the greatest uncertainty.

Fundamental requirements for disposal of radioactive waste (HLW or spent fuel) include reduction of ionizing radiation exposure to humans now and in the future, and environmental protection. These requirements are the basis for rules and regulations promulgated by the ICRP and the IAEA. Probability safety assessments are implied for the post-closure phase of a repository to account for data uncertainty and event probabilities. The ICRP recommends that the risk of health effects to members of the critical group be less than 10^{-5} per year.

The disposal site represents the basis for assuring repository safety during all stages of repository operation and the post-closure period. All geologic repository designs rely on a multi-barrier system consisting of conditioned waste, waste container, backfill, seals, and the geosphere. The containers are designed to function as an impermeable barrier for several hundred or several ten-thousand years. Spatial arrangement and loading density of material that generates heat is governed by host-rock temperature limits (475 K for salt rock and 375 K for granite and clay). Also, formation of critical masses has to be avoided. It is generally accepted that release of radionuclides to the biosphere cannot be avoided in the long term. However, the multi-barrier system must reduce the release sufficiently to satisfy the health criterion.

Long-term safety assessments have been carried out for disposal of vitrified HLW or LEU spent fuel. Safety assessments have not been carried out for MOX spent fuel or vitrified waste with significant amounts of plutonium, but deduction from LEU-spent-fuel assessments leads to the conclusion that safe disposal of MOX spent fuel is feasible. Criticality considerations will probably limit the concentration of plutonium in vitrified waste, whereas heat generation considerations will not likely be limiting. The heat rate from MOX spent fuel will be three times greater than from LEU fuel, thus fewer MOX spent fuel elements than LEU spent fuel elements can be placed in a given volume. The migration of Pa-231, a long-term decay product from Pu-239 and U-235, needs to be assessed as part of post-closure performance.

7.5 Safety

Criticality safety is a major concern in designing the vitrification process. It is required that no single failure could produce a critical accident. Specific criticality analysis of the vitrification process has yet to be done. The issues involved drive the design of the process. In the initial process designs, the primary method of achieving criticality control is by conservatively limiting plutonium mass and its possible geometry. For example, a two-step process is envisoned with the first step a small batch process producing plutonium-borosilicate glass. After homogenization, the potential for criticality is greatly reduced and a larger melter containing higher amounts of plutonium can be used in the second step. Additional neutron absorbers can be incorporated during the second step. As process designs become more definitive more detailed criticality analysis may indicate that a cheaper one-step process is safe.

Criticality safety in the repository also needs to be addressed. The immobilization matrix, plutonium concentration, and neutron absorbers are criticality factors that are directly established in the vitrification process design. Additional criticality factors are established by the repository design. Both designs need to be performed in a system context.

An overview of practices in Russia to prevent plutonium contamination in the nuclear industry was presented. Difficulties with plutonium arise from its chemical nature and its extremely high toxicity (four times higher than most other alpha-decay radionuclides). Plutonium contamination at nuclear power plants results generally only in connection with fuel element cladding failure, and is not a routine concern. Contamination is a continuing concern in research laboratories and in spent fuel reprocessing plants.

Contamination of surfaces in a reprocessing plant begins at the head end during initial cutting of the fuel elements and increases to a maximum in the final product areas. Contamination of surfaces results from adsorption and diffusion processes. Adsorption rates depend on the surface characteristics. Special solutions and conditions are required for decontamination. Initially washing with a medium concentration nitric acid solution that contains a small amount of fluoride is used. This solution permits efficient deposit renewal and reduces the waste that is generated. Specially designed jet sprays are used to reduce the amount of washing solutions.

Numerous advances have been made in techniques for protection of surfaces from plutonium and decontamination of surfaces from plutonium. Means to increase decontamination effectiveness and decrease liquid waste generation continue to be investigated. Significant improvements, especially in new plants, are possible to decrease releases to the environment during operation and in decommissioning.

The practices and basic techniques applied in safeguarding special nuclear material that is in civilian use aim at efficient and effective control by entitled inspectorates. These inspectorates depend on instrumentation development and implementation as well as training being provided by the Joint Research Center Ispra.

The distinction between reactor and weapon plutonium is arbitrary, as either can be used to produce a nuclear weapon. Also, straightforward chemical processes can be used to isolate plutonium from all other elements. The editors note that, although straightforward, the isolation process can require the use of costly facilities. Nevertheless, plutonium has been efficiently controlled for decades under IAEA and Euratom systems of material control and accountancy that are based on well-established material balances periodically verified by physical inventories. The inspectorates must have unlimited access to the nuclear materials. Thus safeguarding plutonium derived from military sources will be intrusive. Solutions for accomplishing the safeguarding and preserving military sensitive information have been proposed. In this regard, the difference in the goals of nuclear disarmament verification and non-proliferation verification must be noted.

Safeguard techniques are based on the existence of a material accountancy system. Computerized means of managing the data including statistical analysis are available. Destructive and non-destructive assays are routinely conducted to make volume and density determinations. Techniques for plutonium measurements rely on neutron coincidence counting, calorimetry, and high-resolution gamma ray spectroscopy. Sealing/identification techniques are effective. Ultrasonic techniques are being investigated for surveillance of PuO_2 in transport containers. Optical surveillance used with data reduction techniques is effective.

Training programs and facilities are in place that provide the capacity to design verification plans. The training provides a means of transfering technology between the laboratory and the industrial measuring environment. Plutonium safeguard practices vary depending on whether the plutonium is liquid or solid, its isotopic content, and whether it is relatively pure or scrap in waste barrels.

The accuracy of measurements varies from less than 1% to as high as 20%. The first requirement for safeguarding excess weapon plutonium in a civilian environment is to accurately assess its initial inventory as it crosses the military/civilian border. Well-defined pure plutonium can be assayed within 1%, while scrap can be assayed within a few percent.

Material protection, control, and accounting (MPC&A) is feasible for immobilization and disposal of excess weapon plutonium. Security of immobilized plutonium depends on a number of barriers. It is not likely that MPC&A would be a determining factor in choosing between MOX spent fuel or immobilization as a means of plutonium disposition. Demonstration of MPC&A is important - new facilities offer an excellent opportunity to incorporate transparency. The need to enhance and demonstrate the protection of nuclear material exists, and could be pursued at a low cost.

The final contribution at the workshop dealt with an overview of contamination from Russian nuclear complexes. Russia has developed one of the largest nuclear weapon complexes in the world over the last 50 years. In doing so, it has also created the world`s largest single inventories of radioactive materials. Although parts of these inventories have decayed to insignificant levels, been stored in engineered systems such as tanks, or been placed into solidified materials following put in storage vaults, tremendous amounts have been released to the environment. The primary contaminated areas are located at facilities that reprocessed fuel from weapons material production reactors and warhead fabrication installations. The potential impacts to the environment and human health from these discharges to surface water systems and underground formations is believed to be one of the most critical disasters in Russia.

Subject Index

GPSR Compliance

The European Union's (EU) General Product Safety Regulation (GPSR) is a set of rules that requires consumer products to be safe and our obligations to ensure this.

If you have any concerns about our products, you can contact us on ProductSafety@springernature.com

In case Publisher is established outside the EU, the EU authorized representative is:

Springer Nature Customer Service Center GmbH
Europaplatz 3
69115 Heidelberg, Germany

Batch number: 10755284

Printed by Printforce, the Netherlands